The Jefferson National Forest

The Jefferson National Forest

An Appalachian Environmental History

Will Sarvis

The University of Tennessee Press / Knoxville

utp

Copyright © 2011 by The University of Tennessee Press / Knoxville.
All Rights Reserved. Manufactured in the United States of America.
First Edition.

The excerpt from "Brier Sermon—'You Must be Born Again'" by Jim Wayne Miller
is reprinted from *The Brier Poems* by permission of Gnomon Press.

The paper in this book meets the requirements of American National Standards Institute / National Information Standards Organization specification Z39.48-1992
(Permanence of Paper). It contains 30 percent post-consumer waste and is certified
by the Forest Stewardship Council.

Library of Congress Cataloging-in-Publication Data

Sarvis, Will, 1959–
The Jefferson National Forest: an Appalachian environmental history / Will Sarvis. — 1st ed.
    p. cm.
Includes bibliographical references and index.
ISBN-13: 978-1-57233-828-9 (alk. paper)
ISBN-10: 1-57233-828-8 (alk. paper)
 1. Jefferson National Forest—History.
 2. Jefferson National Forest—Management.
 3. Forest management—Jefferson National Forest.
 4. United States. Forest Service. Southern Region.
 I. Title.

SD428.J4S27 2011
333.7509755—dc23
2011021566

For my mother,
who endured far away
from her Pacific homeland.

Contents

Illustrations

Preface

This must be one of the more unlikely books ever to reach publication. It is a work begun with the backward process of data in search of philosophy, composed during disjointed periods of unemployment, hindered by the destruction of over eighty boxes of primary sources, financed through sporadic personal savings, and pronounced dead and unrevivable several times over a period now approaching two decades. It is not a procedure I would recommend to anyone. To have the effort finally reach fruition is more than a little amazing to me, in part because this story has a personal aspect.

Perhaps only those who grew up without tangible roots can hold a particular kind of fascination for a certain landscape and wonder at those who are multigenerational products of it. The rooted can always become nomadic, but the nomadic can never become truly rooted; they can only stop moving and accept the locale of their latest transplant.

Blacksburg, Virginia, was the closest approximation I ever had to a hometown. I moved there the first time with my family in 1970, after more than half a dozen moves from coast to coast and points in between. So, for better or worse, I have an unusually personal connection with the subject of this book. I went to high school and college near the old Blacksburg Ranger District of the Jefferson National Forest (JNF) and experienced quite an array of interaction with this federal land. I do not think this experience has unduly biased me in any particular regard to the JNF, despite some of its employees committing minor crimes (slander, defamation of character, plagiarism) against me. Every agency has its rogues. But superseding the unfortunate actions of these individuals was my personal affection for the forest and mountains, where I walked with my dog for many hundreds of miles along portions of the Appalachian Trail and local trails near Pandapas Pond and Mountain Lake, all on JNF territory. It is this personal affection that I have remembered most over the decades, especially when living in the flat Midwest, but also even in the Pacific Northwest with its own beautiful (if very different) mountains and forests.

In 1977 I began helping my great friend Nicholas Farr Nye gather wood from the JNF. We used a "dead and down" permit to collect firewood for sale and mainly traveled Highway 460 and Craig Creek Road looking for supplies.

Other firewood gatherers had scoured the forest uphill from these roads, so often we hustled the wood up steep embankments one stove-length log at a time. I was eighteen years old and still building my strength; often was the time when I fell asleep on the drive home from a day of such hard labor.

Nye had dreams of building a log house. He was trained in architecture and kept a keen eye out for any timber above firewood grade. Selling firewood, in fact, was his way of financing his accumulation of building grade logs. We only had the most rudimentary equipment: a 1948 Chevrolet panel truck and two used chain saws. We utilized the Egyptian technology of levers, fulcrums, and small log rollers to load the timber into the panel truck. What had been theoretical physics in school suddenly became quite real to me.

Nye was also working with the JNF for a more substantial supply of wood. His efforts finally resulted in a forty-two-acre "purchase order" in Craig County, off of the Hall Road, on the old New Castle Ranger District. A purchase order followed a clear-cut timber sale. The professional logger had taken all material he deemed valuable and left the rest. It was our job to cut down all remaining trees, wrist-size diameter and above. We were welcome to take as much firewood as we wanted. The problem with this particular purchase order was its remoteness, lack of good road access, and the years that had passed since the timber sale. We drove up Hall Road from Route 42, then began negotiating what can only be described as an unimproved path through the woods. Huge boulders twisted the truck's suspension and absolutely required the very low first gear (the "granny" gear) to negotiate the terrain at no more than walking speed. The difficult access had left the JNF with no one interested in the purchase order, which paid a measly twenty dollars per acre. The result was that the site had become overgrown in literally millions of tree sprouts, all about four feet tall and thick enough to make walking quite difficult. This was a test. Theoretically, if we finished this site correctly, there would be future sites where we might actually make a modest profit.

We had completed about six of the forty-two acres when Nick Nye was murdered at the foot of Brush Mountain, on June 1, 1978. The JNF surveyed the site and paid $120 to the deceased's estate. Eventually the JNF approached me about completing the purchase order, but in a completely outrageous fashion. This involved timber theft.

Nick and I had been aware that someone was stealing wood near the purchase order. One day when we were working, one of the saws broke, and we were absent machinery noise for a prolonged period while repairing the saw. Suddenly we heard a saw in the distance, coming from the access path to the site. We hopped in the truck, but by the time we got there we found only stumps and sawdust. Later we noticed freshly cut firewood on the porch of a house near the Craig Creek end of Hall Road. We had our suspicions that this might be the culprit, but obviously not even good circumstantial evidence, much less anything more. A certain JNF ranger apparently did not even need that to slander my dead friend and myself.

If I had had a hidden microphone and a half-decent lawyer in 1978, I might have sued the JNF for a fair amount of money and gotten the offending ranger fired. I did not have the microphone, and I could not have afforded the attorney. Basically this ranger (who shall remain anonymous) angrily accused me and Nye of stealing the firewood. It was galling enough for him to do so without any evidence, but to accuse someone like Nye—who had such an acute appreciation for the proper use of timber, who referred to people who cut up saw logs for firewood as "woodpeckers"; who would have cherished George Nakashima's *Soul of a Tree;* who hand built a red oak table that I still own—well, this was too much. The ranger wanted me to finish the purchase order for free, since he believed we had gotten so rich "stealing" half a cord of firewood. It was insulting, slanderous, and outrageous. I simply said I could not afford to lose that much money and left the building. The ranger, detached from reality with his fat bureaucratic salary, had no idea about firewood economics, which turns out to be a money-losing venture even with heavy equipment. I appreciated the economics later. In the meantime, I was living the hand-to-mouth existence characteristic of the more impoverished sort of on-and-off-again college student.

Later this ranger, who had a reputation as an angry racist, was promoted to another position. This "pass the garbage" routine is common in many government institutions, not just the Forest Service. I learned of the ranger's attitude problems from JNF rangers Skipper Frazier and Roy Powell, who oversaw the rest of my work. After getting rid of the problem ranger, I was allowed to finish the purchase order for the previous twenty dollars an acre, then received another sixty-acre purchase order in Montgomery County for twenty-seven dollars an acre. I lost money on both projects, partly because I did not have enough equipment or people to gather very much firewood.

Eventually I realized that there was no economic justification to me for cutting those 102 acres. In the meantime, it was quite an experience, especially regarding the initial 42 acres. The location was so remote that the only way I could break even (if equipment depreciation was ignored) was by camping out on the clear-cut for a week at a time. Someone loaned me their old boy scout tent, and another loaned me a large chest cooler. In advance I would fill three one-gallon antifreeze jugs (which featured thicker plastic than milk jugs) with water and freeze them solid. By keeping the chest cooler in the shade and covered by a canvas tarp, I could keep meat safely refrigerated for about three days. After that the slowly warming cooler would keep cheese and eggs okay, but the last days of the camping episode were always on a meatless diet.

Before departing on these weeklong camping expeditions, I had to go through an enormous checklist of necessities. There were all the usual camping requirements for five days (clothes, sleeping bag, water supply, food), but then there was an entire separate list of tools, spare parts, and equipment for the truck and chainsaws. Since even one trip out during the week would have bankrupted me, it was crucial that I be completely prepared. Without my rent-free residence in a friend's shack during those years, I never would have had the hand-to-mouth finances to pursue such an endeavor. Years later, when I encountered gold-bricking bureaucrats collecting their generous salaries in the JNF supervisor's office, I thought back to those months spent on the mountain.

It was not the most dangerous employment I've ever experienced, but it had its hazards. The most frequent was accidentally walking over subterranean yellow jacket nests. With the noise of the chainsaw and cotton in my ears blocking full hearing, only getting stung alerted me to my misfortune. I would drop the saw and run as fast as possible. Upon returning, the saw would be covered with the wasps (genus *Vespula*), so that necessitated a trip back to camp to retrieve a cup of gasoline. With their permeable exoskeletons, the wasps died instantly upon contact with the petroleum. The remaining aroma dissuaded others from lingering. I could retrieve my saw and continue work.

Other common sights were rattlesnakes and copperheads. I was always surprised that the noise of the saw and the vibration of falling trees did not frighten them away. On Sinking Creek Mountain in Craig County there was one particular copse of a dozen trees or so that I could never cut. Every time I approached I would encounter a coiled poisonous snake. I had concluded that I would have to wait for the cooler days of autumn when one day Ranger Skipper Frazier surprised me with an inspection. I told him about the trees and he laughed. "Don't bother cutting those," he said. "That's the biggest rattlesnake nest on Sinking Creek Mountain." The trees remained standing.

The only time I came close to a predicament was in cutting down a small tree of only a few inches in diameter while standing about six feet in the air atop accumulated slash. This amount of slash was not unusual on the site, but for some reason this tree did not fall as expected and somehow pinned my wrist against its trunk and the slash underneath. It was only mildly painful, though I wondered if it would break my arm with further movement. Without radio equipment I would have had to walk out of the site to the nearest house, about six or seven miles away. Somehow I finally managed to extricate my arm little worse for wear. It did make me think of what would happen were a real accident to come my way. This was my work, for a dollar an hour or less, in 1979–80.

Even people who love the forest and mountains will admit that extended time spent there can become boring. Through-hikers on the Appalachian Trail dread the onset of the "green tunnel," beginning in May in Virginia for many of the northern-bound. Driving on rural roads through a picturesque landscape, a few moments of contemplation at a scenic overlook, or perhaps an afternoon picnic is probably the duration most seek. To stay a week at a time on Sinking Creek Mountain was enjoyable for me because I had an excuse for being there. More than a few loggers have thought about this over the years and find it ironic that their harshest critics tend to enjoy life and work in the suburbs and cities. Like farmers and commercial fishers, they want to be outside. Being in beautiful country offers a great bonus. So even though I was working all the daylight hours in the summer (with a two-hour lunch and siesta in the shade during the heat of the day) there were opportunities nevertheless for gazing across Craig's Creek Valley to enjoy the view of Brush Mountain. In the long run, that may have been the single redeeming feature of the purchase order experience.

A few years after completing the purchase orders, I briefly gathered firewood with a man who employed a small crew and owned a skidder, dump truck, front-end loader, and all sorts of other equipment—but he lost money as well and eventually opened an auto-repair shop instead. The firewood business is simply obsolete as a viable capitalist enterprise. People who want the exercise may enjoy cutting their own fuel wood, but in the marketplace firewood simply cannot compete with cheaper, cleaner, more efficiently delivered, and more convenient fossil fuel heat sources. This was a lesson in economics for me. I still enjoy cutting and splitting all my own wood, and I still heat exclusively with it, but I do not miss trying to make a living selling firewood.

In retrospect, one of the interesting aspects of experiencing the purchase orders (particularly the first, overgrown one) was the empirical observation of all those many thousands of tree sprouts that existed in the years following a clear-cut. This afforded me some insight into the enduring impressionistic ideas of watershed damage linked to clear-cutting. That first purchase order was along the southeastern flank of Sinking Creek Mountain at approximately three thousand feet. About six years after the clear-cut, a conservative estimate would be that the site featured two tree sprouts per square foot. That would amount to over three and a half million sprouts on that forty-two-acre site alone, though natural competition would see those sprouts diminish exponentially over the next decade or so.

This is not to defend clear-cutting, and abusive logging can certainly cause watershed damage. Timber stripping of the late nineteenth and early twentieth centuries largely prompted the 1911 Weeks Act that eventually created eastern national forests like the JNF. But this was more of an emotional and political cause and effect rather than an initiative based in science. Ironically, despite the science of forest hydrology becoming much more sophisticated, the emotional and political reactions have persisted. I doubt many JNF critics have actually walked a clear-cut during a spring or summer following a timber sale.

I used to think clear-cuts were uniformly ugly. They do not bother me so much now, even though I think clear-cutting is mistaken timber policy on eastern national forests in particular. Certainly my favorite place in the JNF (the War Spur Overlook near Mountain Lake) evoked my admiration because it offered an illusion of undisturbed forest landscape. It is an illusion because all the American chestnut trees are missing, because it is a regenerated forest, because excessive ozone has helped kill the old virgin hemlocks near there, and because of other hidden disturbances. Certainly there are no visible clear-cuts from this overlook, and thus it is indeed a grand and glorious woodland vista. Still, in the larger picture, it has become impossible to ignore the impact of distant forces upon the JNF environment.

The greatest deterioration I witnessed all over the JNF was air quality. The Smoky Mountains are but a subset of the southern Appalachian range that features hazy skies in the summer time. This was once purely humidity, and anyone doubting that merely had to wait until winter to watch the vistas grow vividly clear. This was still true during the 1970s. By the late 1980s I began to notice a lingering haze all winter long. It was smog, blown in from midwestern coal-fired power plants, and now the air quality in the southern Appalachians is among the worst in the nation. This factor, perhaps more than any other, forces the JNF into an inescapable global environmental context. My mind was free of such concerns during the purchase order endeavors.

I managed to get through graduate school, and with a fresh master's degree of questionable practical value in hand, I was desperate for employment. The JNF advertised for its first historian to work in the Recreation Division under the forest archaeologist. I applied for and received the temporary position and began working in a frenzied manner with knowledge that I would soon be unemployed again. That occurred eighteen months later. During the interim I managed to work nearly every night and every weekend in the Virginia Tech library and put in somewhere between 1,000 and 1,500 hours of unpaid (and unrecognized) overtime. The insult that followed the injury of being fired was discovering that one of my former colleagues had plagiarized some of my work.

In 1993–94, after months of threatened legal action, I finally had to hire an attorney to warn an editor who refused to print an explanation of the plagiarism. I was trying to get my original work published and, to the untrained eye, it would have looked as if *I* were plagiarizing *the plagiarizer*. In any case, the publication finally agreed to print a mild retraction of sorts, deliberately misspelled my name, and never compensated me for my attorney fees. I let the matter drop.

I relate these tiresome episodes partly because of their irony, for some critics have accused me of being impartially beholden to the Forest Service, particularly critics from the old rival agency, the National Park Service. During the middle and late 1990s, I was doing research on the eminent domain used for land acquisition for the Ozark National Scenic Riverways (ONSR) in the Missouri Ozarks.[1] Some Park Service employees falsely concluded that I was a blind defender and advocate of the Forest Service (once a prospective land manager in what became the ONSR), presumably because of my former temporary employment. This was an unfortunate presumption. In fact, one of the factors that contributed to my getting fired from the JNF was my unwillingness to whitewash certain stories. Pro-agency propaganda is, after all, the elephant in the room of much public history.[2] I was told to my face not to write about clear-cutting. I did anyway, but the eminent domain surrounding the Mount Rogers National Recreation Area proved to be far more controversial and likely contributed to my rapid termination from the JNF. Anyway, obviously I was not cut out for federal bureaucracy, and I blame no one for the brevity of my tenure with the JNF—but certainly I am not biased in favor of any given agency policy. Nevertheless, my brief JNF employment gave me insight that many environmental critics consistently ignore, especially the internal policy disagreements among various Forest Service staff. Also, I witnessed heroic, hard-working, and idealistic Forest Service employees who do the right thing and put in long hours without thought of recognition, extra monetary compensation, or glory. Perhaps the very public emergence of the Forest Service Employees for Environmental Ethics will begin to modify some critics' perspectives.

Acknowledgments

By the time I began this research project, the National Archives in Washington, D.C., had destroyed the primary records they held for the JNF. Oral history interviews became essential in this context, and many JNF retirees (a full list of whom appears in the bibliography) generously shared their memories. During my eighteen months of employment with the Forest Service, various JNF and Forest Service employees, not formally interviewed, also made important contributions to this work through the early phase of its research and composition. Some of them read portions of the most preliminary rough drafts that pertained to their particular specialties and offered helpful suggestions; others supplied much detailed information. These kind people were Gerald Buchanan, Bill Clark, Joyce Crawford, George Freeland, Karen Goode, Skip Griep, John Hinrichs, Tom Heffernan, Cindy Huber, Fred Huber, Gretchen Merrill, Sarah Patterson, Roy Powell, Charles Rozier, Hank Sloan, and Sylvia Whitworth.

Thanks to Bill Cochran for loaning me material relating to his father's career with the JNF; to Shawn Aubitz and the archivists of the Northeast Branch of the National Archives for their very friendly assistance; to Ann Huebner, Thelma Gardner, and Bill Howell, who all helped supply the BLM's payment-in-lieu-of-taxes statistics; and to Pat Fritz, Gloria Bowman, Teresa Love, and Liz Broughton, who all handled many essential daily logistics involving travel or paperwork when I was employed with the JNF. Around 1998 George Washington–Jefferson National Forest Timber Staff Officer Jim Sitton promptly supplied me with extremely valuable and updated timber acreage and cutting statistics that, along with then-recent budget documents, so dramatically illustrated the turn of events in contemporary JNF history.

Jim Loesel, secretary for the Citizens Task Force on National Forest Management, has helped me through several phases of this research and composition. Loesel explained specific, sometimes esoteric topics, particularly the payment-in-lieu-of-taxes scenario. Acting as an informal archivist for an agency naturally preoccupied with current events and future planning, Loesel also supplied me with the invaluable budget documents of recent years, cited in several places of this work and in an appendix. Finally, long after I had left the JNF, Loesel continued—generously, patiently, and enthusiastically—to answer all my questions, giving me direct information or telling me who to ask or where to look. As one of the greatest repositories of JNF institutional memory, Loesel has been an invaluable resource of contemporary history.

Many editors and their anonymous referees have helped publish parts of this work in eight separate articles. I would like to thank Bill Rooney, *American Forests;* Charles Finlay, *Virginia Forests;* David O. Percy and Kevin C. Foy, *Forest and Conservation History;* the late (and great) Hal Rothman, *Environmental History Review;* Hugh Campbell, *Smithfield Review;* Fredrick H. Armstrong, *West Virginia History;* Edward D.C. Campbell and Julie A. Campbell, *Virginia Cavalcade.* These articles are all listed in the bibliography. For help with the Mount Rogers NRA history, I would like to thank, in particular, the following: Charles A. Blankenship, the late Harold Calhoun, William Clark, William Gable, Reginald Kinman, Donald Martindale, Lionel Melancon, Michael J. Penfold, Charles A. Perdue Jr., and George Wolfel.

Years ago William Shands read a premature draft of the entire manuscript and offered very helpful feedback. Harold K. Steen read earlier drafts of the manuscript twice and always offered scholarly and gentlemanly suggestions. Virginia Tech History Department professors Mark V. Barrow Jr., David Burr, A. Roger Ekirch, David Lux, Neil Larry Shumsky, Daniel B. Thorp, Peter Wallenstein, and Young-tsu Wong all offered me generous assistance over many years, some with this project, others with many additional projects, all in a spirit that can only be described as utter dedication to and enthusiasm for the discipline of history. Crandall Shifflett taught a course in Appalachian history at Virginia Tech, and before moving on to Texas, Thomas R. Dunlap taught two courses in American environmental history at Virginia Tech. I thank them both kindly, for these courses turned out to be an especially valuable foundation for this project.

It may be understandable, yet still strange, that some people chose to respond with tremendous rudeness and hostility to my request for research information; especially those self-identifying themselves as "religious" people. But such bad behavior made all the more wondrous the gracious generosity of someone like Thomas Crutcher, who emailed me from his spiritual retreat in Ireland to share his experiences of hiking the Appalachian Trail (an extensive and illuminating quotation from him appears in the chapter on Sacred Land). Some day I hope to have the pleasure of meeting Mr. Crutcher in person, but in the meantime his magnanimity from afar remains very greatly appreciated.

Reverend Bill Gable deserves special thanks for two episodes of great generosity; one in 1992, when he sat for an oral history interview, and the other during several weeks in 2009 when he answered many emails, provided me with a copy of *An Oral History of Konnarock, Virginia,* and (along with Ed

Davis) managed to post the entire transcripts from this project on the Emory and Henry College library Web site. Kind thanks to Reverend Gable, Professor Ed Davis, and Emory and Henry College library for this latter effort.

Thanks to Wendy Probst of the Appalachian Trail newsletter, who kindly posted a free classified ad for me seeking trail stories. Thanks to Antony J. Barrett, who responded to this ad. I also appreciate the help of two people who helped me with the difficult task of trying to gather data from afar: Jim Willis, retired forester, formerly with the Appalachian Sustained Development, and Phil Smith, policy analyst, Virginia Department of Game and Inland Fisheries. Lyle E. Browning of the Archeology Society of Virginia did some very generous and very last minute fact-checking for me.

Very special thanks to Cheryl Oakes of the Forest History Society. Ms. Oakes was there at the beginning of this project in 1992 and, with her usual efficiency and good cheer, continued supplying me with materials right into 2010. Forest historians everywhere hope she will never retire. Kerri Huff and Shai Cullop of the Appalachian Regional Studies Center at Radford University kindly provided the author with manuscripts from this important and growing archival collection. Latter day research for this and several other projects found a friend in law reference librarian extraordinaire Stephanie Midkiff. For all that Don Blackburn taught me about subsurface mineral complications, it was Ms. Midkiff who remembered the notorious "broad deed" from her law school days, which keyword led me to the most pertinent legal sources regarding that topic.

For about eight years now Raymond Bailey has been scratching his head over my outlandish interlibrary loan requests, but he always smiles and makes his best effort. This project and others would be much the poorer without his assistance. From my Missouri days, I thank Leo Drey and Clint Trammel, whose 1998–99 interviews and correspondence made the Pioneer Forest essay possible. These are interesting men who have that rare combination of idealism and pragmatism. All foresters and forest lovers could learn a great deal from their example. And I kindly thank my friend and colleague Joseph Glynn Escobar. Besides being a passionate teacher of history, Joe's wizardry with technology never ceases to amaze me; without his help the visual images in this book would be substantially diminished.

I wish to thank several people associated with the University of Tennessee Press. Referee Will Skelton offered overarching support for the manuscript, which significantly bolstered the entire endeavor just when it needed such support. Another, anonymous reader also offered much support—but very graciously pointed out some shortcomings that needed rectification; in fact, a seemingly minor comment or two from him or her actually became the major beacon that guided me through the last (and most intense) of many revisions. Managing Editor Stan Ivester patiently answered all my questions regarding critical details surrounding the manuscript's final form. Special thanks go to Monica Phillips and Gene Adair for their painstaking and diligent copyediting. Finally, I wish to thank acquisitions editor Kerry Webb, whose initial interest in this work never flagged; eventually I began to realize that her ongoing engagement with the manuscript meant that all my previous years of work were not going to be wasted entirely after all.

Will Sarvis

Eugene, Oregon, 2010

Note on Territorial Designations

In 1995 the U.S. Forest Service combined the Jefferson National Forest with the George Washington National Forest. The James River approximates the former boundary between them. The combined territories now constitute Virginia's only national forest. Other postmerger administrative changes combined the former Blacksburg, Wythe, and New Castle ranger districts into what is now called the Eastern Divide Ranger District. The JNF's former Glenwood District is now the western portion of the Glenwood Pedlar District. This history focuses on the JNF before the merger and will use the earlier ranger district names and their territorial designations. Forest Service parlance refers to activity taking place "on" (rather than "in") a ranger district; that usage is adopted throughout this study.

Southwestern Virginia and surrounding states. Enhanced and modified from a Wikimedia Commons map available at http://en.wikipedia.org/wiki/File:Va_nationalforests.png.

Introduction

The original title I had conceived for this book was "Highland Woodlands," which was derived from the particular Forest Service holdings along the higher elevations of the ridge and valley subregion of the southern Appalachian mountains.[1] The linear valleys are dominated by meadows, farms, waterways, railroads, and major highways. All the large towns are found here, and the lower elevation land is almost completely in private ownership. The linear ridges are overwhelmingly forested, and a great deal of this woodland is managed by the Jefferson National Forest (JNF).

This is a very beautiful land, and now it is easily enjoyed by travelers driving along the Blue Ridge Parkway, Interstate 81, State Route 42, and numerous small roads that crisscross valleys and ridges alike. It was a sacred land to the aboriginal peoples, and it has become a sacred land to successive waves of settlers and visitors. It is also a land that both indigenous peoples and early eastern hemisphere arrivals absolutely relied upon for utilitarian sustenance. For a brief and very intense period industrialists heavily exploited its timber resources, particularly in the higher elevations where the woodlands had already survived many generations of agricultural effort concentrated in the nearby valleys. The story that follows attempts to document how the overall topography of southwestern Virginia has come to feature its mixture of meadows and woodlands, how some of the forest came to be a government-managed commons for everyone, and how various peoples have regarded this land over the centuries.

❧

The Great Valley of Virginia, running southwest to northeast from Tennessee to Maryland, became a major travel route for prehistoric peoples through southwestern Virginia. One valley removed to the northwest from the Great Valley, the natural salines of Rich Valley attracted numerous ancient hunters whose archaeological remains date back to the earliest eras of the North American prehistoric record. And yet no major aboriginal group made southwestern Virginia their permanent home. As the contact period with

Euro-Americans approached, the Cherokee remained based to the south; the Tutelo, Monacan, and Powhatan, to the east; and the Shawnee, to the west and north. Thus, prehistoric southwestern Virginia instead provided early people with a reserve used (if not always peaceably shared) in common. Indigenous peoples used at least one major southwest-northeast travel route, while they appear to have hunted, fished, and gathered food in the area mostly during summer months. There is evidence of occasional and small year-round camps. Overall, prehistoric human influence on the environment remained comparatively mild, though it was intensive at various times in specific areas, especially in proportion to the comparatively small human population.

Prehistoric human activity in southwestern Virginia remains an ongoing interest of the JNF, which is charged with protection of archaeological artifacts. By the 1990s the JNF expended a couple hundred thousand dollars annually in an effort to protect and document this resource. In fact, since the late 1970s JNF archaeologist Mike Barber has become one of the most knowledgeable persons dealing with southwestern Virginia's prehistory.

Through his work with the JNF, his corollary contract archaeology company, Preservation Technologies, and his close association with the Archeological Society of Virginia, Barber has contributed much detailed focus, as well as broader study, to a topic that otherwise would remain mostly obscure. Barber has a long history of successfully extending his archaeological study beyond the narrow confines of artifacts discovered on the JNF itself, since obviously these modern land boundaries arbitrarily demarcate what was, throughout prehistory and history, a cohesive or interrelated territory. Perhaps most notably, the work of Barber, various JNF archaeological staff, and other specialists have focused much attention on the Rich Valley area of present-day Saltville.

Because of mammals' biological need for sodium chloride, the natural salines of Saltville have always attracted animals, and thus some of North America's earliest and subsequent hunters. Even the Smithsonian Institution has taken an interest in this site.[2] Barber and his associates have done other archaeological work in other southwestern Virginia valleys or lowlands conjoining or near actual JNF territory, such as in the Roanoke Valley and near Altavista, beside the Roanoke River. Obviously, Barber and the JNF prehistoric cultural resource management effort has traditionally surpassed merely checking for significant archaeological sites prior to timber sales or road construction, the bare minimum the Forest Service expects from this division.[3]

The following chapter on southwestern Virginia's prehistory is, in fact, largely possible through the JNF cultural resource management endeavor, often drawing upon the work of archaeologists aided by the JNF. Such management applies to historic era resources and topics as well, though this period of inquiry has not received as much funding or attention—mostly because of the nature of cultural resource management being directly linked to ground disturbance rather than, say, structural disturbance. A timber sale and its corresponding road construction requires a preliminary analysis of any possible underground artifacts. Above ground, historic artifacts are generally not threatened with disturbance, and thus the JNF has traditionally studied and documented them on more of an ad hoc basis.

The industrial logging that immediately preceded JNF forest management is an obvious historic cultural resource topic, but other subjects are equally relevant. Preindustrial eighteenth- and nineteenth-century logging, especially as it was associated with charcoal manufacture of salt and iron, pertains to JNF cultural resource management in several ways. JNF land surrounding Saltville was the source, cut many times throughout the eighteenth and nineteenth centuries, for the voracious salt furnaces' fuel. The JNF acquired one former iron mine complex (Fenwick), five old charcoal furnaces (Broce, Catawba, Glenwood, Ravencliffe, and Roaring Run), and vast tracts of land once extensively and repeatedly cut to supply these furnaces with fuel. This is particularly true of the Glenwood, Wythe, and New Castle ranger districts. The old mine and furnaces themselves constitute some of the JNF's richest historic sites, and the repeated cutting that accompanied their historic operation accounts, to some extent, for the successive forests that have grown back in these areas.

Early- and mid-nineteenth-century turnpikes crossing the Glenwood and New Castle ranger districts once carried thousands of tourists traveling to the mineral springs north and northwest of the Great Valley during the Romantic era and America's first age of tourism. The many additional historic cultural resource topics include early-twentieth-century religious education in the southern highlands, which is the historic context surrounding the Konnarock Lutheran Girls School, now part of the JNF's Mount Rogers National Recreation Area. Also in this territory lies the Green Cove train station and part of its associated railroad route. The Clinch Ranger District contains part of the estate of one of the area's historic coal barons, and thus the entire, complicated history of industrial coal acquisition in Appalachia becomes a pertinent JNF historic cultural resource subject. Log cabins and frame houses, and their architectural details, are part of the regional folklore context that has received much study in other locales.[4] Former agricultural sites are part of the broader history of southwestern Virginia agriculture, which once featured the ancient practices of land rotation.[5]

During the seventeenth century, Native Americans and colonists collaborated in one of the first major environmental changes, involving unprecedented intensive white tail deer hunting. Southwestern Virginia, already a part of the Cherokee's hunting grounds, was only one locale that experienced the deer pelt trade's impact. Subsequent Euro-American and Afro-American agricultural occupation included the usual practices of land clearing, annual crop cultivation, and repeated tree felling for fuel wood and building materials. These new farmers also unrelentingly hunted predatory mammals, especially wolves and mountain lions, but also black bears. By the late eighteenth and early nineteenth centuries, charcoal firing of iron and salt furnaces brought about the first industrial woodcutting efforts that denuded second-growth bottomland forests, as well as mountainside woods, whose relatively difficult accessibility became warranted through heightened demand for fuel. Charcoal iron furnaces could consume an acre of forest per day. In both charcoal iron and charcoal salt manufacture woodcutters constituted, by far, the most numerous type of employee.

After the Civil War, a growing resident population increasingly stressed the land's limited agricultural potential. Instead of leaving cropland and pasture to return to woodland, repeated annual plowing and grazing increased soil erosion. Forest habitat reduction impacted all wildlife, and unregulated hunting and fishing drastically reduced game populations. Finally, the southern folk practice of annual or semiannual field burning kept some hillsides bare of any growth beyond rudimentary vegetation.

Railroad construction around the late nineteenth and early twentieth centuries enabled the final and most intensive industrial impact on southwestern Virginia's forests. Development of the coal industry northwest of the Great Valley (in the territory where Virginia, Kentucky, and West Virginia meet) required an enormous amount of wood for railroad ties and mine timbers. Not only did industrial timberers fell more trees over more acreage than previous woodcutters, they did so in areas not previously logged at all. Until the late nineteenth century, the rugged land around Grayson and Washington counties, southeast of the Great Valley, still contained hundreds of acres of primeval trees. The industrial logging that occurred there was what immediately preceded or overlapped with the beginning of the creation of the JNF and other southern Appalachian national forests. Timber and mining companies cut southern Appalachian lands in the fervent and destructive fashion typical of the volatile, unregulated lumber market scenario of the times, when many raced to liquidate their extractive resource stocks before the market grew saturated and prices and profits plummeted. The timber market, like other nineteenth- and early-twentieth-century markets, was genuinely boom and bust.[6] But the resulting environmental degradation and belief that such destruction caused an increase in flooding, both in the mountains and in flatlands downstream, directly led to creation of eastern national forests.

The 1911 Weeks Act became the legal basis for purchasing eastern lands, and purchase they required, for the western public domain created through the Louisiana Purchase and subsequent acquisitions never really existed as such in the eastern United States. Europeans and Americans first claimed southwestern Virginia land through grants from the English Crown, through purchase from colonial land companies, and then through subsequent sales and resales. The JNF acquired every acre of its land through purchases from private citizens or industrial corporations.

American national forests evolved over a number of years through growing public awareness of preservation and conservation matters, public and industrial concerns over timber supply, industrial and governmental concern over timber economics, and, finally, federal legislation designed to address various of these concerns. The early conservation movement developed out of the Progressive era and quickly gathered a diverse, multifaceted, sometimes ideological, and occasionally contradictory life of its own—at once economic and scientific, with urban and rural interests, aesthetic and practical values, and often highly political—and its momentum easily launched it into a major twentieth-century endeavor, as it remains today.[7] The conservation movement of the Progressive era overlapped or quickly followed industrial logging in southwestern Virginia.

JNF history falls into three phases: the late Progressive and Depression eras, the timber primacy of the war and postwar years, and the environmental movement era and its legacies beginning in the 1960s and continuing today. JNF personnel mostly focused on tree planting, fire control, and road building during the first phase. Timber sales dominated the second phase, though wildlife game species also began to attract attention. Finally, values other than commercial timber production began to garner the most public attention.

What began as a public land created amidst late Progressive era ideology of watershed protection and forest fire suppression quickly moved into the tree farm period of Civilian Conservation Corps planters and the postwar harvesters. The JNF most closely resembled the national agency and its timber harvesting priority during this second phase of its existence. Wood products fulfilled a number of military needs, and only shortage of labor hindered the gathering of forest material. The extractive mentality, however, saw new life with war's end. Clear-cutting crashed headlong into unprecedented demands for outdoor recreation.

Postwar public recreation demands preceded the environmental movement in altering the course of JNF management, helping to steer it away from commercial timber sale dominance. The early policy of the JNF's Mount Rogers National Recreation Area (NRA) during the 1960s and 1970s reflected an optimistic attempt to address growing recreational needs. The Mount Rogers NRA also reflected a change in original Forest Service recreation policy by emphasizing facilities construction rather than dispersed, low-key recreation such as hiking, picnicking, or rustic camping. In the meantime, American society underwent the enormous reordering of values of the 1960s and 1970s that changed so many aspects of the nation, including public land management. The changing times certainly helped turn the original NRA plans on their head, as well as inject a number of other concerns into JNF management endeavors. Clear-cutting rapidly became a prominent controversy of national forests, and at about the same time the concept of eastern wilderness gained the status of federal law. Ecological concerns also received unprecedented federal recognition, especially with laws protecting rare or endangered flora and fauna species. The JNF management task was growing complex indeed.

During the 1930s and 1940s, the primary concern for a recovering forest like the JNF lay in letting trees grow back. After mitigating former agricultural and industrial timbering effects on the environment (along with watershed protection and fire suppression), wildlife revitalization within the forest represented the next step toward a more holistic ecosystem recovery. Wildlife management itself underwent dramatic transitions from an initial preoccupation with game species, hunting and fishing, to habitat enhancement and protection of nongame, rare, threatened, and endangered species. As such, botanists, wildlife biologists, and other scientists working with forest plants and creatures came to focus more attention on the greater forest ecosystem and its complex interrelationships.

The JNF's value as a scientific resource remains, perhaps, among its least appreciated aspects. For decades scientists, graduate students, and other experimenters associated with Virginia Tech and the University of Virginia have utilized JNF territory for various studies involving streams, trees, plants, wildlife, and other aspects of forest ecology. Forest Service researchers from the Southeast Forest Experiment Station (near Asheville, North Carolina) have used the JNF as something of a living laboratory for applied science, and other scientists and science students have used this public land for experiments in pure science. Designation of the eleven JNF wilderness areas enhanced this scientific value by ensuring protection (and thus long-term, undisturbed studies) and by providing habitats necessary for the greater encouragement of biodiversity.

In addition to the many different kinds of human action that have affected southwestern Virginia over the millennia, there has also been a remarkable breadth of attitude toward this land. Prehistoric Native American regard will ever remain the most difficult to ascertain, of course. In fact, no comprehensive description will ever emerge, given the nature of source materials at hand. What does seem apparent is an intimate, sometimes sacred regard for the land that incorporated both animism and skillful exploitation. But whatever the details of their regard for the natural environment, certainly it was a very close one, borne of lives closely dependent on it.

Even the vaguest indication of the indigenous worldview provides marked contrast to the historic period immediately following Indian occupation. Euro-Americans brought many cultural preconceptions toward wild, forested lands, including contradictory notions of mythic natural world paradise and "untamed" territory conducive to satanic temptation. But they had to grapple with these concepts in new ways once they were faced with actually *living* in the wilderness as opposed to regarding it hypothetically or from a distance. Their initial incursion grew preoccupied with "conquering" the wilderness and taming it into agricultural productivity. Hostile or potentially hostile Native Americans, of course, had to be removed.

During the interesting interim between the sliding time line of European contact and permanent agricultural occupation of southwestern Virginia, the heightened white tail deer trade radically disrupted whatever cosmological regard the Cherokee and other Indian groups had had for the natural environment.[8] The very stakes for overall survival of specific tribes were raised as alliances and warfare eventually pitted Native Americans against each other in unprecedented levels of conflict, as they were caught up in colonial extensions of European national conflicts (mainly the French and Indian War and Revolutionary War).

After the eighteenth century, an interesting duality arose in regard to mountainous Virginia. On one hand, an understandable utilitarianism pervaded among permanent residents who struggled to survive, mostly through agricultural endeavors. But as the human population grew during the first decades of the nineteenth century, so did the area's first commercial transportation network in the form of turnpikes. And along these turnpikes scores of travelers—the area's first tourists—flocked to various mineral springs, scenic landscape designations, and to the many points of accommodation along the various mountain roads. Among these travelers were writers and illustrators who sojourned into the highlands where the landscape, now inspirational, gave rise (self-consciously or otherwise) to their various creative works. Thus southwestern Virginia participated in the Romantic movement that swept the nation and paralleled Europe. But the Romantics were a leisurely sort who were prosperous enough to enjoy the mountains without having to consider the difficult economic challenges posed by living in such a topography.

From the 1860s into the 1920s, the spectrum of practical regard encompassed a new wave of industrialists increasingly exploiting the land's natural resources to those eking out an agricultural existence in a limited habitat stressed by a swelling residential population. With the rise of railroad networks elsewhere in the United States, train tourism and significant numbers of visitors bypassed the region, for only the Virginia and Tennessee Railroad ran down the Great Valley during the postwar nineteenth century, and by then old mountain turnpike companies were increasingly defunct. Either way, economic survival or economic profit preoccupied most of the resident (permanent or temporary) people interacting with the area. But overlapping and partly inspired by such a pervasive view, the Progressive movement and its concern for woodland conservation helped give rise to a conservation era out of which the JNF itself came into creation.

From the end of the Romantic movement, it took about another century before anything approaching a "sacred" regard for the landscape again pervaded the American consciousness. When it did, it arose as part of a growing "ecology" or environmental movement that actually had much older beginnings. For southwestern Virginia, this era would, again, involve a notable intermixture of local and visiting (or transplanted) people. That admixture remains today, and JNF management has increasingly been in response to these various "publics" and their widely differing expectations and demands upon the forest environment.

The demographic shifts of the past twenty years in southern Appalachia have continued to affect JNF management considerations. Private forest owners have changed in America as the population has shifted from a predominantly rural to urban lifestyle. With the demise of small family farms, teachers, doctors, lawyers, and other white-collar professionals have bought many acres of rural lands. These people generally favor the aesthetic qualities of forests and oppose timber-cutting. During the past thirty-five years, the trend has reached JNF territory, with white middle-class urbanites escaping cities and wishing to slam the door on development in the region they've "discovered." Conflicts with "natives" have especially involved trespassing, where newcomers demand a kind of privacy previously alien to most local people.[9]

Returning to the land, getting back to nature, trying to recapture a mythical Garden of Eden paradise, returning to a simpler way of life, agriculture's romantic appeal, urban industrialism's depressing pollution—all have been themes of varying intensity and frequency at different times and places in Western culture, but of special pertinence since the industrial age began. Prior to industrialism, humans regarded environments not substantially altered by themselves in various but generally ambiguous ways. While there are numerous examples of viewing undisturbed nature as idyllic, industrialism gave such views new urgency. This pervasive emotional trend very much figures into public regard for the JNF and other eastern national forests today.

Several features distinguish the JNF and other eastern national forests from their western counterparts. All eastern national forests consist of land acquired from private interests and were not originally reserved from the public domain, as with western national forests. National forests in Arkansas began with a little public domain, reflecting the geographical progression of frontier land acquisition during the early years of the conservation movement. From their earliest history, eastern national forests have been in close proximity to the majority of the U.S. population. Constituting only 6 percent of forested lands east of the Mississippi in 1976, eastern national forests continue to face much higher overall recreational demand than their western counterparts.[10]

Eastern national forests still consist of smaller, often patchwork tracts interspersed with many acres of nonfederal lands, called inholdings when completely or nearly surrounded. A forest like the JNF actually manages far less than half of the acreage within its gross proclamation boundaries. Congress established or modified the proclamation boundaries at various periods beginning in 1911, and an act of Congress remains necessary to alter these greater boundaries. In the meantime, the JNF has regularly bought, sold, and traded various tracts within their proclamation boundaries, often with an eye toward consolidating certain tracts. This kind of land management approach, at the very least, necessitates a high degree of cooperation between federal, state, county, industrial, and private citizen interests.[11]

The balancing act the JNF has performed during the past forty years or so has reflected a dynamic intermingling of the democratic process and the agency's bureaucratic elites. A high degree of effective grassroots democracy has often characterized the public participatory aspect of the process, despite bureaucratic growth and accompanying torpor. Nowhere was this kind of public influence demonstrated more dramatically than in the reversal of concentrated recreation plans for the Mount Rogers NRA. The process of this radical change in management occurred over the better part of a decade and was often highly controversial, but in the end the JNF proved (sometimes grudgingly) receptive and responsive to public mood. Effective public influence has manifested itself in other enduring and important ways. The creation of almost 60,000 acres of wilderness on lands managed by an agency formerly much more utilitarian and extractive-resource oriented presented a monumental shift in management attitude and technique. The Virginia Wilderness Committee and other wilderness advocates successfully promoted the preservation of these lands amidst a new atmosphere of general public regard for the scenic, recreational, and less disturbed possibilities for the JNF. But opponents of wilderness designation have made their wishes known and have helped prevent designation of some wilderness areas.[12]

Timber harvesting continued on the JNF, even in the era of ecosystem management and wilderness preservation. After all, it is a national forest (not a national park), and the tremendous amount of public-influenced evolution the JNF has undergone has not reached the extent of rewriting the Forest Service's mission statement, which would virtually require an entire restructuring of federal bureaucracy, perhaps taking the Forest Service out of the Department of Agriculture and placing it in the Department of the Interior (which Secretary of the Interior Harold Ickes and others attempted during the 1930s).[13] Nevertheless, the JNF is hardly recognizable if today's agency is compared to the one that managed southwestern Virginia lands fifty or even forty years ago.

In 1995 the JNF merged with the George Washington National Forest. Of the combined acres (1,751,254), only 689,600 are slated for possible future timber harvesting. Of JNF's postmerger lands (690,254 acres), slightly less than half (339,600 acres) are deemed suitable for possible future timber harvesting.[14] Sixty years earlier JNF staff viewed practically the entire forest as future logging lands. Such a shift from utilitarian timber extraction to a more holistic ecological approach seems nothing short of revolutionary.

Chapter 1

Prehistoric Southwestern Virginia

Native American land and resource use in southwestern Virginia spans almost the entire scope of North American prehistory and includes evidence of early Paleo-Indian hunters, all the way up to tribal peoples who met the impact of Euro-American occupation. Southwestern Virginia prehistory shares much in common with archaeological patterns in the southern Appalachians and greater southeastern United States, but, like any subregion, remains ultimately unique. In fact, some archaeologists argue that the western Virginia highland presents a singular version of an intermontane prehistoric culture distinct even from its neighboring mountain aboriginal cultures. One important feature of pre-recorded human occupation in southwestern Virginia includes less human-induced environmental change compared to surrounding lowland areas or even the proto-Cherokee highlands to the south.

Long before the Virginia mountains became a Euro-American frontier on the edge of advancing agricultural settlement, it had functioned as something of a "natural reserve" for cohesive native groups (later designated "tribes") situated in comparatively larger numbers surrounding, but not intensively occupying, the area. While such groups traveled and temporarily camped in the mountains for hunting, fishing, resource gathering, or trading purposes, certain smaller groups of aboriginal peoples also lived in southwestern Virginia for prolonged periods. All left behind archaeological evidence.

During the past 20,000 years, both natural and human-induced causes have rendered dramatic changes in the environment of the Virginia highlands. The end of the Wisconsin glaciation, 18,000–12,000 BC, ended the Pleistocene period and introduced the Holocene with radically modifying effects. The end of the Wisconsin glaciation also raised sea levels and submerged the land bridge between Siberia and Alaska, which archaeologists commonly believe was a major route ancient Asian peoples used in populating North and South America. As the earliest hunters began traversing the new continents, a spruce forest gradually evolved into one dominated by hemlock, which in turn gave way to a deciduous forest. Oak species became prevalent between 3000–2500 BC, and a warm and dry climate encouraged the development of subsequent deciduous combinations involving oak, chestnut, and hickory trees.[1]

While few archaeologists or anthropologists argue for exclusive environmental determinism, much evidence in the southern Appalachians suggests that land forms naturally had significant influences on prehistoric human behavior. In general, low-lying areas associated with rivers and flood plains became the most conducive for the most elaborate cultural development, while rougher topography proved the least.[2]

From the earliest days of the Paleo-Indian era, rivers became natural corridors for travel among the mountains. The New, Clinch, Tennessee, Holston, and Powell rivers all flowed toward the greater Mississippi drainage area and provided natural travel routes into and through the westernmost section of Virginia. The Roanoke and James flowed toward the Atlantic Ocean and offered travel routes into the mountains from the east. As prehistoric Indian lifestyles became progressively more sedentary, the flood plains of these rivers became the most common sites of permanent and semipermanent villages and horticultural activity.[3] River valleys also became the routes through which various Indian groups interacted and exchanged culture. In this sense, local cultural traditions expressed through ceramic variations and trade goods arose in association with various waterways, and thus even the Dan, Shenandoah, and Potomac rivers influenced southwestern Virginia's easternmost prehistoric culture.[4]

Along with rivers, geological sources of salt figured very importantly in southwestern Virginia's prehistory. Ancient clays embedded with salt in the present-day Roanoke area and a more concentrated salt formation in Rich Valley (modern-day Saltville) attracted mammals who in turn lured hunters of mastodon, giant sloth, and later creatures such as white tail deer, eastern elk, and black bear.

As America's first humans evolved from a highly nomadic lifestyle to a more stationary one, their culture grew more complex, characterized by the gradual rise of such activities as ceramic manufacture, increased plant cultivation, and eventually the crafting of ritual goods and burial items from materials such as mica and copper. Archaeologists have distinguished three broad stages to delineate these changes in culture. The Paleo-Indian, Archaic, and Woodland eras all distinguish themselves with unique cultural attributes in southwestern Virginia and other parts of North America. The Mississippian tradition arose as sort of a hybrid and ultrasophisticated version of the Woodland. While centered along the Mississippi Valley itself, this tradition also coincided and overlapped with the Woodland tradition in the areas contingent to the Mississippi Valley. In this manner, the Mississippian made its mark even as far away as the upper Tennessee valley and southwestern Virginia. Recent scholarship suggests that the western Virginia highlands were home to at least three major Mississippian sites.[5]

Paleo-Indian Period (ca. 9500 BC–ca. 8000 BC)

During the past century, archaeologists have found numerous Paleo-Indian artifacts throughout eastern North America. The basin areas of the Mississippi, Ohio, Cumberland, and Tennessee rivers have especially divulged rich finds in Paleo-Indian projectile points. In fact, despite the early fame of Folsom and Clovis finds in New Mexico and the American Southwest, present-day Kentucky and Tennessee and the bordering halves of all adjacent states have yielded the very richest sources of Paleo-Indian points in North America. The southern Appalachians, including all of southwestern Virginia, were part of this early hunting period, and, remarkably, the Paleo-Indian finds of eastern coastal areas of Virginia and other states have revealed notably fewer projectile points than the highlands themselves.[6] But despite extensive evidence reflecting transient hunting practices, archaeologists studying southwestern Virginia and other parts of the southern Appalachians in West Virginia, eastern Kentucky, and western North Carolina have yet to find evidence of larger Paleo-Indian settlements, such as the Flint Run or Thunderbird sites in eastern Virginia.[7] Thus, the evidence to date—though incomplete and partially destroyed through looting, past indiscriminate artifact gathering, and earlier road and building construction—would suggest that nomadic hunters traversed all of southwestern Virginia. Particularly indicative are the numerous Paleo-Indian projectile points found in Rich Valley, where animals roamed in high numbers seeking the natural salines.[8] About 10,000 years ago many of these megafaunal species became extinct. Why they died remains mysterious and controversial, and archaeologists and paleontologists have pondered a number of possible causes, including climate change, disease, natural evolution, human predation—or a combination of some or all of these factors, or from additional causes not yet known.[9] In any case, their demise, and the effect of their demise on early humans, brought the Paleo-Indian era to a close.

Archaic Period (ca. 8000 BC–ca. 1000 BC)

Archaeologists distinguish the Archaic Period from the Paleo-Indian Period by both natural changes and modifications in human behavior. The final stages of the climatic transition from the Pleistocene arrived, creating remarkable environmental changes and thus interrelated human behavioral changes. During the early Archaic, a pine and oak forest began to replace natural grasslands, the mastodon became extinct, and bison numbers decreased.[10] The climate followed a general warming trend. Somewhere around 5000 BC, the basic flora and fauna of the modern era's eastern United States established itself and provided Indians with vast new food sources, such as shellfish, acorns, chestnuts, and wild turkeys. Again, the broad riverain area associated with the Mississippi—including the Ohio, Cumberland, and

Tennessee rivers—became a concentration of Archaic culture.[11] Early people altered their behavior from primary reliance on hunting to a more diverse, less itinerant way of life that involved a heavier use of plant foods, both wild and cultivated. Indians in west central Illinois apparently began domesticating gourds and squash as early as 5000 BC, and such horticultural practice seems to have reached the Tennessee and Kentucky area by 2500 BC.[12] Such plant cultivation apparently spread into southwestern Virginia as well, and by the end of the Archaic Period, aborigines in this area seem to have used a wide range of the area's natural food resources.[13]

Prehistoric sites and artifacts from the Archaic Period in southwestern Virginia mostly reflect tool making, tool maintenance, and the hunting and food processing practices associated with them.[14] Indians developed new tools to fulfill new functions, such as stone milling equipment used for greater utilization of plant foods. Archaic people in southwestern Virginia used manos, a handheld globular rock, to grind food substances against flat rocks called metates. Indians also began to make more refined types of stone tools resembling axes and adzes. Additionally, they manufactured stone weights fitted for a new weapon called an atlatl, which utilized leverage through a simple yet highly effective hinged device for throwing spears. Finally, they sometimes developed specific tools in particular areas which reflected local uses, such as nut harvesting, fish weir production, seed processing, and forest clearing.[15]

Around 3000 BC, Archaic people in various parts of the eastern United States began practicing a new subsistence pattern involving annual migrations between summer and winter camps, thereby utilizing seasonal resources.[16] The Grayson County area of southwestern Virginia reflected a regional variation of this new development in which flood plains and uplands grew interrelated. Here, Indians established their more substantial hunting camps in low-lying areas, from which they traveled into the mountains to smaller, more transient camps.[17]

The oldest known aboriginal occupation in southwestern Virginia occurred between 8240–7440 BC at the Daughtery Cave in Russell County. Indians living at this rock shelter appeared to have been transient hunters.[18] As mentioned, shellfish represented a new food source during the Archaic Period, and the Daughtery Cave site has revealed the earliest known evidence of shellfish consumption in the western Virginia highlands. A growing reliance on shellfish may have contributed to the progressively stationary behavior that characterized the Archaic Period.[19] Certainly more settled activity and accompanying cultural development occurred around 1000 BC, distinguishing an entirely new cultural tradition, commonly called the Woodland.

Woodland Period (1000 BC–ca. 1607, European Contact)

As the Archaic Period evolved into the Woodland Period, Indians developed North America's most sophisticated prehistoric culture. This culture reached its climax in certain locations with the Mississippian tradition, which began around AD 700 or 800 and also continued until European contact. The Mississippian cultural tradition itself became complex enough to generate local variations, and several of these traditions, including the Pisgah, Dallas, and Fort Ancient, directly or indirectly influenced prehistoric culture in southwestern Virginia. Where Mississippian influences were *not* felt in eastern North America the relatively less elaborate Woodland tradition persisted. Thus, depending upon the location, the first Europeans encountered either a Woodland or Mississippian people, with numerous unique local cultural idiosyncrasies.

After about 1000 BC, the eastern woodlands saw the rise of numerous politically autonomous and economically self-sufficient groups that rose in population, engaged in more intensive horticulture, and began to exchange tools, pottery, ornamental items, and other trade goods with neighboring groups. Eastern Native Americans, particularly around the Mississippi and Ohio River valleys, settled along floodplains, where they began an unprecedented level of food cultivation, especially the much-noted triad of maize, beans, and squash. Horticulture and its associated sedentary behavior inspired profound cultural changes, both locally and in regard to exchange with other Indian groups. Indians made more use of pottery, traded "wealth items" (copper-ware, shell, and mica jewelry and ornaments), developed new tools, used tools more extensively, performed fairly elaborate burials, and developed (in some cases) relatively complex political systems, commonly called chiefdoms.[20]

Archaeologists have found sites reflecting Woodland Indians' hunting, gathering, and plant-cultivating Woodland culture throughout the western Virginia highlands.[21] As in other places in the eastern United States, palisaded villages, circular dwelling places, flexed burials, triangular projectile points, various distinctive ceramic styles, and horticultural evidence characterize Woodland Period occupations in southwestern Virginia.[22] Woodland sites range in size and complexity from single family units to larger, more diverse sites that reflect a high degree of cultural interaction with neighboring peoples.[23] A crucial development of this era in the eastern United States lay in a much more sophisticated pottery manufacture involving the first fired ceramics, which were capable of withstanding high temperatures and drastic temperature changes. Such a capability allowed Indians to engage in more sophisticated cooking activities, particularly the cooking of starchy seeds and the leaching of acorns. In eastern Kentucky, such new food processing technology revolutionized Indian use of the forest cover, from about 15 percent in 2000 BC to more than 80 percent a thousand years later.[24]

The first fired ceramics appeared in most of Virginia by about 1200 BC. This new technology likely arrived in eastern Virginia from the Indian peoples of coastal South Carolina and Georgia. Evidence of this newer pottery, however, does not appear in southwestern Virginia until three hundred years later, when Swannanoa ware, a sand- or grit-tempered pottery arises, apparently through exchange with southern peoples.[25] Archaeologists categorize and attempt to trace subsequent ceramic types through their differing surface designs and their tempering medium. Indians of the southern Appalachians used devices such as fabric, cords, and carved wooden paddles to decorate pottery, as well as a variety of materials, including shells (mostly periwinkle and mussel), sand or grit, and limestone to temper their ware.[26]

Ceramic types help archaeologists determine different time periods, cultural groups, and possible exchanges between groups. As Sebert Sisson wrote in regard to the Pot Rock Cliff Shelter in Carroll County, Virginia, "the relatively great amounts of pottery, with the evidence of type changes through time, prove that the shelter was extensively used during Woodland times."[27] Pot sherds found in southwestern Virginia help reveal much about the complex story of cultural interchange, particularly in regard to localized peoples and influences from neighboring groups, or the lack of such cultural interchange. For instance, certain sites in Lee County have revealed pottery types distinctly associated with the Dallas and Pisgah cultures (local, distinct Mississippian cultural variations) to the west and south. On the other hand, the nearby Crab Orchard Site in Tazewell County divulged almost only indigenous-type pottery remains, reflecting little or no interaction with Dallas or Pisgah peoples.[28] The Brown Johnson Site in Bland County produced similar conclusions where the ceramic evidence was relatively meager but nevertheless clearly reflected native pottery as opposed to outside influences.[29] On the other hand, the Flannery Site in Washington County contained a varied collection of ceramic artifacts that indicated early indigenous Indian occupation, followed by later outside influences.[30] Obviously, pot sherds are among the most important artifacts that archaeologists find.

During the Middle Woodland Period (ca. 500 BC–ca. AD 900), the use of Indian corn or maize spread throughout the eastern United States, bows and arrows replaced spears, and the first socially stratified cultures arose among aborigines, particularly in two cultural areas. Beginning around 500 BC, the Adena culture developed primarily in southern Ohio but extended into all adjacent states until around AD 700. Indians of the Adena culture built the famous burial mounds of the Ohio Valley region and beyond that remain in evidence across the landscape today. Around AD 900 the Hopewell cultural tradition also arose in southern Ohio, overlapping to some extent with the

Adena but extending over a much broader area of eastern North America in what is sometimes called the Hopewellian Influence (or Interaction) Sphere. Within only a couple of hundred years the Hopewellian Interaction Sphere reached as far north as Montana and Michigan, as far south as the Gulf Coast, and as far east as the Appalachian Mountains of North Carolina and Virginia. Increased production of maize and other cultigens introduced from Mesoamerica as well as indigenous domesticates provided an economic basis from which Indians all over eastern North America began interchanging ideas, raw and manufactured goods, and various cultural practices.[31]

The Middle Woodland Period in Virginia reveals characteristics typical of a broader eastern North America trend as well as unique local traits. In general, Indians of the eastern forests continued to settle into an increasingly settled lifestyle accompanied by population increases and a greater degree of social stratification. Various groups became more territorial and, by extension, began to develop regional cultural characteristics.[32] Indians in southwestern Virginia manufactured a distinct style of ceramics that apparently reflect some degree of cultural interaction with other southern mountain aboriginals. Around the Blue Ridge area, however, ceramic traits seem to resemble those found to the east. Social characteristics probably ranged from nonstratified societies typical of earlier eras to the more sociopolitically complex relationships that were developing among other eastern Indian groups of the period. Additionally, southwestern Virginia's Middle Woodland sites reveal explicit differences in artifacts depending on elevation. Thus, the distinction between valley settlements and highland hunting camps continued.[33]

Horticultural practices intensified during the Late Woodland Period in Virginia (900–1607), when Indians began employing slash-and-burn techniques, including the girdling of trees. As with fired ceramics, horticulture arrived somewhat later in the western Virginia highlands compared to the eastern piedmont and coastal areas. Archaeological evidence reflects a late, rather than early, Woodland horticultural configuration in the Virginia mountains. But during this late Woodland Period, floodplains and associated plant cultivation became the sites of major, semipermanent occupations.[34] This situation is reflected quite prominently in the Crab Orchard Site in Tazewell County near the headwaters of the Clinch River. Indians built the 400-feet-long, palisaded Crab Orchard village around the year 1500. Inside the palisade, archaeologists discovered circular homes arranged in rows, about 180 burials, and various storage pits. They surmised the population of the village at about 400 people. Outside the palisade, the Indians had a large, semi-subterranean council house, and beyond this complex, along the Clinch River, they cultivated food.[35]

To date, the Crab Orchard Site remains unique in southwestern Virginia, for no other site displays its particular arrangement of circular dwellings, palisade, council house, and mixture of two distinct ceramic types (shell-tempered, plain-surface, and Radford Series). The Crab Orchard Site becomes especially significant when considering, as in much of southwestern Virginia, the absence of indigenous inhabitants during the subsequent Contact Period (that is, after 1607). With such a dearth of documentary evidence, the Crab Orchard Site is so far the sole and most important resource for understanding late prehistoric lifeways in the Tazewell County area.[36] The Crab Orchard complex also to some extent reflects part of a broader prehistoric highland culture in Virginia.

During the late Woodland Period, distinct natural areas developed in Virginia based on the various zones of geography, such as the coastal plain, piedmont, Blue Ridge, and Appalachian plateau regions. Distinct natural areas arose when the particular traits of an area's natural features began to distinguish markedly the local Indian population, now increasingly based in specific locales. In addition to this growing indigenous Indian population, southwestern Virginia became an area traversed by neighboring groups (also growing in cultural distinction), particularly along river ways. Indians from eastern Virginia used the James, Dan, and Roanoke rivers to reach southwestern Virginia, often for hunting purposes and concurrent weapon manufacture. Indians from various Mississippian cultural areas to the north, west, and south traveled into southwestern Virginia along Kanawha, New, Tennessee, and Holston rivers.[37] It was up from the Tennessee River that Virginia received its most significant Mississippian influences.

Beginning around AD 700, Mississippian culture began to replace or evolve out of the Woodland culture in the valley areas associated with the Mississippi, Illinois, Tennessee, and Ohio rivers. An enhanced strain of corn, the addition of beans, and an overall increase in horticultural production contributed to an increase of population among Mississippian peoples. Mississippian culture and its horticultural food dependence initiated the earliest known effects on Indian social hierarchy by creating stratified social and political patterns and may well have fostered a class-ranked society in areas as distant as southwestern Virginia.[38] As the core of Mississippian culture developed in the lowland areas, populations on the fringes of these areas moved into higher, more remote river valleys, such as the Powell in far southwestern Virginia. Since the more remote valleys were smaller, had less arable land, a shorter growing season, and generally offered fewer natural resources, Indians living there tended to develop communities of somewhat less elaborate social and political structure compared to the core Mississippian locales.[39] Furthest southwestern Virginia, around Lee, Scott, and Wise counties, experienced just such a later developing influence.

Southwestern Virginia's Mississippian traits, as seen in items such as ceramics, reflected distinct elements of localized Mississippian traditions that had developed immediately to the south, west, and north. Around AD 1000, the Pisgah variation of the Mississippian culture arose in western North Carolina, as did the Dallas cultural tradition in eastern Tennessee and the Fort Ancient tradition in southern West Virginia and eastern Kentucky. The Fort Ancient influence possibly entered southwestern Virginia through the Kanawha and New River Valley route, and the Dallas and Pisgah definitely did via the Tennessee River.[40]

The Mississippian mound builders represented some of the most advanced of late prehistoric peoples, and without question Lee County's truncated pyramids represent Virginia's most outstanding Mississippian sites. Lucien Carr, of the Peabody Museum, excavated the Ely Mound in Lee County during the early 1870s and was among the first archaeologists to identify southwestern Virginia's burial mounds within the greater Mississippian cultural complex.[41] Carr discovered various burials, projectile points, Indian corn, pottery, horn implements, "small disks of stone, pottery and hematite [and] shells of *Melania*, converted into beads." Two remarkable Ely Mound artifacts included a chunky stone associated with an Indian game and a weeping eye ornamental shell pendant. The Ely Mound burials generally corresponded with similar customs of the later Cherokee and Chickasaw tribes.[42]

Where the Dallas variation of Mississippian culture spread of its own accord to the area around the Ely Mound, other Indians of southwestern Virginia borrowed various aspects of the Pisgah variation. Therefore, where Lee County reflects "true" Mississippian cultural, biological, and ethnic traits, Washington and Smyth counties, further to the east, reflect an indigenous non-Mississippian population adopting Mississippian traits. A number of rock shelters found in far southwestern Virginia also indicate contact with Pisgah peoples through trade items such as marine shell and mica fragments.[43] Since these Indians voluntarily borrowed and interacted with the Pisgah culture and people, the interchange did not entail an invasion from an outside group and thus represented gradual and elective cultural development.[44] Exchange among Indians participating in the Mississippian culture also involved the Southern Cult phenomenon.

Archaeologists and anthropologists have identified the Southern Cult by an array of religious, ornamental, and other types of artifacts that many southeastern U.S. Indian groups traded among themselves. The climax of this cultural exchange phenomenon seems to have occurred sometime during or shortly after the 1100s. Native Americans in southwestern Virginia definitely participated in the Southern Cult, and artifacts recovered in the Rich Valley area appear to reflect interchange or influence from both the ancestors of the Cherokee to the south and the antecedents of the Siouan Indians to the east.[45] The proto-Cherokee immediately south of Rich Valley would have been part

of the Pisgah version of Mississippian culture, while the late prehistoric eastern Siouan Indians would have practiced a Woodland culture. And for all the
magnificence and prominence of the Mississippian cultures in the southern
highlands and to the west, the peoples of the Atlantic coastal and piedmont
areas certainly developed significantly during the final stage of prehistory.
Their influences also figure into southwestern Virginia's aboriginal story.

The easternmost area of southwestern Virginia involved a notably different prehistoric cultural complex focusing on the Roanoke and James rivers.
Here, too, rock shelters constitute some of the most significant prehistoric
finds. Geology fundamentally dictated local stone tool manufacture, and
where western rock shelters reflect the surrounding sedimentary geology
and yield almost only chert artifacts, eastern rock shelters' proximity to the
igneous Blue Ridge render mostly quartz and quartzite, and some jasper artifacts.[46] It was precisely stone that the noted Powhatan Indians of extreme
eastern Virginia lacked in any sufficient local amount, and thus they had to
rely upon western rock sources—either directly or through trade—for most
of the stone tools they wished to make or use. The Blue Ridge area abutting
the foothills and piedmont also served as a hunting and fishing area for Indians living to its east, likely the Siouan peoples of the Monacan alliance who
periodically rivaled the Algonquian Powhatans for control of piedmont and
foothill territory. Within the cultural exchange, eastern sources provided
southwestern Virginia with Southern Cult artifacts, while exotic ornamental items such as mountain lion claws ended up among eastern Indians and
probably ultimately derived from the highlands.[47]

Obviously the exchange of goods traveled in many directions among
many groups of Indians through a number of direct and indirect channels
that superseded local political antagonisms. Southwestern Virginia's prehistory reflects numerous attributes during various periods of the distant past.
Aborigines traversing and living in southwestern Virginia selected and developed particular cultural traits and created a unique hybrid culture in itself.[48] Singular cultural traits aside, Indians living in the western Virginia
mountains impacted their natural environment, as all humans must. This
impact generally strikes a contemporary observer as minimal. A relatively
sparse population combined with a worldview not oriented toward concerted
natural resource exploitation largely explains this minimal impact.

Environmental Prehistory

Prehistoric southeastern U.S. Native Americans may have contributed to the
extinction of various animal species through their hunting practices over
very long periods of time, a time period that also experienced greater climatic
changes that naturally altered ecosystems. In certain locations, Indians certainly made significant changes to the natural environment, particularly

through deliberate burning of grasses or forests or through massive bison kills via stampeding over cliffs (a practical method for obtaining meat before the horse arrived). As Shepard Krech has reminded us, it remains important to resist re-creating Rousseau's romanticized version of Native Americans as idyllic and idealized children of nature.[49]

Given the situation in the greater southern environment, southwestern Virginia in particular probably remained one of the areas least affected by prehistoric human activity. As a fringe area of mostly seasonal hunting and fishing throughout most of prehistory, and apparently supporting only a few permanent or semipermanent camps fairly late in the prehistoric record, it stood to experience some of the most dramatic transformation following Euro-American occupation. Considering this radical alteration, the prehistoric Indian environmental worldview becomes all the more intriguing, for it involves both a particular landscape and a regard for it, both altered or lost long ago.

Since the 1970s certain writers have created an image, in their own environmentally correct likeness, of prehistoric Native Americans as the first "ecologists."[50] This most recent version of the Noble Savage myth has unfortunately obscured what little may be concluded about prehistoric Indian attitudes toward and practices affecting the environment. This obfuscation is doubly unfortunate when considering the actual contrast between Indian and Euro-American attitudes toward their natural surroundings during the sixteenth and seventeenth centuries, not to mention the environmental conceptual difference between early European migrants and successive American generations.[51]

Native American interaction with the natural surroundings in the southeastern United States constituted a mixture of pragmatic manipulation and carefully ritualized reverence, respect, and awe.[52] As with all earthly creatures, the Indians' survival depended upon exploiting their environment, though this exploitation sometimes highly impacted the land in proportion to aboriginal population sizes. Hunting, fishing, and cultivating food sometimes involved setting deliberate fires, girdling trees, and utilizing natural poisons. Certainly, prehistoric and especially historic-era Indians contributed to the decimation if not outright extinction of certain animal species through overhunting.[53] For both Euro-Americans and aborigines, modern concepts of overhunting appear to be reactions to the concerted animal killings of the seventeenth through nineteenth centuries.[54]

Long before Native Americans in the southeastern United States began cultivating food, their hunting and fishing practices had impacted the environment. As previously mentioned, the Archaic Period began with a dramatic change in fauna, perhaps partially attributable to human hunting practices. In the later prehistoric and early historic eras, of course, Indian subsistence hunting continued. A sense of survival combined with a religious respect for the natural world stemming from animism and totemism generally characterized fishing and hunting practices. Indians ingeniously employed a number of sophisticated methods for obtaining wild food.

In addition to spearing, hooking, netting, and other common methods of catching fish, Native Americans used organic poisons derived from various indigenous plants to stun the creatures for easy capture. Buckeye nuts, walnut bark, and devil's shoestring (*Tephrosia virginiana*) for poisoning or stunning fish were available in the Virginia mountains, and Indians may well have used them there for this purpose.[55] Southeastern Indians also commonly used fire to flush game into a killing ground.[56] Around the southern Appalachian region, aboriginal fire—both as a hunting device and as part of slash-and-burn horticulture—had the most drastic effect on the environment. Indians also girdled trees to create more enduring cleared areas that attracted game. The slash-and-burn practices associated with Woodland Period horticultural activity intensified aboriginal use of fire, and a general increase of Indian population throughout the prehistoric era also intensified human impact upon the environment.[57]

Even though fire obviously dramatically altered an area's ecology, it did not do so in a completely destructive manner. Fires caused by lightning long preceded human-induced fire, and various flora evolved survival mechanisms that actually came to *depend* upon fire for survival. So, as scientists working during the past half century have increasingly appreciated, fire serves important ecological functions and facilitates the growth of certain plant (and by extension consumer-animal) species, even at the detriment of others. Indians readily employed fire as a practical tool.

The overall low aboriginal population and their relatively conservative hunting and fishing practices put their use of fire closer to the natural lightning-induced-fire side of a spectrum, the other extreme of which came to be defined by nineteenth- and twentieth-century Euro-American practices of deliberate burning in association with more intense agricultural activities. Smaller population numbers would probably mitigate the environmental impact of almost any human group. So the more interesting aspect of a particular group involves specific cultural orientations that include conscious decisions regarding the surrounding world and its resources.

Prehistoric Environmental Worldview

Beyond the available details involving aboriginal environmental interaction, the much trickier question arises concerning Indian *conceptions* of their environment—their "environmental philosophy," if the term may be used. The fact is, exactly how prehistoric Indians of the southern Appalachians felt about the land will never be known. The closest approximation of their perspective may only be approached through several filters, where time (in itself a culturally loaded concept), evolved tradition, and various unavoidable subjective interpretations modify ancient perspectives.[58] Through contemporary Euro-American observations, modern anthropology, James Mooney's late-nineteenth and early-twentieth-century anthropology, and latter-day Indian mythology (gathered from old people in a language other than their native tongue and through a medium, the written word, novel even among the Cherokee, with their Sequoyan syllabary) plenty of speculation, imagination, guesswork, fantasy, and romanticism may be generated. In the case of southwestern Virginia, an additional geographic barrier arises in that no prominent cultural tradition comparable to the Cherokee actually permanently occupied the territory in question, at least into the historic era. But given the proximity of the Cherokee, whose ancestors probably traversed and lived in southwestern Virginia at various times during their prehistory, some extrapolation of their environmental worldview seems worthwhile. Similarly, the Shawnee and Tutelo, who also had some involvement with southwestern Virginia, may offer important variations upon an approximation of the aboriginal perspective. Finally, despite all the historic era's modifications that affected later versions of Indian traditions, it seems reasonable to expect a certain amount of continuity stemming from a fundamentally distinct regard for the environment ultimately rooted in prehistory.[59]

Amidst all their activity, Indians were intimately aware of their natural surroundings, as could only be expected from a people who lived in such daily close proximity to it, and whose daily subsistence depended directly upon it. Like other nonliterate peoples whose intellectual faculties are used to other ends (such as memorizing literally hours of detailed oral tradition), the southeastern Indians were experts on the details of their landscape. They drew excellent maps and could recount intricate details such as individual trees next to specific bends in a particular river, sometimes hundreds of miles from their home base.[60]

The aborigines' environmental intimacy contributed to a view of their world steeped in natural forces and phenomena, such as weather features and animals. In many ways, their perspective was typically animistic and similar to other animistic cultures, such as the Shinto of the Japanese. Animism entails the regard of all objects—plants, animals, rocks, water—as possessing spiritual qualities.[61] But contrary to recent romantic stereotypes, the spectrum of southeastern U.S. Native American worldview ran the gamut from deep veneration to fierce hatred. Among snakes, for instance, the Cherokee greatly revered rattlesnakes but absolutely despised spreading adders and copperheads. They would not eat birds of prey or any carnivores (omnivorous black bears excepted), based on concepts of cleanliness and the idea that animals that ate meat were unclean.[62]

The Cherokee outlook was also generally anthropomorphic in that Indians assigned human qualities to nonhuman entities, such as arranging animal groups in totems resembling human family or clan groupings.[63] Thus Little Deer became a deity of sorts and acted as chief of the deer tribe. They believed that other animals, such as bear, were really human underneath a guise of animalness. What might be called the Indian conservation ethic was obviously interwoven with their animistic outlook.[64] To a certain extent, it was out of religious respect for a deity such as Little Deer that the Cherokee would avoid wanton killing of the deer species. John Lawson encountered a similar ethos in 1700 among the mountain Indians northwest of High Point, North Carolina. He wrote, "All the *Indians* hereabouts carefully preserve the Bones of the Flesh they eat, and burn them, as being of Opinion, that if they omitted that Custom, the Game would leave their Country, and they should not be able to maintain themselves by their Hunting."[65] (That aborigines would soon participate in the industrial deer hide trade reflects how powerful acculturation forces can be.)

The particular caution that the Cherokee associated with killing wolves mixed their need to eliminate a competing predator with their special respect for the wolf and thus required a specialist properly ordained for such an act. On the other hand, their general aversion for killing snakes fell more purely into the religious realm and represented an interestingly obverse taboo compared with the Judaic-Christian fear of serpents.[66]

A generally animistic outlook certainly extended far beyond animals and encompassed seemingly every aspect of the world around them. Rivers figured centrally in this worldview. Beyond the obvious facility as transportation corridors and sources of fish and shellfish, southwestern Virginia's rivers might be considered from an Indian's spiritual perspective. The Cherokee assigned anthropomorphic qualities to rivers, thinking of them as giant men whose heads lay high in the mountains and whose feet stretched down into the lowlands. Daily purification in such waters became profoundly important. James Adair, a trader and resident among the Cherokee from 1736 to 1743, wrote that they were "strongly attached to rivers,—all retaining the opinion of the ancients, that rivers are necessary to constitute a paradise."[67]

In some ways the Cherokee regard for rivers captured their entire environmental perspective, with all its multifaceted aspects of utilitarianism, animism, and religious purification. Beyond this, something might be said for the "energy" surrounding bodies of water that has always captivated all peoples in one form or another. In ancient religions, this led to what scholars have come to call "numina." Numina (plural for "numen") are the deistic extensions of animism, wherein people associate a divine being or force with some natural phenomenon, such as rivers, trees, or rocks in the forest, and certainly mountains.[68] Numinous energy, of course, supersedes aquatic biology, the distinct aromas that arise from such ecosystems, or even the physical details of such environments. This phenomenon, perhaps more conveyed to human instinct or emotion rather than human intellect, has long been the domain of mystics and artists, and really remains impossible to pin down logically. But this limitation does not or should not detract from its importance. Perhaps the most that can be said is that what many contemporary Americans might now sense in admiring rivers, the Indians sensed at least in equal measure and, with all romanticism or idealization aside, probably to a significantly greater degree.

If aborigines associated numina with rivers, they certainly did so with mountains. For the Cherokee the highlands were the domain of bears (a previously human race who had departed for a different existence), two races of spirit peoples (the Nunne'hi and Yunwi Tsunsdi), and Tsul'kalu, the "Slant-Eyed Giant" who was keeper of the game.[69] All had supernatural powers. The bears, the Cherokee believed, were a race akin to themselves who had chosen to live in the forest. The spirit peoples were fond of music and dancing, would aid lost Indians in the mountains, and even help the Cherokee against other Indian enemies encroaching in the highland territory.[70] Bald mountains free of trees held special significance for the Cherokee, who often dreaded them as taboo places where shamans would banish diseases. Mountain gaps also carried significance as the temporary abodes of disease in the process of banishment, and a series of mountain gaps signified progressive stages in the banishment ceremony.[71]

Prehistoric southeastern Indians, finally, did not share the uniquely post-seventeenth-century Western attitude of "progress" in regard to their environment or anything else. Despite such dramatic innovations as the gradual adoption of plant cultivation over a strictly hunting, fishing, and foraging lifestyle—or the invention and utilization of such weapons as the atlatl spear thrower or bow and arrow—Indians in and around southwestern Virginia nevertheless continued to live a highly diverse and thus ultimately less environmentally disruptive existence.[72] Their behavioral modifications, therefore, did not resemble linear change as much as it did lateral change. In this sense, the Native Americans shared a generally nonlinear outlook commonly found among many non-European peoples.[73] Thus, in terms of worldview, the American Indians probably could not have encountered a people more diametrically opposed to them than the Europeans. These contrasting peoples' differing actions toward the natural environment and the ultimate results of these behaviors reflected, in part, this greater cultural clash.

Shepard Krech makes some important points, both specific and general, about the Rousseauian renaissance in idealizing Native Americans and their interaction with the natural environment.[74] Aside from the rebirth of Noble Savage romanticism among white American middle-class Protestants, another misleading influence likely arises from native religion itself. Here a gulf often exists between the differing approaches in presenting mythology and history (or prehistory). Mythologists like Mircea Eliade or Joseph Campbell tend to describe religion in its conceptually pure form, whereas historians and archaeologists are likely to look for actual behavior, and behavior as it contrasts from the pure form. Conceptually, aboriginal religion seems indeed designed for perfect harmony with the natural environment. But just as a naive student might be tempted to conclude that medieval European Christians were all pacifists who accepted poverty voluntarily, so a student of Indian religion might be tempted to conclude that all natives were flora and fauna lovers living in harmony with nature. Prehistorically, how far Indians deviated from their worldview in its pure conceptual form will likely remain mysterious. In many ways, the fact that they were intimately involved with and aware of their natural environment simply appeals to common sense.

Chapter 2

Agricultural Settlers

From the Atlantic shore various Euro- and Afro-Americans and their descendents tended to follow the Great Valley of Virginia in a west, southwest direction toward Tennessee. The Great Valley and some of its parallel linear valleys in southwestern Virginia offered the easiest travel along well-established footpaths that animals and aborigines had traversed for millennia. The valley floors were also the most conducive to agriculture, and thus its forests were the first to fall.

At this point named Indian groups begin to emerge.[1] The groups that bear mentioning in relation to southwestern Virginia were, in order of significance, the Cherokee, Shawnee, Tutelo, and Yuchi. From the sixteenth through the early nineteenth centuries, the Cherokee Indians constituted one of the largest and one of the most enduring tribes in what became the southeast United States, with permanent villages established from northern Georgia to western North Carolina. They roamed southwestern Virginia and eastern Tennessee as part of their traditional hunting grounds. And as the southeast's greatest tribe, the Cherokee easily posed the most substantial resident in proximity to southwestern Virginia.

At the outset of the historical era, the Cherokee numbered around 20,000 and lived in concentrated villages spread over an area of about 15,000 square miles. Their vast hunting grounds, of course, encompassed much more land, including southwestern Virginia as far east as the New River, most of Kentucky, eastern and central portions of Tennessee, and parts of northern Alabama. Other tribes, such as the Creeks and Shawnees, also used various parts of this territory.[2] Before the pelt trade with Euro-Americans arose, Indians tended to hunt in the western Virginia highlands only seasonally. Also, considering the Indians' relatively small population numbers and their animistic respect for mammals and other living entities besides themselves, their predatory approach generally existed in a less disruptive fashion in regard to greater biological and naturalistic forces. These hunting activities had tended to sustain the aborigines century after century without extinguishing animal species. All this drastically changed as Native Americans acquired firearms and horses and began hunting animals year round to supply Euro-Americans with pelts.[3]

The Shawnee influence on southwestern Virginia is more difficult to ascertain. By the early historic period, the Shawnee appear to have lived around the Cumberland River in Tennessee. With the rise of the deerskin trade in the early eighteenth century, the Cherokee and Chickasaw drove the Shawnee north to the Scioto River in Ohio. It was from this northern base that the Shawnee periodically raided southwestern Virginia and the Cherokee territory during the years surrounding the French and Indian War. They appear frequently in the colonial record enumerating Indian raids on southwestern Virginia's white settlements and, indeed, helped temporarily reverse agricultural settlement for a time prior to the conclusion of this noted conflict. Around midcentury some Shawnee unsuccessfully tried to reestablish themselves around their former Cumberland River home before banding together with the Creeks.[4] Much more itinerant than the Cherokee, they appear to have represented a relatively more sporadic, though sometimes quite intense influence on southwestern Virginia. Being a highly mobile people, the Shawnee did not have a sedentary sphere of influence comparable to that of the Cherokee. They may have been among southwestern Virginia's prehistoric hunters and fishers, and during the early historic era southwestern Virginia seemed to function as something of a buffer zone between the Cherokee and Shawnee, as the Shawnee sphere of influence then centered around part of Kentucky and the Ohio Valley.[5]

In addition to the famous Cherokee and Shawnee Indians were lesser known groups who occupied the area around the Roanoke and New River valleys on the eastern edge of the focus area, and along the western edge into the mountains of Tennessee. The tribe of Tutelo or Totero were a small group of Indians who apparently abandoned their late prehistoric period home on the eastern edge of Virginia's mountainous region during the first half century following European contact.[6] The Yuchi Indians apparently lived in eastern Tennessee only after the historic era began, and then only as Cherokee and/or Creek groups subsumed them into their larger, more powerful groups, perhaps sometime during the Woodland/Mississippian prehistoric era or, at the latest, during the early historic period. Certainly by the late seventeenth century they lived in eastern Tennessee with the large Creek tribe, and from that locale they may have fished and foraged into southwestern Virginia, as well as participated in the deerskin trade.[7]

Heightened Indian hunting and increasing Euro-American treks into southwestern Virginia characterized the century following Batts and Fallam's 1671 journey, the first recorded far westward trek into Virginia's mountains. Such exploration and trade motivated early whites, who used rivers and Indian trails to journey into the frontier.[8] For the time being, however, the Indians remained the principal travelers and hunters in the southern highlands. The Cherokee were the principal deer hunters in southwestern Virginia during the bonanza years of the hide trade, from about 1730 to 1750. But the early- and mid-eighteenth-century deerskin trade transformed southwestern Virginia into a major killing territory as lands closer to the Cherokee villages soon lost their substantial animals populations. Already adept predators, the Cherokee escalated their efforts in order to trade deerskins for guns, ammunition, cloth, and ornamental objects. The Euro-American middlemen traded and bought hundreds of thousands of hides from the Cherokee alone and hauled them by wagons and mule and horse packs to the coast (mostly to Charleston, South Carolina), where they shipped the product to English and continental European markets.

Between 1700 and 1715 these shipments included nearly 900,000 dressed and undressed deerskins. Beaver pelts figured a distant second place at 9,841, with 8,159 fox furs coming in third. The Indian–Euro-American trade also supplied elk, mountain lion, otter, moose, and raccoon pelts.[9] The deerskin trade, which began during the latter half of the seventeenth century, had reached such a frenzy of activity by the early eighteenth century as to precipitate numerous conflicts between Indians and Euro-Americans, as well as among Native American groups themselves. Rival tribes intensified their traditional enmity within this new commercial context, and tribes contending for the same hunting territory may have even developed new rivalries over competition for skins.[10]

The southeast's pelt trade peaked around the middle of the eighteenth century, by which time the process had devastated the area's deer population and irrevocably altered the Indians' culture. In the meantime, European-introduced diseases such as measles, typhus, and especially smallpox cut the Cherokee population in half. The smallpox epidemic of 1738–39 was particularly devastating. One consequence of reduced Cherokee population numbers became a loss of dispersed activities and a greater concentration of Indians in core village areas, the northernmost of which were located in eastern Tennessee and western North Carolina.[11] The Cherokee not only survived the smallpox epidemic of 1738–39, but also apparently rebounded in a major way during the following years. By this time, however, their interaction with colonists had moved from involvement in the animal skin trade to direct competition for lands. Between 1768 and 1775, the Cherokee ceded to the English all their remaining claims to western Virginia lands.[12] By 1701 southwestern Virginia's Native Americans had disappeared through introduction of European diseases, outward migration, or a combination of both. Indians of eastern Tennessee seem to have experienced a similar disappearance.[13] With the arrival and endurance of the new peoples came a new phase in environmental history.

As early as the 1650s, Virginia settlers had moved as far west as the fall line of the James and other eastward-flowing rivers. Traders regularly ventured beyond these settlements to do business closer to the mountains, with Indians such as the Monacan and Nottoway. During the last quarter of the seventeenth century, Virginians also participated in the greater southeast Indian trade, but did so somewhat haphazardly via the famous Occaneechi path, by which route they had to contend with sometimes hostile intermediary Indians. The Occaneechi path traversed the Virginia and North Carolina piedmont parallel the Blue Ridge Mountains in a southwest-to-northeast direction that stretched from modern-day Petersburg, Virginia, to Charlotte, North Carolina.[14] The first recorded exploration into Virginia's mountains coincided with this piedmont trade. The Euro-American invasion of Virginia's western lands had begun.

For the next several decades, traders and the first few agricultural settlers began pushing west into the Virginia wilderness. Frontier immigrants had to clear land, raise livestock, and plant crops in order to claim land, and this policy continued to define settlement in southwestern Virginia throughout the eighteenth century. By the 1720s farmers had established themselves above the James River Falls. During the following decades a more concerted settlement effort began to emerge. In 1748–49, the Council of the Province of Virginia authorized the creation of both the Ohio and Loyal land companies, which subsequently surveyed almost 2 million acres in southwestern Virginia for the purposes of agricultural settlement. By 1768 settlers had reached the valleys surrounding the Holston River headwaters.[15]

The end of the large-scale, organized Indian wars in the southern Appalachians in 1763 eventually spurred further settlement when the King of England proclaimed thousands of acres in land grants throughout southwestern Virginia for war veterans. The 1768 Treaty of Fort Stanwix (New York) made the Kanawha River the western demarcation of European settlement, and other treaties with various Indian tribes followed, constantly pushing the frontier westward.[16] Despite continuing periodic conflicts with Indians, Euro-American agricultural settlement soon escalated after the French and Indian War. The first people of African descent began to trickle into the mountains as slave laborers. Most newcomers traveled down the Great Valley to settle unclaimed bottom lands or push on to what would become Tennessee and Kentucky.[17] Pioneers settled the Valley itself earlier than surrounding mountains, coves, hollows, and more remote valleys. After homesteading the Great Valley, settlers followed increasingly smaller valleys deeper into mountainous terrain, and in this way they gradually penetrated the more remote regions. Beyond the Great Valley, major settlement patterns followed the New River Valley south toward North Carolina and north toward West Virginia. In Giles County, near the present-day West Virginia state line, pioneers followed Walker's Creek Valley west toward Kimberling Creek and eventually to the Clinch Valley before crossing over into Kentucky. Advancement of settlements covered this territory over a period of fifty years, until early pioneers established the first outpost in the Tazewell County area by 1771.[18]

The first white settlers in southwestern Virginia cut wide tracts of valley forests, introduced foreign flora and fauna, and initiated the first concentrated slaughter of wolves. Food crops varied with geology, topography, and changing transportation facilities, but generally bottomlands proved conducive to cultivation of corn, wheat, rye, and oats. Hillsides provided mostly grazing, and livestock husbandry became a very significant component of southwestern Virginia agriculture. The mast-filled forest of acorns, hickory nuts, and especially chestnuts contributed to making the southern Appalachian region one of the most important hog-producing areas in the United States, as domesticated swine roamed the woodlands in the manner of livestock foraging on free range.[19] Farmers also raised numerous cows and sheep. The general advantage of raising livestock lay in their mobility. Stock could be sold and driven to market without benefit of improved roads. Difficulty in transportation also encouraged distillation of grain, since an acre of rye or corn transformed into liquor was comparatively easy to move.

Prosperous settlers regularly offered rewards for wolf heads.[20] In addition to professional bounty hunters, farmers regularly killed wolves. By the late nineteenth century, few, if any, wolves remained, and mountain lions had also become scarce. And though not strictly predators, black bears also plummeted in population with the arrival of white settlement and agriculture. Other animal populations suffered tremendous losses, such as the wild turkey and the Carolina parakeet, the latter eventually hunted into extinction. Where passenger pigeons once darkened the sky with their enormous numbers and broke tree branches with their collective weight, only a few decades of Euro-American occupation all but eradicated them from their former habitat. Hunters also eliminated eastern elk and bison completely. Wild animal populations continued to diminish in general, some completely disappearing, until twentieth-century conservation efforts began helping to reestablish wildlife and their habitats.[21]

The new people radically altered the area's flora with the introduction of new plants, in what Alfred Crosby called "probably the greatest biological revolution in the Americas since the end of the Pleistocene era."[22] Euro-Americans introduced to North America the white potato, bloodwort, wormwood, peach trees, and many other species. New flora related to raising livestock particularly altered North American ecology, as the new farmers found native grasses unsuitable for grazing, and therefore introduced Kentucky bluegrass, timothy, redtop, white clover, and other grasses. Other new pastoral plants included daisies, yarrow, dandelions, and buttercup. With the proliferation of new pollen-bearing plants and the curtailing of forest land, honey bees—native to Europe but not North America—became very common in the southern highlands by the late eighteenth century. Tobacco from South America (*Nicotiana tabacum*), of course, revolutionized the colonial economy in Virginia and the Carolinas when found more suitable for smoking than the original "Indian" tobacco (*Nicotiana rustica*).[23] The introduction of new flora coincided with the destruction of other flora, not the least of which were trees. Euro-American agricultural settlement of the area during the rest of the eighteenth and into the early nineteenth century caused the removal of much of the forest, especially in bottomlands.

During the nineteenth century, southwestern Virginia charcoal iron manufacture figured prominently in state and sometimes regional industrialization. Iron production began in southwestern Virginia shortly after agricultural settlement, when charcoal furnaces and forges supplied farmers with plows, stoves, horseshoes, nails, and other agricultural and household implements. Ironmasters established the Poplar Camp furnace in Wythe County as early as 1778, and by the turn of the century they had built furnaces and forges in Craig, Grayson, and Carroll counties, and as far west as Washington and Scott counties. The Civil War brought southwestern Virginia's iron industry to its peak when the Tredegar works in Richmond became the South's major iron producer.[24] Tredegar relied heavily on iron manufacturing in the upper James River area, which became Virginia's most important iron source during the last half of the nineteenth century.[25]

A charcoal iron furnace could consume as much as an acre of woodland every day to meet its fuel requirements. By the 1870s iron manufacturers in the upper James River basin area began to decline partly because of no available fuel wood. Ironmasters had already cut the forest in that region several times over. By 1880 Virginia's other important iron-producing center, around the Wythe County area, had experienced a hundred years of intense charcoal iron production, leading to serious forest destruction in southern Wythe and northern Carroll County.[26] Southwestern Virginia's charcoal iron industry also involved some of the most frequent instances of industrial slavery. Most of these slaves were woodcutters, and woodcutters easily constituted the most numerous workers in any charcoal-fired operation.[27] Similar patterns of woodcutting labor and fuel wood consumption surrounded nineteenth-century salt manufacturing in and around Saltville, Virginia.

As early as the late eighteenth century, Saltville area residents began manufacturing salt by boiling deep well brine in kettles, and later furnaces, fueled by wood. During the nineteenth century Saltville produced millions of bushels of salt, generally ranging from 75,000 to 450,000 annual bushels, with peak output during the Civil War year of 1864 at 4 million bushels. Every bushel of Saltville salt required about six cubic feet of hardwood to render it. In the long run, this voracious wood consumption contributed to limiting Saltville's production. Around 1800, trees still covered the hills surrounding the saltworks, but by 1829, with only one well in operation, salt maker Francis Smith burned twenty cords of wood daily and transported wood from two or three miles distant. By the summer of 1838, the Reverend H. Ruffner reported "all the country near the [North Fork of the Holston] river for twenty miles, was stripped of its fine forests," and that the salt business "was beginning to languish for want of fuel." Forests grew back, of course, but the wasteful designs of the Saltville furnaces tended to deplete it more rapidly than it could regenerate. By midcentury the output of 300,000 annual bushels required the use of twenty-five teams of horses that helped haul the year's required 10,000 cords (1.8 million cubic feet) of fuel wood. At that point, the salt manufacturers' woodcutters rendered about fifty cords per acre from second- and third-growth trees that had sprung back from lands previously cut during the 1820s. In this particular wave of local deforestation, salt makers had cut about 3,000 acres of the surrounding forest and available fuel wood was again growing scarce, as it had during earlier waves of forest clearing.[28]

During the Civil War, the Virginia and Tennessee Railroad hauled enormous quantities of wood into Saltville. For example, during just two and a half months in 1864, between mid-August and the end of October, the railroad hauled almost 8,000 cords to the saltworks. But despite such shipments, the train simply could not meet Saltville's tremendous fuel demands during the Civil War. Local legend has it that lines of wagons waiting to receive salt were loaded with much-needed wood for firing the furnaces, and, in fact, the salt makers required a wagonload of wood as part of the payment. Cash alone would not suffice. Certainly the Virginia and Tennessee Railroad hauled many thousands of cords of wood to Saltville and left equally laden with salt. In 1870 the Holston Salt & Plaster Company kept some 18,000 cords on hand and continued to use wood fuel into the 1880s. In fact, in 1874, Saltville's seven furnaces consumed sixty-five cords of wood daily. In 1880 alone, Virginia salt making (which Saltville then dominated) consumed 540,448 cords, at which time Saltville salt manufacturers reported some $36,000 in annual fuel wood expenditures. Naturally, at this point the railroad brought in almost all this wood from eastern Virginia, as most of the forest around Saltville had long been cut, some of it many times over.[29]

Chapter 3

Turnpikes and Romance in the Mountains

As the American wilderness began to diminish, the American people began to cherish it. With Indians subjugated and predatory animals greatly reduced, the wilderness no longer seemed so threatening. However, it remained rural and comparatively undeveloped and thus greatly appealing to those who lived in the Atlantic seaboard's growing cities. The southern highlands also appealed to the southern plantation elite. Growing miles of turnpikes and hotel accommodations, as well as stagecoach and horse maintenance facilitated America's first wave of tourism.[1] This was what scholar Edward Halsey Foster described as Americans' desire to establish a "civilized wilderness," replete with travelers' facilities and a dynamic social component.[2] Nineteenth-century travelers even created a tri-part categorization of rural scenery, which they described as either beautiful, picturesque, or sublime. Beautiful scenery was pastoral lowlands or hills, while picturesque and sublime were varying degrees of mountainous country. Tourists sought sublime lands most of all, for they had the power to unite humankind with eternity.[3]

Romanticism's appeal during the early nineteenth century grew as a reaction against early industrialism and represented a departure from the Enlightenment's heavy emphasis on logic and reason to focus on aesthetics and emotion. But without the long literary, architectural, and artistic traditions of the Old World, the Romantic era in the United States grew intimately associated with natural beauty. As early as the 1820s–40s Thomas Cole and the Hudson River School of landscape painting captured this mood in their art that glorified the American wilderness.

Romanticizing the mountains and the forest was far more than entertainment; from the outset it carried religious overtones.[4] As Cole himself wrote, "Poetry and Painting sublime and purify thought, by grasping the past, the present, and the future—they give the mind a foretaste of its immortality, and thus prepare it for performing an exalted part amid the realities of life."[5] The art that Cole referred to was, of course, oriented toward forested mountains not yet seriously altered by encroaching human activity. "Prophets of old retired into the solitudes of nature to wait the inspiration of

Illustrations such as this one epitomized the Romantic movement in southwestern Virginia. From *Virginia Illustrated* (1871).

heaven," Cole wrote, and "that voice is YET heard among the mountains! St. John preached in the desert;—the wilderness is YET a fitting place to speak of God."[6] Cole also described the simple human enjoyment of beautiful, natural scenery and the important respite such lands offered people in an age of "meagre utilitarianism" and relentless toil.[7] Cole singled out mountains as the most prominent landscape feature deserving praise and enjoyment, but also praised America's rivers, forests, lakes, and even her skyscapes.[8] Finally, Cole concluded that Americans were "still in Eden; the wall that shuts us out of the garden is our own ignorance and folly."[9]

Early colonists had carried a paradox of fear and idealization toward the American wilderness that had its cultural roots in the Judaic-Christian tradition. Biblical attitudes toward the wilderness included both reverence for wild lands left over from mythic visions of the Garden of Eden, but also fear of wilderness as a source of satanic temptation.[10] This cultural baggage, along with the realities of possible starvation, the threat of large carnivores and potentially hostile Indians, became posited in the early American mentality. A new phase in American environmental regard arose with the first elements of the new nation's economic prosperity.[11]

Beginning in the late eighteenth century, privileged Europeans and the first group of Americans prosperous enough to travel in the back country at their leisure, then return to eastern cities and write about their experiences, initiated the first romanticization of Virginia high country. This paradoxical aspect of humans in relation to the environment has, to some extent, characterized American environmental history ever since. A certain amount of economic prosperity generated, in part, by natural resource exploitation gives rise to conservation of (if not reverence for) those very natural resources. During the nineteenth century, these natural resources involved the mountainous eastern United States in general. But before travelers could begin romanticizing the hardships of earlier generations, they had to await the construction of hundreds of miles of turnpikes, hotels and inns, mineral springs facilities, and stagecoach manufacture.

Stagecoach Tourism

The turnpike network that arose in western Virginia between the 1830s and the 1850s provided the means of transportation that became crucial for travelers to reach the area.[12] Early visitors felt an emotional regard for the landscape and wrote dramatic and melodramatic accounts of Natural Bridge, Natural Tunnel, the Peaks of Otter, Mountain Lake, and other prominent features. Another group of visitors traveled into the Virginia mountains mainly to reach the famous mineral springs situated in and around Greenbriar, Monroe, Bath, Montgomery, and Allegheny counties. Some went to the springs for health reasons, but many went to participate in a popular social interchange that involved polite mingling, business and political discussions, and courting rituals among the southern elite. The ultimate result was an early, concerted exchange between an inland Appalachian region and the wider world. The fortuitous coinciding of turnpike construction, the expansion of stagecoach manufacture, mineral spa development, and the Romantic movement all gave the Virginia mountains its first era of tourism.

Between 1830 and 1860, a turnpike network arose among three major regions of central and western Virginia; the piedmont, the upper Shenandoah Valley, and the more rugged area of the Allegheny Mountains north and northwest of the valley. As settlement progressed westward throughout the eighteenth and nineteenth centuries, various turnpike builders constructed new routes or improved upon existing routes over the Blue Ridge and into the Allegheny region.[13] From the east, usually by way of Richmond, travelers reached the mineral springs by one of two major routes. One ran through Charlottesville, over the Blue Ridge toward Staunton, and on to Warm Springs. The other went through Lynchburg, over the Blue Ridge to Fincastle (and later Buchanan), and on to White Sulphur Springs or Sweet Springs. By the 1850s travelers from Lynchburg could cross the Blue Ridge at several different gaps, including the scenic route by the Peaks of Otter.[14] The sheer logistics of moving so many people through the rugged topography would have been impossible without adequate roads, and travel would have remained more limited to horseback rather than what came to be the favored mode of conveyance, the stagecoach.

Between 1820 and 1845, stagecoaches dominated mail carriage and overland transportation in the early republic, preceding railroads throughout the western frontier, and persisting in mountainous areas, such as the southern Appalachians, long after railroads crisscrossed flat lands. By 1820 an entire economic network had developed around stagecoach travel in the United States, including company proprietors, stage agents and drivers, manufacturers of coaches, as well as wheelwrights and blacksmiths. Hotels also benefited from stage travel, as did boardinghouses and taverns. The very construction of turnpikes and subsequent stage traffic gave direct rise to certain roadside ordinaries and inns.[15] All of these kinds of transportation and economic activities involved themselves with western Virginia's tourist trade.

Local hunters, fishers, and farmers had always provided fresh food for travelers, but by 1832 they catered to a more structured group of local taverns

Before the advent of paved highways, pilgrimages into the mountains sometimes became arduous. From *Virginia Illustrated* (1871).

and inns. Two hostelries arose where the Sweet Springs and Price's Mountain Turnpike forded Pott's Creek in Allegheny County, which served the turnpike's horse, wagon, and stage traffic.[16] At this early interval, tourism was already enhancing the local economy and, in fact, pushing the entire area out of its previous, more strictly agricultural mode into the first stage of a developed (though preindustrial) economy.[17] The town of Fincastle acted as sort of a "gateway" community for tourists venturing further into the mountains and reflected various aspects of the tourist economy between 1825 and 1860, featuring as many as three hotels, three taverns, ten boardinghouses, and various wheelwrights, saddle makers, wagon makers, and blacksmiths who found a viable trade with the passing traffic.[18]

As journeys into the mountains from lowland plantations began to resemble annual pilgrimages, some travelers must have visited friends along the route and perhaps methodically patronized the same hotels and inns each year. During peak summer months, when the spas' quarters overflowed capacity, residents living along the springs' routes accommodated travelers with room and board. As an 1834 visitor to White Sulphur Springs recalled, "almost every house for miles on the roads leading to the springs, was thronged with persons who had been turned off at the hotel."[19]

Numerous stage lines arose in Lynchburg and Lexington offering passage over the Blue Ridge and Allegheny mountains to the mineral springs. As proprietors and the Board of Public Works formed new turnpike companies and built roads over the Blue Ridge, new stage lines and special "accommodation" coaches began offering additional routes over the mountains, tours to local scenic attractions, and passage to the mineral springs.[20] Concurrent with increased turnpike travel, local stagecoach manufacturers produced more vehicles and a greater variety of models.[21]

Despite some instances of monetary success, even eastern Virginia turnpikes traversing flat land proved unprofitable in the long run, and mountain turnpikes were certainly doomed. Inadequate capital, high maintenance costs, "shunpikes" (which skirted tollgates), poor management, canal and railroad competition, overall low traffic, and tollkeepers' salaries contributed to the turnpikes' unprofitability and eventual demise. Stagecoaches themselves were notorious for paying variable tolls, and consistently among the lowest of tolls. Virginia turnpikes collected relatively low tolls compared to northern and western states to begin with, and the special rates allowed stagecoaches often created problems.[22] The sporadic finances for maintaining turnpikes faced the deteriorating forces of vehicular traffic and, especially, inclement weather. The accounts of the Fincastle and Blue Ridge Turnpike reflect the typical problems that steep roads of the era faced. Heavy rain storms, such as those of 1843, damaged the steep sections of the road to the extent that the company had to deduct $469 from that year's tolls of $697. The company wrote to the Board of Public Works in November 1843, "The mountain portion . . . has been repeatedly washed by the heavy rains of the last summer which has rendered it indispensible [*sic*] that blasting of rock should be done . . . and other additional expenses incurred—thereby materially affecting the dividends both to the state and stockholders."[23] Nineteenth-century accounts of turnpike conditions varied considerably, and writers denounced or praised road conditions depending upon their state of repair. Travel on the western Virginia turnpikes could be quite uncomfortable or very smooth.[24]

In any case, with the advent of rail travel the old stage network began to break down. By 1854 the *Lynchburg Daily Virginian* recognized at least seven mineral springs in close proximity to the new Virginia and Tennessee Railroad. Coiner's Sulphur Springs, near Bonsack's Depot, lay within 250 yards from the track. Other springs were within half a day's coach ride from the railroad, and stage companies modified their business to accommodate the new traffic patterns.[25] Virginia turnpikes peaked in 1851, with some 3,000 miles in use, but only seven Virginia turnpike companies remained operative by 1866.[26]

In addition to the thousands of actual travelers themselves, who hailed from many parts of the nation and abroad, transitory stage drivers and sta-

The Romantic Movement coincided and overlapped with the regional stagecoach pilgrimages to various mineral springs in the Virginia mountains. From *Virginia Illustrated* (1871).

tionary local innkeepers also became cultural agents. The stage driver grew distinguished among the tavern keepers, hotel proprietors, farmers, and other people along his route as a messenger and authority on the outside world. Stage drivers delivered merchandise, interpreted the landscape and communities for his travelers, and bore city and national news for the local people. In all these capacities he acted as something of a liaison between people along his stage line and the wider world. Local innkeepers also held prominent positions as repositories of mail, merchandise, newspapers, and foreign visitors. Many acted as postmasters, and some held political offices.[27] At the very least, the dynamic early tourist trade of the Virginia mineral springs amply debunks the negative isolation myth that later arose concerning southern Appalachia.[28]

With this traffic between mountain and flat regions lies a conspicuous absence of negative depictions concerning the local people or culture. Any complaints that visitors registered in their diaries could have been the common travelers' complaints levied anywhere in the United States, and certainly anywhere on the frontier. In fact, the overwhelming descriptions that travelers left of western Virginia either focused on the springs' society or on the landscape itself and, in the latter category, mostly reflected acute admiration for the mountains. Given their limited economic contributions and their failure to alleviate sectionalism, the western Virginia turnpikes' role in this cultural phenomenon became, by far, their most positive attributes. So it was the post–Civil War color writers—not the earliest tourists—who created the negative stereotypes of the southern mountaineers.[29]

Romance

Thomas Cole naturally focused most on the scenery of his native northeastern United States, but other Romantics followed his lead and flocked to all parts of the eastern high country. The art they rendered was not always as magnificent as Cole's, but they evoked a similar religious reverence for the beautiful land. A gentlemanly tradition arose among writers (some of whom were probably women, writing anonymously) in eastern cities who ventured inland to the Adirondacks or Appalachians, where they described natural beauty in the most romantic terms.[30] Southwestern Virginia and the Virginia mountains in general became very much a part of this tradition.

Nineteenth-century America only presented the latest touring grounds in a well-established Western cultural tradition. Well back into the eighteenth century Europeans had been embarking on the Grand Tour of continental Europe and the Tour of Britain. The northern United States incorporated this tradition into the Fashionable Tour of the early nineteenth century.[31] A tour of the northeastern United States could include the wilderness of the Adirondacks, the romantic landscape of the Catskills and Hudson River Valley, the

Scott County's Natural Tunnel was one of southwestern Virginia's landscape attractions during
the Romantic Era. From *Virginia Illustrated* (1871).

pastoralism of the Connecticut countryside, and the magnificent spectacle of
Niagara Falls. Some travelers ventured south to experience the underground
cathedral-like experience of Kentucky's Mammoth Cave.[32] At one level or another, western Virginia offered all of these kinds of attractions in a more con-
densed locale, in something like a southern version of the Fashionable Tour.

The landscape surrounding Virginia's mineral springs fulfilled the de-
sires of those seeking a romantic, sometimes mystic communion with nature.
The upper Shenandoah Valley offered the pastoral beauty of farmland, while
merely a few hours' stage drive took passengers deep into completely forested
mountains. Hawk's Nest and the Peaks of Otter became designated lookouts
for these landscapes. Natural Bridge offered a spectacle for those seeking the
exotic, as did Weyer's Cave and Natural Tunnel for those seeking an "underground cathedral" experience. During the course of a single journey, several
or all of these aspects of the Virginia backcountry appealed to visitors. The
area's pastoralism particularly appealed to Dandridge Spotswood when he
wrote in 1848, "my eye surveyed the green meadows, the lovely residences,
the golden groves, the hills and mountains piled one on the other, [and] I

dispaired [*sic*] of language to delineate the spectacle."[33] In 1834 a Philadelphia doctor described the area's forested mountains with special appreciation:

> These mountains are the striking features of the route, and excite strong interest by the extensive views from their summits, of adjoining mountains, which rising in the distance, and covered with a luxuriant growth of forest trees, resemble a fine undulating grass sward, from the uniformity and thickness of the trees upon them which conceal completely the surface of the soil. The dark foliage of the pines, as you examine the distant mountains, contrasts finely with the lighter tints of the oak, chesnut [*sic*] and hickory, and is seen clustered in distinct groves in the midst of the other trees.[34]

Some pilgrims to the wilderness, and they often thought of themselves as such, made the leap from aesthetic admiration to outright religious conversion. As an anonymous visitor wrote of his experience at the Peaks of Otter:

> The dullest soul must have been impressed by the wonderful show [of scenery]. My own susceptibilities, ever alive to the glories of nature, were in a condition of unusual excitation, produced by other causes. And the great solitude and silence around me as I stood alone upon the topmost rock of the summit, enhanced my emotions. I was enraptured. My enjoyment increased to the highest point of enthusiasm, as I communed with nature. Soon the thought of the infinite majesty of the Creator of all I looked upon . . . entered my mind and overpowered me. Solemn awe took possession of me.[35]

The landscape did not evoke such a heightened mystical response from all observers, of course, though many poured out their admiration in the flowery, even syrupy prose fashionable for the era and experience. An anonymous writer of 1837 wrote a typical example:

> Behold those majestic mountains piercing the clouds! compared with which, Pelion, Ossa and Olympus piled on each other, would appear a dwarf beside a giant. Ascend, and if you cannot thence mount the throne of Jupiter, you can at least survey a world beneath you. Pierce now into that romantic glen, through which the Dryads might love to wander; slake your thirst that limpid brook, rippling down from the fern and moss-covered rocks; or recline under the projecting crag, enjoying that delicious calm.[36]

Around 1834 "Peregrine Prolix" traveled from White Sulphur Springs to Salt Sulphur Springs and wrote, "the scene for miles is wild and romantic, being laid in the heart of an ancient forest, flanked at intervals by mountain spurs terminating in lofty promontories of rock." From Salt Sulphur Springs to Red Sulphur Springs he declared the land "a mountainous and woody region which grows wilder and more romantic as you proceed."[37] Edward A. Pollard, writing about far southwestern Virginia's Natural Tunnel around 1870, even turned geology into a metaphysical contemplation. "Power, winged with all the winds of heaven and browed like the thunderbolt," he wrote, "that has *battered* its way through the solid rock . . . scatters dismay around and leaves behind it the voiceless uninscribed [*sic*] monuments of a sublime and inscrutable wonder!"[38] Later in his journey Pollard ascended Salt Pond Mountain in Giles County (now associated with Mountain Lake) and discovered the most exquisite scenery of his journey. He exaggerated in his claim to see five states from its summit, then waxed most romantically in his description of the vista: "The scene was hung with strands of raveled clouds; a variable purplish tint rested on the mountains; the sun was throwing across them the last lengthened javelins of the day. It was a scene of surpassing width and grandeur. . . . savage grandeur of mountain scenery was spread around us and lifted up into the sky."[39]

Even a somewhat oddly shaped sycamore (incidentally, one of the world's oldest tree species) captured the interest of Romantics seeking the picturesque. This one was located in the Montgomery-Giles counties area of the New River. From *Picturesque America*, ed. William Cullen Bryant (1872).

Around midcentury, John Esten Cooke found himself enraptured by the mountains, but even found inspiration in the forest, especially as the autumn colors developed. Cooke revealed some knowledge of tree species in the process:

> The trees rose like mighty monarchs, clad in royal robes of blue and yellow, emerald and gold, and crimson; the forest kings and little princely alders, ashes and red dogwoods, all were in their glory. Chiefly the emperor tulip-tree, however, shook to the air its noble vestments, and lit up all the hill-side with its beauty. The streams ran merrily in the rich light—the oriole swayed upon the gorgeous boughs and sang away his soul—over all drooped the diaphanous haze of October, like an enchanting dream.[40]

Cooke went on to distinguish oak, dogwood, red maple, "yellow hickory," and chestnut trees.[41] For all the focus on landscape, some nineteenth-century writers also began to note (and romanticize) the area's fairly recent history. In the opening lines of his 1920 history, *Conquest of the Old Southwest*, Archibald Henderson wrote, "The romantic and thrilling story of the southward and westward migration of successive waves of transplanted European peoples throughout the entire course of the eighteenth century is the history of the growth and evolution of American democracy. Upon the American continent was wrought out, through almost superhuman daring, incredible hardship, and surpassing endurance, the formation of a new society."[42]

Since the writers themselves enjoyed stagecoach travel, the services of hostlers and innkeepers, they were in awe of the pioneers who had roughed it a few generations before them. On their way to far southwestern Virginia's Natural Tunnel in 1870, Edward A. Pollard wrote of his group, "Riding through the beautiful and remarkable scenery . . . we are lost in very different meditations. A single remark conjures up all the resources of romance. *We are traveling precisely the same road that Daniel Boone traversed a hundred years ago, when he moved into Kentucky.*"[43] Pollard went on to ponder the warfare between whites and Indians that had permeated the settling of the frontier and, consistent with his romantic mood, sympathized with the natives. "What tales of blood yet cling to these mountains!" Pollard wrote. "What calamities and trials come fresh in remembrance in the midst of these scenes! They compose a story as yet but scantily written, and one that cannot be written entirely to the advantage and honor of the white man, when we remember that ruthless warfare sometimes made by the settlers."[44]

Not everyone was a romantic. Some travelers, while clearly appreciative of the beautiful landscape, described their observations in relatively sober terms. Such seems to have been the disposition of Anne Royall, famous traveler of the early nineteenth century. After leaving the Virginia mineral springs she described the Kanawha River in mostly detached, descriptive language.[45] One of Royall's traveling companions, a man from Philadelphia, was romantically captivated by the mountain scenery. But Royall thought more about the harsh realities of living in the mountains; how the residents there struggled against harsh winters and poor agricultural productivity. Because of this consciousness, she wrote, "I confess I cannot admire mountains as I hear many do."[46] No doubt Royall was not alone in these sentiments, though by nature of evidence it was the romantics who left such voluminous records of their impressions.

This juxtaposition continues in the Virginia mountains today. Visitors and transplants to the region—who are often people of comparatively prosperous means—want to enjoy the luxury of romanticizing the mountains, while residents often struggle with difficult economic circumstances. Attitudes toward extractive forest resources sometimes fall into two camps, purely aesthetic and at least somewhat utilitarian. A return to selective cutting, or all-age timber management, presents an alluring compromise in this scenario (more on this later). But even those trying to make a living from timber today must acknowledge the folly of earlier industrial logging, which created a short-term economic boom followed by bust. This was the environmentally devastating activity that immediately preceded JNF management of the highland woodlands. Industrial timber stripping, and what it inspired afterward, is worth careful consideration for all who judge national forest management in the southern Appalachians today.

Chapter 4

Industrial Logging Discovers Appalachia

Toward the end of the nineteenth century, when timber cutters had already denuded most areas of the eastern United States of their old growth forests, large areas of southern Appalachia remained untouched by industrial logging. Virgin timber covered literally hundreds of thousands of acres.[1] Steep, inaccessible terrain, not yet penetrated by railroads, had prevented large-scale, commercial lumbering operations. In 1880 parts of Wise County still featured huge poplar trees, some of which were six and eight feet in diameter. White oak predominated in many areas, such as Pulaski, northern Wythe, and Washington counties, and the Iron and White Top Mountain area still featured mostly primitive forest.[2] Around 1881 Charles Boyd toured southwestern Virginia and upon visiting Russell County reported that "it would astonish any one from the great timber and lumber markets, where everything of the kind has its value, to see the quantities of fine trees destroyed annually in clearing lands. Very frequently the log heaps are composed of walnut, hickory, white oak, and poplar, ready for the torch. It is the only way the people have of getting the timber that they do not want for fencing purposes out of the way."[3] As late as 1884, Charles Sargent declared that the timber lands of mountainous southwestern Virginia and neighboring territory in North Carolina, Tennessee, Kentucky, and West Virginia contained "the most valuable hard-wood forest remaining on the continent." He accurately attributed this preserved forest to the absence of adequate industrial transportation through either a railroad network or navigable waterways.[4] This delayed exploitation, however, only contributed to the impending timbering frenzy, for once railroads and logging companies reached southern Appalachia, large-scale lumber cutting began.

This intensive cutting was enabled not only by railroads but also by the greater mechanization of the logging industry in general. The invention of the Shay locomotive during the 1870s, and the subsequent use of it and other gear-driven engines, especially facilitated logging in steep terrain such as the southern Appalachian mountains. The Shay, Climax, and Heisler were slow but very powerful machines, and their tremendous torque could transport

heavy log trains over grades as steep as 14 percent.⁵ Steam yarders and load-
ers enabled faster log loading, and large band mills sawed the giant trees
faster than earlier circular saws. The most rugged areas of the southern Appalachian mountains even witnessed the occasional employment of incline
logging, which utilized an anchored steam winch that pulled lumber cars
up mountainside rails far too steep for even the geared locomotives.⁶ Industrial interests largely from the northeast utilized all these technologies. They
moved into the southern mountains well after diminishing timber supplies
in their own area.

The Douglass Land Company of New York formed in 1893 specifically
to exploit 113,000 acres in the Washington County area of southwestern
Virginia. Soon several timber companies, including the Laurel River, U.S.
Spruce, Damascus, and Grayson lumber companies, leased or bought this
land and industrial logging commenced. Between 1902 and 1928, Luther
Hassinger, owner of the Hassinger Lumber Company from Pennsylvania,
claimed to have cut about 3.5 million board feet of lumber from 30,000 acres
in Washington County, Virginia. Another lumber company, W. M. Ritter, cut
3 billion board feet of southern Appalachian timber between 1890 and 1940,
in Wise, Dickenson, and Giles counties, Virginia. From 1890 to 1900, logging and timber milling establishments in West Virginia more than doubled,
from 454 to 950, and ranked second in state industry only after iron and steel
manufacture. Virginia reflected a similar growth, rising from 663 "lumber
and timber products" establishments in 1890, to 1,341 in 1900, third in industrial rank after tobacco and flour-and-grist milling. At this point, in 1900,
southern Appalachian timber accounted for 30 percent of the nation's hardwood lumber production. Since lowland timber had long been cut or permanently cleared, this remarkable rise in industrial timbering clearly reflected
a highland phenomenon. By 1901, farmers, loggers, and other woodcutters
had cleared 46 percent of the relatively accessible 711,872 acres in the New
River Basin. But in the less accessible 233,280 acres constituting the southern
tributaries of the Holston River's south fork (including Green Cove, Damascus, and White Top), 77 percent of the land still remained in forest.⁷

Both West Virginia and Virginia lumber production reached their peak
in 1909, at the climax of Appalachian logging. In 1909 Virginia sawmills cut
more than 2 billion board feet of lumber, while West Virginia cut nearly 1.5
billion. That same year, 33,287 Virginia wage earners worked in 2,617 lumber and timber products operations, which accounted for 46 percent of all
the state's manufacturers. West Virginia reflected a similar scenario, featuring 18,643 laborers working in 1,016 lumber and timber establishments,
accounting for 47 percent of all manufacturers. In both states, logging and
timber milling establishments far outnumbered all other industries. At the
height of Appalachian industrial logging, the region produced 40 percent of

Yellow Poplar Lumber Company splash dam, Russell Fork, Dickenson County, Virginia, early twentieth century. Virginia Department of Forestry photograph.

the nation's lumber. At that point Virginia sawmills cut more than 2 billion board feet of lumber, ranking Virginia sixth in national production. In 1910, the National Lumber Magazine reported more lumber sawn in Washington County, Virginia, than any other U.S. county. Even as late as 1920, timber companies cut about 500,000 acres a year in Appalachia. But the boom began to wane, and by the early 1920s, when only stumps covered many mountain acres, Virginia lumber production fell to sixteenth in the nation, producing less than 800 million board feet. By the early 1930s timber companies had stripped more than 60 percent of the territory that constituted the JNF of 1990. A few, very small stands, such as the few hemlock trees near Mountain Lake, or the old hemlocks in Washington County near Bear Creek Camp, proved too difficult to reach or were considered unmarketable and thus survived. After they had finished logging, the Hassinger operation and other lumber companies gladly sold their holdings to the National Forest Reservation Commission in order to escape paying property taxes on land they had purchased primarily for standing timber.[8]

Potts Mountain Incline Railroad—A Case Study

Between 1892 and 1932, a remarkable combination of economics and technology came together to give southern Monroe County, West Virginia, and northern Craig County, Virginia, an experience in heavy industrialization. Prior to the arrival of the Potts Valley Branch of the Norfolk & Western Railway, most of the people living in these areas practiced the small scale agriculture that was then typical in many parts of mountainous Virginia and

West Virginia. The railroad, fully operative in 1909, provided the first large commercial outlet for timber products. Enough marketable timber existed in the Johns Creek Valley of Craig County to entice a Pennsylvania logging company to temporarily wed its interests with the Potts Valley Branch.

Regardless of political boundaries, the topography of the Big Stony and Potts Creek valleys link the northern portions of Giles and Craig counties (Virginia) with the southern portion of Monroe County (West Virginia). The Eastern Continental Divide separates the heads of these two valleys, but with such a mild peak that early railroad and road builders recognized the area as a natural route. As early as the 1860s various industrialists, farmers, and promoters of iron and wood industry, agriculture, and recreation all eyed the area with future plans in mind. Famous mineral springs resorts abounded in the area, as well as Mountain Lake, later the site of a resort.[9]

On December 9, 1910, the Potts Valley Branch officially began operation.[10] Norfolk & Western Railroad sought access to various iron ore deposits throughout the Potts Creek Valley, such as the "Cornfield Mine" near Paint Bank, where Virginia Iron, Coal & Coke Company had built a siding, store, and housing for their miners, and where they planned to construct a tipple—all in anticipation of the railroad.[11] Initially, N&W considered timber as only a secondary economic concern, even though the Potts Creek Valley contained significant tracts of virgin forest. For example, in 1905 N&W Chief Engineer Charles Churchill reported just one of several substantial timber tracts in Monroe County as containing an estimated $350,000 worth of timber.[12] In the long run, despite N&W's original iron ore ambitions, timbering became the valley's major industry.[13]

The majority of woodcutters along the Potts Valley Branch were local, small-scale tanbark and railroad tie gatherers and manufacturers. In August 1908 N&W Chief Engineer Churchill reported tanbark and railroad ties stacked at Waiteville, West Virginia, in anticipation of the coming railroad. By November, Churchill estimated the Waiteville tanbark storage at some 2,000 tons.[14] Regular service began to Waiteville on August 16, 1909, and finally provided an outlet for tanbark, which valley residents had previously spent three days hauling by wagon over Peter's Mountain, further into Monroe County to places like Gap Mills and Ronceverte, in Greenbrier County.[15] Throughout the early 1910s, the Potts Valley Lumber Company in Waiteville advertised themselves as buyers of tanbark, cross ties, and lumber, which they in turn shipped out on the railroad.[16]

As important as the railroad became to local, small-scale timber cutters and tanbark gatherers, one industrial logging company called Tri-State linked its entire Virginia / West Virginia operation to the Potts Valley Branch. Like numerous other timber companies that began operating in the southern Appalachians during the late 1800s and early 1900s, Tri-State was a northern company—in this case, from Pennsylvania. In 1912 Tri-State bought the timber rights to 28,139 acres in Craig and Giles counties, Virginia.[17] This timber lay in the Johns Creek Valley on a tract measuring about fifteen miles long and seven miles wide, which supposedly contained 100–150 million board feet of ancient white oak, chestnut oak, red oak, poplar, hickory, basswood, and hemlock. In addition to this sawtimber, the tract also contained various amounts of potential tanbark and pole timber. At its uppermost perimeter, this land lay within 1,600 feet of the Potts Valley Branch—but on the opposite side of Potts Mountain.[18] With no railroad into Johns Creek Valley, the Potts Valley Branch became the transport route for Johns Creek Valley timber. But to reach the main railroad required construction of an incline haulage system.

Throughout the 1910s and 1920s, as the more-accessible areas of the mountains grew progressively denuded through regular forest railroads, logging companies operating in the southern Appalachians grew more interested in incline technology. The Tri-State operation atop Potts Mountain, straddling Craig County, Virginia, and Monroe County, West Virginia, began in the midst of this period and typified many such operations during the early twentieth century.[19]

Incline technology had its origin in the old yarding donkey that yanked itself around the forest via a horizontal steam boiler mounted on skids, powering a rolling drum around which were wrapped many yards of steel cable. Anchoring the yarding donkey to a tree or large stump allowed loggers to drag timber to a central landing, after which the yarder could pull itself to a new area. Permanently anchoring the yarding donkey atop a mountain crest, then adding a rail track up and down the mountain flanks, gave birth to the first and simplest incline system, the one-way or single-line incline. The Yeon and Pelton logging company employed just such a type of incline as early as 1905 or 1906 in the mountains of Oregon.[20] Improvements to technology included the use of two-way or counterbalance inclines, which raised empty cars as loaded ones were lowered. Such systems required dual tracks for part of the distance on the delivery side of the mountain so cars could pass alongside one another.[21]

Southern Appalachia was the only region in the eastern United States where loggers used incline technology. Loggers working in the region utilized one of the earliest incline operations in West Virginia as early as 1909.[22] After Tri-State completed construction of their incline system, they used horses to drag logs down from ridges, and workers loaded them onto the narrow-gauge cars. When loggers finished cutting the timber in a certain hollow, track laborers pulled up the track and relaid it in a fresh cutting area. The railroad carried the logs to the mill pond, at the foot of Potts Mountain, where sawyers used a band mill to make lumber.[23] The workers then loaded the sawn lumber back on the railroad and used geared locomotives to haul it as far up Potts Mountain as the grade would permit. At this point, the workers disconnected the locomotive, attached a steel cable, and thereafter the incline system transported the lumber up the south side of Potts Mountain and lowered it down the north slope. Like other incline operations, by using narrow-gauge rails throughout their logging network, Tri-State avoided transferring the load from their locomotive cars to their incline cars.[24] At the main railroad, however, workers had to reload the lumber onto the standard-gauge cars of N&W's Potts Valley Branch. From there, N&W carried the lumber toward the New River and markets beyond.

Tri-State's logging operation involved two distinct sets of workers: railroad builders and loggers. As in many southern Appalachian logging and railroad operations, segregated Italians and blacks built the railroads, and various Caucasian workers performed the actual logging. Italian railroad workers predominated in the northern United States, while African Americans predominated in the South. In 1905 more than 30,000 track laborers, "largely Negroes," worked in Virginia earning $1.05 a day.[25] While the Italian railroad workers involved with West Virginia operations reflect this larger picture, those prevalent around the logging operations of mountainous Virginia would seem to present something of an exception to this situation.[26]

The Italian railroad builders who worked for the Tri-State operation were but a few of the many thousands of immigrants from southern Italy conspicuous on track gangs in the eastern United States during the late nineteenth and early twentieth centuries. Gangs of workers typically toiled under a bilingual straw boss who directed them at construction, maintenance, and repair of rail lines, including the preparation of the roadbed, laying ties and track, tunneling, and later, replacing ties.[27]

Italians became but one group of late-nineteenth- and early-twentieth-century immigrants exploited by industrialists operating in the Southern Appalachian region. Company agents typically used false or exaggerated promises of wealth to lure workers, especially to coal mines in West Virginia.[28] Longtime Waiteville resident Herman Harry recalled growing up in Potts Creek Valley and hearing how Italian (and possibly Greek) railroad workers toiled for six months without pay, merely to compensate company agents who had paid their passage to America. By the time the workers had finished paying off their passage, they had also finished building the Potts Valley railroad and had to search for employment elsewhere. The railroad certainly did not hire them as permanent employees.[29]

Over in the Johns Creek Valley, loggers lived in houses of three to five rooms, while Italian railroad builders (called "Tallies" by the local people) lived in a segregated village comprised of one-room shanties. The Italian railroad workers involved with Tri-State, however, met with a somewhat more prolonged labor situation than their N&W counterparts, for Tri-State kept the Italian workers busy laying down new line, then picking it up and laying it down again in a neighboring hollow as timbering progressed.[30] Tri-State probably employed Italians toward the very end of their operations and possibly had them remove the final track after finishing the Johns Creek cut.[31]

Working around logging and railroad equipment was certainly highly dangerous, and in 1916 Tri-State met with their worst accident, involving an overturned locomotive that severely scalded four men, two of whom died. Apparently E. E. Bitners, Bob McMasters (the two fatalities, both from Pennsylvania), Ed Tolley, and Bob Atkins (both from Waiteville) were all riding a single Shay locomotive through a multitrack area near the band mill in Johns Creek Valley, in the process of shifting some cars onto a siding. They apparently failed to release a cable coupling connecting their locomotive with the cars they were moving, and as the locomotive and cars diverged at a widening point on the dual tracks, the heavier cars caused the locomotive to flip.[32]

The Tri-State operation faded amidst the end of the southern Appalachian industrial logging boom and declining national lumber market. After 1909 the industry steadily dwindled in both Virginia and West Virginia.[33] Tri-State's operations certainly never achieved the "twelve to fifteen years" duration as originally anticipated and only operated its band mill "sporadically for five or six years" before portable circular mills took over lumbering activity.[34] By September 1919, at the latest, Craig County court records described Tri-State's premises as "demised."[35]

Given the sharply curtailed timber industry and the abandoned iron ore industry, it seems rather amazing that the N&W's Potts Valley Branch continued to operate for another decade. But between 1913 and 1917 freight haulage seriously declined, and N&W began to make plans to abandon the line. In July 1932 the Interstate Commerce Commission gave permission to remove the track, and N&W authorized abandonment that November. By 1933 the Potts Valley Branch had disappeared.[36]

With the demise of Tri-State's logging operation and the Potts Valley Branch, the brief historical phase of heavy industrialism ended in both Johns Creek and Pott's Creek valleys. This presented a modest variation as far as industrial histories go, but it stood in marked contrast to any economic activity or environmental impact experienced by the region before or since. In 1910, just before Tri-State arrived, farmers overwhelmingly predominated in Johns Creek Valley, with a few working in building trades, such as carpentry or masonry, or in light industry such as saddle-making or blacksmithing. No one listed their occupation as logger, sawyer, or railroad worker.[37] Without question, farming continued to dominate the work scene for the next decade, but for the first time professional loggers and track layers worked in Johns Creek Valley. By 1920 they were gone, and the two Virginia-born men who

Timber stripping followed by massive slash burning, near Konnarock, Virginia, 1926. This later became part of the JNF's Holston Ranger District. This was the sort of devastation often left behind by early-twentieth-century industrial logging all over the southern Appalachians. U.S. Forest Service photograph.

listed their occupation as "woodchoppers" were far outnumbered by their farming counterparts.[38] Local woodcutting continued, but on a much smaller scale and with lighter equipment.[39]

The brief but devastating era of early industrial logging in southwestern Virginia and other parts of southern Appalachia coincided with the first applications of scientific forestry efforts in the region. Such forestry science came out of German techniques of selective cutting, or all-age timber management. Later such science supported highly controversial clear-cutting, or even-age timber management. Many have argued that economics far more than science characterized the latter. But any sort of timber management contrasted starkly from the timber stripping of the early industrial era, when loggers cut all marketable trees and cut railroads and skid trails through the woodlands with utter disregard to environmental consequences.

Scientific Forestry

In 1915 the Forest Service established experimental plots on the Biltmore Estate in western North Carolina for studying the effects of thinnings. Other work focused heavily on fire prevention and the effects of fire on the forest environment, but also examinations of individual tree species and regeneration experiments. In 1921 the USDA created a Forest Service experiment station in Asheville with E. H. Frothingham as its director. With its proximity to the Biltmore Estate, Frothingham kept in close contact with their forestry experiments and conducted his own.[40] From the outset, the role of science in the Appalachian Mountain and other Forest Service experiment stations clearly emphasized applied (as opposed to "pure") science. As Frothingham wrote in 1922,

> The solution of the technical problems concerned with the complete utilization of forest lands will fall to forest research. They will include such things as the exhaustive study of the characteristics and requirements of individual species, the classification of forest soils (or "sites") in terms of their timber producing capacity, the proper degree and frequency of thinnings, the encouragement of young reproduction by freeing it from competition and overhead shade, the planting of seedlings and the sowing by hand of tree seeds, and the big subject of forest measurements. Investigations of some of these subjects have already begun.[41]

Obviously pragmatism dominated early Forest Service scientific work. "How long will it take yellow poplar to grow to a diameter of 15 inches?" were the kind of questions facing Frothingham and other Forest Service scientists.[42] Even with the utilitarian bent, Frothingham had to contend with a prevailing shortsightedness in the logging industry that had yet to realize the very practical ends of much forestry experimentation. In 1923 he wrote, "there is, in fact, some reason to question whether, in plunging so entirely into the solution of 'practical' problems, we are not sacrificing opportunities for considerably greater usefulness later on." Pioneers in the field of scientific forestry had to struggle with such practical demands at the same time they were beginning to realize just how little they knew about the forest ecosystem. Frothingham defended his work and the work of other scientific foresters by writing, "as owners of forest and cutover lands lumbermen should be directly interested in these studies of regeneration, protection, growth, and management of the future second-growth forest."[43] Early scientific forestry also included designating experimental forests on national forest acreage, the purpose of which was "to make permanently available for silvicultural, range, products, and other related forest research carefully selected areas as fully representative as possible of conditions in important parts of forest regions and large enough to meet present and foreseeable future needs."[44]

Forestry schools arose concurrently with the advent of federal forestry, and Congress supported these efforts in 1928 as part of the McSweeney-McNary Act, which helped fund forestry schools and state-level scientific efforts. From this early beginning, federal forestry research grew to become the world's largest endeavor of its kind.[45] An early form of timber survey characterized some of the Forest Service's earliest scientific efforts in the southern Appalachian region. Scientists and timber surveyors had to describe the forest before more extensive experimentation and study could begin.[46]

Preliminary observations of the earliest JNF tracts began just before and following the 1911 Weeks Law, when the National Forest Reservation Commission bought the first purchase units in southwestern Virginia. The JNF's own localized timber surveys began in the 1930s and coincided with the Forest Service's first comprehensive survey of the southeastern United States. Forest scientists began giving unprecedented attention to the woodlands and eventually the greater southwestern Virginia forest ecosystem.

Chapter 5

JNF and the Rise of National Conservation

While early industrialism and spreading farm acreage continued to alter southwestern Virginia's environment, Romanticism was to some extent morphing into national events that were coalescing around a new environmental conservation ethic.[1] After the 1860s, preservation and conservation of public lands began to gain greater public appeal. A significant fear of timber shortages preoccupied certain conservation circles. Related to timber shortage fears, but even more captivating in regard to human loss, were catastrophic forest fires that helped draw national attention to the need of some kind of public intervention in industrial logging practices.[2] Nebraska held the first Arbor Day in 1872, the same year Congress established America's first national park, the Yellowstone. In 1876 Congress passed an appropriation act that initiated examination of the nation's forests, with plans for reforestation and conservative harvesting.

In 1881 Franklin B. Hough became chief of a new agency, the Division of Forestry (later called the Forest Service), established in the Department of Agriculture. By 1886 scientifically trained German forester Bernhard E. Fernow operated the young Division of Forestry. Continuing along the dual paths of preservation and conservation, Congress created the Yosemite National Park in 1890 and passed the Forest Reserve Act in 1891. The Forest Reserve Act enabled the president to retain forest lands from the public domain, and thus the infrastructure for designating national forests began. Further federal conservation legislation followed in 1897 with the Organic Act, which specified that the new federal reserves were "for the purpose of securing favorable conditions of water flows, and to furnish a continuous supply of timber for the use and necessities of citizens of the United States."[3] In 1898 Gifford Pinchot became the Division of Forestry's chief, and in 1905 the Department of Agriculture took control of the forest reserves. By 1907 the Department called them national forests.[4]

Gifford Pinchot advocated several tenets that remain essentially unchanged in the Forest Service today: a certain amount of permanent public forest ownership, forest fire control, and professional and scientific management of public forests that would, theoretically, influence private forests along similar conservation guidelines.[5] German forestry practices very much influenced Pinchot, with their emphasis on state involvement and scientific management. Pinchot himself originally sought to emulate such an approach to American forestry in something of a dictatorial mode, where a scientific elite would operate free of congressional (or public) interference. However, the 1908–10 political conflict that arose between the executive and legislative branches over federal forestry eventually forced Pinchot and his supporters to compromise their exclusive approach by appealing to the public sentiment of the Progressive era. In something of a contradiction, public sentiment operated largely around the misconception that industrial forestry practices of the time were all destructive and irresponsible, and therefore salvation awaited in federal forestry practiced on public woodlands. In reality, many industrial foresters of the time, for purely economic reasons, originally supported Pinchot and other advocates of a scientific approach to forestry. The Germans, after all, were at the vanguard of the day's most sophisticated forestry practices, and thus the American wood products industries were highly interested in those trained in the German school.[6] Ironically, the southern Appalachian region where Pinchot and the "cradle of American forestry" began was about to witness precisely the kind of uncontrolled industrial cutting against which scientific forestry reacted.

Early Legislation

Much of the early legislation pertaining to the nation's forests specifically concerned lands in the Appalachian Region. Even before the 1890–1920 southern Appalachian industrial timbering era had run its course, eastern groups such as the Appalachian Mountain Club and larger organizations like the National Academy of Science supported protecting or preserving Appalachian land.[7] In 1885 two doctors, Henry O. Marcy and Chase P. Ambler, also began promoting Appalachian forest preservation. In 1899 they helped form the Appalachian National Park Association (later called the Appalachian National Forest Reserve Association) in Asheville, North Carolina. They advocated preserving a large forested region for economic reasons, aesthetic values, and what they believed were the region's natural health-inducing qualities. North Carolina Senator Jeter C. Pritchard supported their effort and persuaded Congress to investigate. By 1901, with congressional backing, President William McKinley agreed to the need for Appalachian forest reserves. The next year, Secretary of Agriculture James Wilson submitted a report concerning the Appalachian region. Among his conclusions, he stated:

> The rivers which originate in the Southern Appalachians flow into or
> along the edges of every State from Ohio to the Gulf and from the Atlan-

tic to the Mississippi. Along their courses are agricultural, water-power, and navigation interests whose preservation is absolutely essential to the well-being of the nation. . . . The regulation of the flow of these rivers can be accomplished only by the conservation of the forests.[8]

That same year, in 1902, the Senate approved $10 million for land purchases in the Appalachian Region.[9] Similar efforts in New England's White Mountains soon followed, and by 1905 congressmen from Minnesota, Missouri, New York, Texas, and West Virginia advocated forest reserves in their states. Clearly a movement had begun, and special focus fell upon the eastern mountains stretching from Maine to Alabama.

During its early years, the Forest Service utilized diverse propaganda in its various arguments for federal forestry in the Appalachians. The focus of Forest Service employees' official publications ranged from timber protection, fire prevention, inland water navigation, and local wood products economic stimulation. When the Forest Service foresaw a possible crisis in future hardwood timber supply, protected forest lands in the Appalachians became the answer. The idea of year-round navigation on such rivers as the Kanawha, upper Tennessee, and Yadkin rivers could be possible, according to the proponents, only if the federal government regulated the Appalachian forests. U.S. Geological Survey workers H. B. Ayres and W. W. Ashe linked potential hydrological power with forest protection.[10] Watershed protection seized many conservationists' imaginations, and thus the core of the ideology that lay behind much of the movement for creating eastern national forests developed.[11] On July 23, 1909, Massachusetts Congressman John Weeks introduced his historic bill, which would become the Weeks Act two years later, on March 27, 1911.[12]

The Weeks Act provided the first major legislation for purchase of private and corporate lands that began making up various eastern national forests. The dominant motivation behind the creation of these early federal forest reserves centered around the belief that widespread deforestation in the Appalachians contributed to increased flooding, which killed or endangered the lives of local people, destroyed or damaged their property, and wreaked disaster upon the lives and property of people living downstream. Among the worst floods, the 1907 Monongahela River deluge caused over $100 million in damage in West Virginia and cities downstream, especially Pittsburgh. The Monongahela flood caused Leonidas Glenn to begin his study of the southern Appalachian watersheds for the U.S. Geological Survey, and he concluded that deforestation was one of the major culprits in the recent flooding. Ideas generally linking forests or lack of forests with climate variations had actually been around for decades, including the theory that deforestation caused increased flooding.[13] The extent of the effects of deforestation on erosion and increased precipitation runoff, however, was highly controversial at the time and for many following years.[14] Although intensive timber cutting could exacerbate flooding in some instances, the ideology became part of the gospel of the Progressive conservation crusade, overwhelming the more complex scientific truth, which had only begun to emerge amidst new hydrology studies.[15]

Certain parties, whose priority was a reduction in mountain deforestation, were perfectly willing to use the most readily available plausible argument. Given the state of industrial logging in the southern highlands at the time, a propensity to bend the facts or draw profound conclusions from partial evidence seems understandable. Scientific findings take much time to establish, and during the early twentieth century every month saw new acreage stripped of its trees. Conservationists, such as the American Forestry Association, focused heavily upon the need for watershed protection. Their concern began to manifest itself in the political and public consciousness right around the peak of Appalachian logging.

The Weeks Act created a National Forest Reservation Commission (NFRC), involving the Departments of War, Interior, and Agriculture, and four congressmen to authorize land acquisition. The NFRC's early acquisition agenda began moderately, with the ultimate purchase objective being no more than 5 million southern Appalachian acres. During the first year they bought only 31,876 acres and took an additional two years to expend their first annual budget. But the NFRC soon built momentum, and as the southern Appalachian timber boom began to wind down, they accelerated their acquisitions. By 1914 the Weeks Act enabled the NFRC to buy about 260,000 acres of Virginia lands in five purchase units.[16] The JNF's purchase boundaries already concentrated on the upper elevations of the highlands, but the secretary of agriculture could go even further to concentrate national forest holdings. NFRC survey crews or other parties could inform the secretary that certain lands within the purchase boundaries were more suitable to farming, and thus the secretary could exclude their purchase.[17] It is little wonder that the JNF ended up with a distinct geography of highland woodlands interspersed with private, relative lowland, inholdings.

By 1922 the NFRC was progressing with Weeks Act purchases of more than 2 million acres in eleven eastern states. Motivation for federal forest lands acquisition in the southern Appalachians began to expand from an initial watershed protection and fire prevention emphasis to a greater focus on timber management. In 1923 the Senate appointed a select committee to explore forest regions not already covered by the Weeks Act. The select committee investigated lands linked with watersheds, but not necessarily related to stream navigability, as the earlier legislation had specified. When the select committee returned, they recommended more federal land purchases, but this time they emphasized timber production in addition to watershed protection. The Clarke-McNary Act of June 7, 1924, authorized these further purchases,[18] and by 1925 the NFRC had completed the purchase of almost 2 million acres in southern Appalachia alone. Between 1911 and 1932, the NFRC spent more than $23 million for some 5 million acres comprising twenty-four eastern national forests.[19] The NFRC purchases coincided with the waning southern Appalachian timber boom and a glut in finished lumber supplies that depressed timber prices. The accelerated NFRC purchases were, in part, encouraged by an enthusiasm among industrial interests to divest themselves of former timber lands.

Early JNF Land Acquisitions

On March 21, 1902, the Virginia Legislature unanimously agreed to give the federal government permission to buy land and create "purchase units." Purchase units were designated areas generally 100,000 acres or more, or smaller tracts deemed worthy of economic management. If a purchase unit grew or was combined with one or more other purchase units, the land transferred to national forest status. Immediately following the enactment of the Weeks Act, the NFRC approved the 255,027 acre purchase boundary for the White Top Purchase Unit in Virginia and Tennessee.[20]

The "purchase boundary" was an important geographical designation still widely misunderstood in the eastern United States. A typical road atlas offers the illusion that eastern national forests are consolidated blocks of land when, in fact, these green areas on the map actually represent maximum purchase boundaries. Private inholdings abound. But the Forest Service can readily trade and purchase lands within purchase boundaries without congressional approval. Expanding the purchase boundary requires congressional approval. Thus, the highland woodland holdings of the current JNF reflect a historic geographical intent.

The contemporary JNF's purchase boundaries reflect a long process of accumulating and consolidating purchase unit acreage. By fiscal year 1912 the NFRC expanded the White Top boundary to contain 132,986 acres in Tennessee and 186,639 Virginia acres in Grayson, Smyth, Washington, and Wythe counties. The NFRC 1915 annual report described the area: "The original forest consisted of oak, chestnut, and poplar, with [a] large mixture of white pine of very excellent quality. The bulk of this timber had been cut. What remains is mostly low-grade old timber for which there is now a market and which should be cut in order to make way for the young stand."[21] The White Top unit included 13,450 acres bought from the Douglas Land Company, which had figured prominently in earlier regional timber exploitation. Much of the old Douglas Land Company holdings later became part of the JNF's Holston Ranger District (now part of the Mount Rogers National

Recreation Area).[22] This reflects a classic case of timber companies divesting their cutover holdings. Their motives likely included initial short-term profits, avoidance of property taxes on land that would be unproductive for decades, and eventually (during the 1930s) a need to liquidate any and all sorts of property amidst the dismal lumber prices of the Great Depression.

During the next few decades the JNF slowly formed from numerous land acquisitions all over southwestern Virginia.[23] There were the several purchase units that later became ranger districts and, very briefly, small Virginia national forests called Natural Bridge, Shenandoah, and Mountain Lake.[24] All of these lands eventually merged into what is now Virginia's single national forest, the George Washington–Thomas Jefferson. At the regional level, the Forest Service was also shifting regional administrative districts and their holdings around, which reflected similar land purchases in New England,

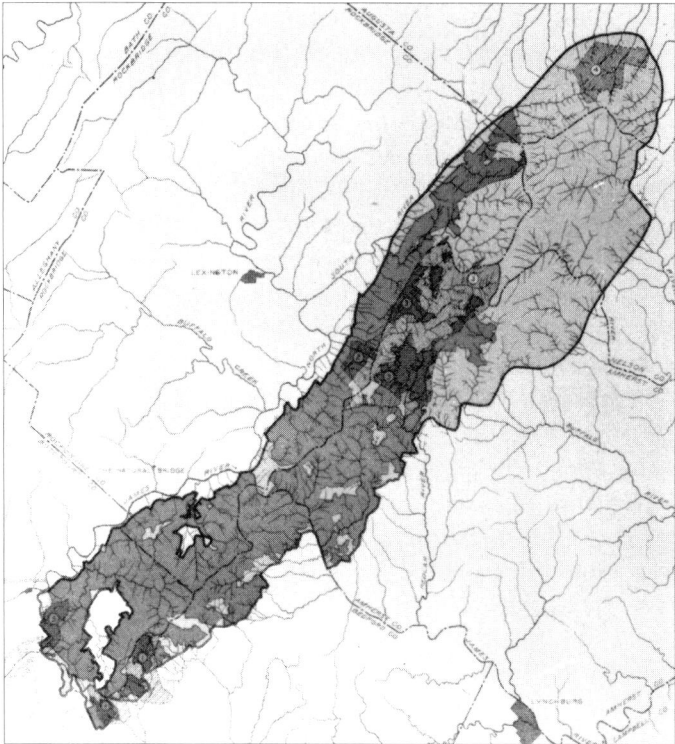

Natural Bridge purchase area, 1910s–1920s. This map shows some of the purchase boundaries established by Congress, as well as early purchases made by the National Forest Reservation Commission. The JNF and other eastern national forests remain patchwork holdings with scores of private "inholdings." U.S. Forest Service map.

Pennsylvania, and the southern Appalachians.[25] The somewhat complicated administrative history involving all the purchase units and small, temporary national forests reflects the piecemeal manner in which the Forest Service generally acquired eastern national forests. The land surveying and legal work involved in these endeavors was formidable and entailed many thousands of hours on the part of field crews and title searchers. The file cabinets filled with the resulting paperwork constitute some of the most important primary records of eastern national forests and remain viable legal documents in ongoing land purchases and/or exchanges.

On July 1, 1937, Secretary of Agriculture Milburn L. Wilson traveled to the Clinch Ranger District to dedicate the Jefferson National Forest as part of the annual Rhododendron Festival held at Norton, Virginia. Approximately 15,000–20,000 people attended the festival, including the governors of Virginia and West Virginia and community leaders seeking to promote their area for tourism.[26] Wilson took an audience up to High Knob, where he read his dedication speech. "By request, President Roosevelt has named it the Jefferson," Wilson said of the new Forest, and he proceeded to illustrate some of Thomas Jefferson's ideas concerning conservation: "Jefferson was a conservationist, for he knew that man's welfare depended in large part on use of natural resources . . . recognizing that problems of conservation cannot be solved merely by preservation; that instead we must in the main produce, use, and renew. And to do this we must, of course, treat wisely our soil, water, and all plant and animal resources, cultivated and wild."[27] Wilson went on to praise forests for both their practical and aesthetic values, stressing their ability for renewal. The occasion marked a small event as far as New Deal affairs went. And yet the creation of the JNF presented a momentous occasion in southwestern Virginia's environmental history, in its land use evolution, and with all the accompanying social, economic, and cultural ramifications. The JNF came into creation in the wake of intensive industrial logging, in one of the area's hardest hit by the national depression, and through a somewhat convoluted administrative reshuffling that coincided with a quarter century of patchwork acquisition. These circumstances, along with the first generation of Forest Service employees, gave the JNF its early character.

Continuing Land Acquisition and "Friendly Condemnation"

The advent of the Forest Service in southern Appalachia during the early twentieth century significantly contrasted any of the numerous large outside entities that introduced industrialism or a federal presence into the region. Railroad, timber, and coal interests, as well as the Park Service and the Tennessee Valley Authority, all invaded Appalachia in quest of land, resources, or recreational or industrial facilities. The effects of these various outside entities upon the area's people involved traumatic, often detrimental economic, social, and cultural results. Contrary to romantic images of self-sufficient pioneers, much of Appalachia's history began with absentee land speculators and numerous tenants.[28] In some ways, the lack of fee simple landownership helped foster the southern Appalachian custom of using the upper slope woodlands as open range or a de facto commons.[29] Some of this land passed into small holdings afterward, but during the late nineteenth century land speculators and early capitalists launched another wave of mass land acquisition. They acquired much of this land through questionable or unscrupulous means, such as surveying land of uncertain or lost title, then claiming it for themselves. Many mountain people, who did not realize the national market value of their timber and coal, unwittingly sold the rights to these resources to speculators.[30] Poverty tended to accompany or follow industrialization, and economic booms in Appalachia mostly benefited outside capitalists rather than the indigenous population. Concurrently, Appalachia's population grew beyond the numbers that the preindustrial era's agricultural lifestyle had supported. The absentee ownership of land that preceded and accompanied industrialism, of course, exacerbated the problem of a growing population in a region of limited agricultural possibilities by removing even more land from possible agricultural cultivation.[31]

The Forest Service ultimately became far too minor a land manager to make any drastic improvement upon the absentee landownership legacy, but they played a markedly different role than had timber and mineral companies, especially in regard to long-term commitment and sustained resource yields. As in the rest of the nation, the advent of national forests introduced Appalachia to an entire new concept of land and resource control. But certainly absentee ownership, as some writers have charged, did not truly apply to Forest Service holdings in Appalachia.[32] With ranger stations scattered throughout a national forest and Forest Service personnel at least somewhat accountable for their actions and policies, the Forest Service was never truly absent to begin with. Indeed, many native people have historically worked for the Forest Service, and personnel transferred from other places traditionally strove to become an intricate part of their new communities. Also, unlike timber and coal companies, profit did not drive the Forest Service. And unlike the Park Service and the Tennessee Valley Authority, the Forest Service did not attempt or even realistically desire absolute and inclusive land acquisition—and thus it never displaced the thousands of families these other federal agencies forced off their land.[33] As previously described, holdings in eastern and even Midwestern national forests continue to resemble a patchwork quilt rather than an all-encompassing blanket, despite potentially misleading road atlases and other general maps. In addition to flexible land acquisition agendas, the Forest Service historically accommodated local communities and their economies through timber sales, and at least some economic benefits derived from recreation and tourism. All of these factors provide marked contrast to other outside entities who had acted upon Appalachia during the past century.

The 1930s saw much corporate land pass into JNF management from depression era liquidation interests, or from an inability or unwillingness to pay property taxes. Corporations sold the JNF its larger tracts. These land-holding companies were not always specifically timber companies but also mineral companies that had harvested or sold the valuable timber before divesting themselves of land. The businesses were naturally reluctant to pay property taxes on cutover land they had originally purchased only for timber, or on land that had lost its mineral value.[34] In 1937 the Clinchfield Coal Corporation and Consolidation Coal Company sold the JNF 130,489 acres in the Breaks area of Pike and Letcher counties in Kentucky, and Dickenson, Buchanan, and Wise counties in Virginia. Naturally, this land did not include significant coal beds. Other companies who sold tens of thousands of acres to the JNF included the Pocahontas Coal Company, Hassinger Lumber Company, and the Craig-Giles Iron Company.[35]

Land acquisition did not come about without periodic conflicts. The NFRC reported forty-three condemnation cases among all Appalachian national forests during fiscal 1914, and clearing titles took a process averaging seventeen and a half months.[36] Acquiring land presented early JNF staff with one of their first public relations challenges. Instead of eminent domain for acquisition, disputed boundary lines and uncertain land titles—sometimes complicated by lost or destroyed county court records—immediately involved the JNF in "friendly condemnation" lawsuits. In contrast to regular eminent domain proceedings, land sellers and the JNF used the procedure here as a legal device. The JNF inherited title conflicts and old boundary disputes that had gone undetected or unresolved for decades, even centuries.[37] Resulting lawsuits could be confusing, and in 1938 the New Castle District's local newspaper reported that "a great deal of misunderstanding" had surrounded court summons involving condemnation. The newspaper reported, "The Forest Service acquires land by purchase from those who are willing to sell at the prices offered. Due to imperfections in the title of some tracts, it is necessary to place land in condemnation, a friendly court procedure intended only for clearing the title to lands already under purchase agreement. The Forest Service cannot purchase land that does not have a perfect title."[38]

Such a situation befell future JNF land in Lee County, Virginia, that the Virginia Iron and Coal Company wanted to sell to the NFRC. The NFRC approved direct purchase of 19,213 acres but had to condemn a remaining 759 acres that did not carry a safe title.[39] In 1935 the NFRC directly purchased

10,792 acres in Craig and Botetourt counties for the JNF from the Virginia Timber Corporation. Another 2,256 acres, however, fell under condemnation. Of these remaining acres, USDA abstractor Draper W. Phillips gave the following explanation in 1936:

> These laps [lapses; i.e., disputed tracts in question] cannot be acquired by direct purchase because of the numerous claimants of the property. Apparently none have established title by adverse possession and each lap [tract] is covered by two or more claimants or sets of claimants, and in instances the title to the mineral and other rights alone are divided four ways or more. These claimants in many instances are unknown heirs at law of former record owners. The title is further complicated by the fact that there are numerous defects in the recorded instruments and there are breaks in the title, making it possible that others might assert interests.[40]

The NFRC routinely settled such suits through condemnation. In one of the most celebrated cases, 16,168 acres of the Hagan Estate in Scott and Wise counties fell under condemnation, involving $9,600 in delinquent taxes and fourteen separate tracts of land, ranging from 4 to 10,760 acres.[41] As many as thirty heirs claimed title to land on Walker Mountain that the JNF eventually added to the Wythe Ranger District. When the courts finally settled the claims and gave the JNF clear title, the JNF directly paid the court, which then determined how the money would be distributed among the heirs.[42] Charles Sexton remembered a six-acre tract in Wise County on which his father paid property taxes for over thirty years, only to discover, according to JNF survey, that he did not really own the land. Apparently, the Hagan Estate had sold him land that actually belonged to the Virginia Iron and Coal Company, from which the JNF made its purchase. The senior Mr. Sexton decided not to pursue this matter in court,[43] but other cases languished in lawsuits for decades, and to this day some people feel the government cheated their ancestors out of land.

All in all, acquiring land could prove a delicate task. William Campbell, who did much of the early JNF survey work, recalled one of his experiences on the Clinch Ranger District. He remembered, "Well, I used to be real diplomatic. The madder they got, the calmer I got. I never really had any trouble with them. I had one guy threaten to stomp me in the middle of the road one day, but I told him, 'I didn't come down here to fight; if we can settle it, we'll settle it; if not, well, I'll leave.' By the time I left everything was settled."[44]

Resolving these type of landownership disputes and clearing titles through "friendly condemnation" clearly should not be confused with the forced displacement of other Appalachian residents that took place during the 1930s. These land seizures under the Fifth Amendment took place in Cades Cove, for the Great Smoky Mountain National Park,[45] for Virginia's Shenandoah National Park,[46] and the Norris Basin area in eastern Tennessee for the Tennessee Valley Authority's first hydroelectric impoundment.[47] Thousands of people lost their homes, farms, and cemeteries through strong arm strategies that often involved unfair property payments, shady legal proceedings, as well as an absence of legal representation among property owners, and ultimately the forced physical removal of residents. In all these situations, government agencies, private recreational promoters, or both, dominated area residents and little grassroots democracy prevailed.[48]

During the same period as these massive displacements, the JNF and the Forest Service in general faced a significantly different management task that did not mandate acquiring a consolidated area within demarcated boundaries. To this day the JNF does not manage anywhere close to all the lands within its purchase boundaries, nor has this ever been a goal (with the brief and uncompleted exception of the Mount Rogers National Recreation Area, discussed in a following chapter). The Forest Service in the eastern United States in general enjoyed the pleasure of acquiring most of its land through willing sellers.[49] The difference partly arose from the Forest Service's mission, unique among federal agencies, which dictated that national forests should satisfy many different groups interested in the various forest resources, such as timber, minerals, recreation, or wildlife.

During the World War II and immediate postwar era, the Holston Ranger District reflected the wider JNF policy. The Holston Working Circle desired acquisition of an additional 5,000 acres, but strictly on a voluntary basis, and then only when sellers were asking "reasonable" prices. In fact, the JNF had already turned down many tracts because of prices they perceived as too high. They were willing to pay above average prices for land whose owner had avoided timber stripping and practiced fire prevention. They also sought right-of-way easements across private land in order to reach land already under JNF management.[50] But by 1950 they maintained the conclusion that "a concerted acquisition program is not warranted."[51] In acquiring timber producing lands, they specifically sought to avoid infringing upon lands they deemed better used for agriculture or recreation.[52] This approach illustrates the broader philosophy that eventually rendered the entire JNF as mostly a highland woodlands.

During this earlier period the JNF did not displace residents and instead, in the Forest Service tradition, offered lifetime residency rights to those who wished to both sell and retain occupancy.[53] Such an approach certainly engendered conflicts from the outset but also allowed an important element of flexibility and adaptation to special circumstances. The Forest Service found it in everyone's interest to accommodate local people in regard to residency on federal land. As historian Harold K. Steen wrote of *The U.S. Forest Service Use Book,* published in 1905, "Significantly, the first regulation assured all legitimate occupants of the forest reserves that the Forest Service would protect their rights even to the extent of giving them preferential treatment. . . . The next seven regulations dealt with permit procedures and free use permits, both detailing the needs and rights of those living on or near the reserves."[54]

The JNF naturally followed this tradition and, as an alternative to outright sale, issued special use permits that enabled tenants to live on the national forest. The residents who engaged in such special use permits sold their land to the JNF, escaped property taxes, and retained lifetime residency rights. William Campbell remembered one such elderly lady tenant who lived in Currin Valley, above Marion on the old Holston District. After she died, some of her younger relatives tried to stay on in the house, but District Ranger Ockers "aggravated them so much that they finally left," and he and Campbell immediately razed the farm's structures to prevent the relatives from moving back. For legal tenants, the JNF would actually do repair work on residences, such as the roof they reshingled with split-shakes on a house located on the New Castle District near Newport.[55] But the JNF did not have a perfect record in regard to tenancy. In 1938 an in-house inspection criticized the JNF for neglecting severely impoverished tenants on the Clinch Ranger District and for focusing too much on recreational development at the cost of these indigent people.[56] Despite such mistakes, the JNF generally advocated and practiced a flexible approach to land acquisition and, indeed, interaction with local populations that gave the agency a good grounding in public relations. To a great extent, JNF employees became part of the community.

Harmonious Public Relations

Since the national forests developed at the very heart of rural America, the Forest Service sought to integrate its personnel into the existing sociocultural setting. This became especially important in eastern national forests, where land once privately occupied became public, in sort of a reversal of the historic American trend of making public domain available for private homesteading, farming, industrialization, or other development. Thus the Forest Service encouraged their rangers to move into local areas and become part of individual communities.[57] As the *National Forest Manual* read in 1929, "The district ranger should by all means consider himself a member of the community in which he is located and take part in community affairs to the fullest extent compatible with his duties and the legal and departmental limitations on political activity. He should be content to raise his family in the village or isolated locality where the headquarters of district rangers are often necessarily located."[58] Such was largely the case upon the early JNF.

William Campbell described some of his interaction with the local people: "If I was downtown on Saturday, these people would come to town. [If they saw me they'd say,] 'Hey, I need a wood permit.' And I'd go up to the office and get them a wood permit—take about five minutes and, you know, good public relations."[59]

The main idea was education and interaction. The Forest Service wanted its rangers to diffuse knowledge about forestry, diplomatically interpret official government policy under local conditions, and generally perform all official duties with a pervading sense of goodwill. "Information will be given tactfully and violations prevented by friendly advice rather than by offensive warnings," the Forest Service manual stated in 1929.[60] The early JNF typified this approach, and rangers and ranger assistants regularly ingratiated themselves with the local community through informal social interaction and professional public relations.

Early JNF employee Brad Clark, a native of Scott County, walked from house to house during the 1930s, introducing the new national forest and disseminating fire prevention information. Clark and others tried to calm fears that "the government" planned to seize land whether the owners liked it or not. They also asked many personal family questions, such as the numbers and ages of children. This latter query was part of the JNF's attempt to assess its community profile.[61] Like other national forests, the JNF began sponsoring "show me" trips in an effort to educate and entertain the public. An early example, undertaken on the Holston District in November 1937, involved local residents from Abingdon, Marion, and Wytheville who went on a tour that covered practically every facet of national forest management. They visited the Backbone Rock recreation area and the Feather Camp fire tower, ate lunch at a CCC camp, and observed a private timber sale near Troutdale.[62]

In the wake of timber-stripping in the Mount Rogers area, the Holston Ranger District followed the broader JNF agenda. During the 1940s, the priority was commodity timber production of the most marketable tree species for local, small operators. The Holston Ranger District accurately saw itself as a "stabilizing influence" following the industrial logging frenzy.[63]

JNF personnel also tried to work with their communities by employing local people, ranging from temporary hires during firefighting season, to full-time employees who spent thirty-year careers working for the JNF. During the late 1930s New Castle Ranger J. N. Van Alstine employed many local men in extensive telephone line construction.[64] Ernie Karger, who worked for the JNF during the 1930s and 1940s, recalled what a valuable asset Scott County native Brad Clark was to the early Clinch District. Clark not only knew most of the local people but, more important, their nuances and range of attitudes toward the Forest Service. As a neighbor and sometimes a relative, Clark could approach area residents about sensitive issues, such as incendiarism or poaching, often with far more success than outsiders such as Karger.[65] Firefighting in particular required local crews, and the JNF regularly hired local fire wardens. Before World War II a massive campaign against wildfire dominated the entire Forest Service, including the JNF, and such a campaign obviously required the trust and cooperation of area residents.[66]

In the meantime, broader trends within the greater agency and, indeed, the nation continued to drive the Forest Service toward its multifaceted and increasingly complicated land management endeavor. Arthur H. Carhart, Aldo Leopold, and others in the Forest Service advocated preserving land within national forests as early as the 1910s and 1920s. In 1924 the Forest Service set aside its first wilderness area in New Mexico's Gila National Forest. The Forest Service continued to reserve various lands for special protection and in 1930 designated its first roadless areas. Other early Forest Service considerations for undisturbed land included "research natural areas."[67]

In 1934, near the JNF's New Castle District, professors and scientists began teaching summer biology courses at the Mountain Lake Biological Station, and research during the 1930s and 1940s included limnology studies and native fish surveys.[68] The station functioned more as a summer camp for biologists during this earlier era, when biology involved careless introduction of nonnative species, and the collection and even overcollection of other interesting species. Early biologists tended to focus on description and classification, species' life histories, and geological association of various flora.[69] But even in the JNF's earliest era, when utilitarian resource concerns (or at least their custodial management) preoccupied its initial endeavors, already beginning was the effort for preserved land, pure science, nongame species protection, and other considerations that grew over subsequent decades.

This has remained a constant Forest Service theme, the balancing or juggling of national policy or greater public sentiment under local conditions with consideration for local people—but in all cases the striving for excellent stewardship of the forest environment. The ambition of such a multifaceted endeavor has, in some ways, been both the consecration and curse of the Forest Service. The versatility deserves much admiration but also draws criticism from advocates who emphasize only one of the many national forest resources. The JNF largely escaped this kind of controversy during the early years of its existence, mainly because the Forest Service was in a recovery mode in this part of the nation. Plenty of controversies over management and policy would arrive later. But, initially, JNF rangers, land managers, and other personnel fostered a great deal of harmonious interaction with the local residents and visitors. Such harmony in itself involved quite varied resources, involving nearly all aspects of the Forest Service mission.

Chapter 6

The Depression Era

The JNF arose out of an exceptionally dynamic period in American conservation history. The New Deal's social programs, particularly the Civilian Conservation Corps (CCC)—Roosevelt's "tree-planting army"—accentuated the creation and maintenance of the JNF. Indeed, the 1930s were something of a culmination of at least a hundred years of a developing conservation ethic that grew throughout the nineteenth century until reaching early fruition during the Progressive era.[1] But by the time FDR officially established the JNF, most of Appalachia was in even worse economic difficulty compared to the rest of the nation. In 1936 almost half of southern Appalachian people accepted federal relief. Along with the Great Depression, Appalachia now suffered from a previous half century of intensive natural resource exploitation, prevalent absentee landownership and, concurrently, a surging local population that contributed to a growing demand on shrinking hillside farms. In many ways, the people of southwestern Virginia and other southern Appalachian locales were caught in a vicious circle.

By the early twentieth century increasing population and a shrinking per capita land base had rendered the traditional "land rotation" type of highland agriculture untenable. With a relatively low population base, previous farmers had practiced land use somewhat similar to (but more protracted than) the medieval European "open field" system—only the open land did not remain fields at all, but rather were allowed to return to forest, likely through natural seeding by abutting woodlands. The disruption of sustainable population to land ratios pressured mountaineers to violate their traditional practices, which meant not leaving land fallow long enough, which in turn increased erosion and resultant fertility loss.[2]

In any case, unprecedented intense hillside farming and annual burning caused the most severe soil erosion on the steep lands constituting most of the JNF.[3] Putting such land back into forest was often the best choice, and amidst the Great Depression, the Forest Service was one of the few entities financially and ideologically equipped to do so. Urban growth, its economic attractions, environmental degradation caused by coal strip mines, and general agricultural decline encouraged an exodus from the mountains. Even before the Great Depression (and continuing into the 1960s) cleared farmland began to diminish in the southern highlands.[4]

The JNF helped rehabilitate large tracts of damaged land and, in the role of caretaker, began to play an intimate role with the remaining local people and their economy. The early timber program initially entailed cruising the districts to compile inventory estimates and volume tables for potential growth. Most early timber sales involved only a few select trees to local cutters. Where timber companies had provided a very large but temporary economic boom, the JNF began regularly sustaining small, portable mill operators and supplementing the income or resources of local farmers who engaged in small-scale timbering or firewood gathering during their off-season.[5] As the woodlands recovered, they once again became potential wildlife habitat, and game managers eventually restocked streams with trout, and the forest with deer and turkey. Additionally, the JNF began to feature a number of outdoor recreation sites, many fostered by Civilian Conservation Corps construction. All these activities presented much interesting interaction between a large federal agency and the local population, whose habits and culture evolved along with the new national forest.

JNF employee Joe Glass and his motion picture equipment, touring southwestern Virginia in 1937 to educate the public about forest fire suppression. Fire fighting was one of the main preoccupations of early JNF personnel. U.S. Forest Service photograph.

Clinch District fire-fighting crew, 1937. U.S. Forest Service photograph.

Fire Control

By the time the Forest Service established the JNF, they had long won the battle for protecting or nurturing forests back from a cutover state in the name of watershed protection. The other ideology central to the original Forest Service mission, fire prevention, followed a somewhat more convoluted route. Initially, of course, the Forest Service began with the adamant curtailing of all fire. JNF rangers found themselves in a protective role, with fire prevention in the slash-filled, cutover woods their major concern.

The advent of Forest Service fire control marked a distinctly new era in environmental impact. Long before human activity took place in North America, certain flora species had adapted to lightning fires, which still account for 10 percent of the fires in the United States today. Native Americans broadcast fire, mostly for hunting purposes, and Euro-Americans settling in the southern United States evolved a particular folk culture involving regular burning of wood and agricultural lands. Consequently, the South came to dominate fire incidence in the United States, and fire control dominated Southern national forest history during the early twentieth century. By 1901 some 4.5 million acres of the Southern Appalachian region had experienced the effects of light fire, at least, which consumed the forest litter and killed many smaller plants and some trees. Over 78,000 acres had received severe fire damage that had killed most of the trees.[6]

State firefighting preceded federal fire suppression in southwestern Virginia, and as early as 1915 Virginia's state forester, Chapin Jones, used Weeks

Apple Orchard Lookout Tower, Glenwood Ranger District, probably 1960s. U.S. Forest Service photograph.

Act fire funds to oversee fire patrols in Smyth, Grayson, and Washington counties. Various lumber, land, and coal companies built some of the earliest fire towers on JNF or adjoining territory. Both Marion's Douglas Land Company and Konnarock's Hassinger Lumber Company helped finance early fire protection, and by 1916 two fire wardens patrolled their holdings. In 1917 Wise County's Virginia Iron and Coal Company built a wooden fire tower near Big Stone Gap, and Clinchfield Coal Corporation and W. M. Ritter Lumber Company contributed three wooden towers during the same period. But the huge fire control effort arose during the 1930s when the CCC performed much fire prevention work on the JNF, including road, trail, fire tower, and telephone line construction, as well as direct fire suppression.[7]

The original tower men could see about ten miles in any direction and scanned the horizon every fifteen minutes during daylight hours. During a bad fire season, they took only short naps throughout the night and rose every hour to watch for fire. Early JNF fire-prevention technology included the Osborne firefinder, which coordinated a circular dial with a map of a given tower's territory (zero indicating due north) and enabled the tower man to radio or telephone a fire's location to a dispatcher. The dispatcher, in turn, alerted area fire wardens. Then the tower man would continue to monitor the fire with the Osborne firefinder, notifying the dispatcher of any changes.[8] Fire

JNF ranger Lewis Smith, Glenwood District, using the Osborne Fire-Finder, probably 1950s. U.S. Forest Service photograph.

danger stations also contributed to early firefighting efforts by monitoring forest conditions. The station employed an ingenious device for measuring ground moisture: rangers simply calibrated thin, absorbent basswood strips, whose weight significantly changed with minute moisture fluctuations. Three daily readings were necessary during fire season, when relatively safe forest conditions could turn highly dangerous within only a few hours. By 1940 the JNF had nine of these stations, and during the height of spring fire season, between March and May, rangers would check them three times a day.[9]

As the southernmost national forest in the Forest Service's old Region 7, the JNF became a hot spot in fire control. In 1936 the Jefferson and Cumberland (now the Daniel Boone) national forests accounted for 80 percent of Region 7's fire problems. JNF Supervisor John McNair recorded more than 400 acres burned in forty-one fires during the forest's first operating month (the May fire season) alone. Debris burners and smokers caused about half of these fires; lightning, accidental, industrial, and recreational fires accounted for another fourth. Arsonists made their early mark, deliberately burning almost 100 acres. The JNF began requiring a campfire permit after October 1936 in an effort to diminish accidental fires, but nevertheless had contended with 114 fires of various origin by the year's end. In 1939 alone 120 fires burned some 3,000 JNF acres.[10]

Fire problems tended to accompany newly acquired territory, and once JNF personnel began working with local residents, fire incidence declined.[11] Rangers used a variety of materials and techniques to convey their antifire message, including posters and signs, public lectures, radio messages, newspaper and magazine articles, motion pictures, exhibits, and visits to area schools.[12] "You loafed around country stores," former JNF Holston Ranger District employee William Campbell remembered. "That was part of fire prevention." Much of the Forest Service's success depended upon its rangers' ability to become part of local communities, as well as Forest Service employment of local fire wardens. A 1931 Forest Service Fire Handbook illustrated this important aspect and almost read like a study in psychology:

> The bulk of the action will have to be taken by human beings through other human beings. Much of it will be man-to-man work. Of prime importance we have the task of getting close to our local people, of knowing them as intimately as possible, of studying ways and means of directing their thoughts into right channels. Such a task requires the best of tact, diplomacy, and fellowship. It demands a keen insight into the workings of a man's mind, a study of the facts or fancies which cause him to reason in certain ways.[13]

Indeed, during 1938–39 the Forest Service sponsored a psychological study of the traditional southern folk burning in an effort to better understand, and therefore combat, this prevalent problem.[14] Such an approach proved quite important when involving arson. Several motivations lay behind arson, including revenge against the Forest Service, simple pyromania, or the economic incentive provided by employment on firefighting crews. In dealing with arsonists, the Forest Service again advocated rational investigation: "The incendiary. An individual who is hard to reach. Study people. Most incendiaries are local folks and they set fire to the woods for some specific reason. Find out that reason. What is the incendiary's motive? If known, are we not in a better position to make an attempt to convince him of the error of his ways? Meanwhile, watch him."[15]

The employment motivation in arson knew great extent. Local firesetters seeking work with the Forest Service sometimes rigged time-delayed incendiary devices in the forest, such as a slow burning "rope punk," or a magnifying glass positioned to focus the noon sun upon a pack of matches. Then they would joke about it, predicting when and where they were going to have a fire that day—and they would.[16] The Forest Service sometimes countered this tactic by employing only firefighters from outside the immediate locale, thus discouraging the local job incentive.[17] But again, the Forest Service believed psychology was an important factor concerning fire investigation:

Good interviewing requires ingenuity and hard thinking; no two cases can be handled exactly alike. A man who is reluctant to talk can often be brought to it by directing the conversation first along lines in which he is personally interested, even though this at first has no connection with what you want him to talk about. A little flattery also is often effective. Reluctance to talk, however, may arise from a fear that you are trying to implicate him in the case. Such a suspicion should be guarded against when it is unfounded. Antagonism can often be avoided by stating to a witness that you have been requested by headquarters, or are required by regulation, to get the facts in the case, and will appreciate it if he can tell you anything about it.[18]

All in all, catching arsonists proved very difficult and often impossible. "You almost had to see them light the match," William Campbell recalled. As a young general district assistant, he wrote up one man he suspected of torching the woods, only to have the case thrown out of court. Months later he learned that the judge depended upon the arsonist for bootleg whiskey, so after that Campbell only concentrated on educating the public and extinguishing fires once they started.[19] In another case, on the New Castle

Dispatcher J. N. Jefferson tracking forest fires, Clinch Ranger District, 1937. The poster in the background reads, "Your Forest, Your Fault." U.S. Forest Service photograph.

District, local firefighter J. H. Watson Smith remembered finding a suspected arsonist standing beside the last of a long string of small fires that Smith and other firefighters had followed through the forest. But the man denied setting the fires and claimed he had started the one beside him—not yet out of control—only to warm his hands. In court, the man escaped conviction because of an absence of, technically, anything but circumstantial evidence.[20]

Despite the arson, most early JNF forest fires were accidental. Careless brush burners, hunters, lumber mills, and campers all contributed to accidental fire, as did open-flame mash-burning moonshine operators. Before the days of chemical pesticides, tobacco farmers burned their fields every year to kill weeds and insects, and a fire in a field abutting the forest could easily spread. Also, before the days of barbed wire, forest fires spurred farmers to light their own backfires in an effort to protect their chestnut rail fences—no longer replaceable with the blight's destruction of that tree. Sometimes these backfires got out of control and burned much more than the intended boundary territory, and in general burned much of the forest that might otherwise have escaped damage.[21]

During the early years of the campaign against fire, the Forest Service sought punishment more as an educational measure than anything else. Former JNF Clinch Ranger District employee Brad Clark recalled the general legal policy surrounding accidental fires on the early Clinch as a four-part process: the first offense elicited a friendly letter from the ranger. The next violation earned the careless burner a ticket, which the local judge would usually suspend with a warning. The third offense resulted in a moderate fine, and the fourth, a heavy fine of fifty to seventy-five dollars during the 1930s.[22] Former JNF employee Ernie Karger recalled a similar approach, wherein the Forest Service wanted to educate "firebugs" more than punish them. Karger and various JNF rangers would approach a judge prior to a trial and explain their desire to befriend, rather than alienate, incendiaries. As a result, a judge tended to hand down a suspended sentence at first, accompanied by a stern lecture, with the promise that a second offense would incur actual and additional punishment combined with the original suspended sentence.[23]

Timber and Other Forest Products

The two great thrusts of early JNF land management, watershed protection and fire prevention, came right out of original Forest Service advocacy and generally succeeded quite well. Directly related to these efforts, of course, was the propagation of timber. In the minds of early conservationists, trees constituted the central resource of national forests. Before timber management could begin, JNF rangers had to determine just what kind of forest they had. Therefore, they focused much of their early effort on timber survey. Such a massive endeavor required large numbers of men, and the JNF immediately utilized CCC labor. During the JNF's first month of operation in 1936, Glen-wood Ranger Lewis Smith used CCC crews out of Camp F-17 at Big Island to run baselines and cruise timber. During this same month, JNF personnel gave 1,543 hours of practical and forestry-oriented instruction to 127 enroll-ees from Camp F-13 at Arnold's Valley. A year later, the timber survey pro-gram began in earnest. By April 1937 "one of the most ambitious programs yet undertaken on the Jefferson National Forest" involved surveying the entire Holston District's timber, covering some 95,000 acres. The proposed project would take a year to complete and would use the labor of ninety CCC enrollees.[24] This timber survey constituted the first step of placing the terri-tory's timber on a sustained-yield program.

In addition to timber survey, early JNF timber management concen-trated on mostly small timber sales, small-scale cutting of selected individ-ual trees or species, tree planting, combating the white pine blister rust, and coping with the devastating chestnut blight. Additionally, early JNF foresters engaged in timber stand improvement (TSI). TSI involved careful thinning or cutting to eliminate damaged, diseased, or disease-prone trees in order to fa-vor straight trunk trees or certain commercially valuable trees. All early JNF foresters marked their timber selectively, for all-age management was still the Forest Service standard. The Holston District allowed woodcutters to har-vest various hardwoods like birch and hickory for mine props. Early Glen-wood sales tended toward pine pulpwood, railroad ties, chestnut poles, and "extract" (also called "extractwood" or "acidwood"), which entailed mostly chestnut (and some chestnut oak) from which leather manufacturing com-panies derived tannic acid. The Glenwood District and New Castle districts sold wood to the Lynchburg and Radford extract plants, Covington's West Virginia Pulp and Paper Company, West Virginia's coalfields in need of mine props, local railroad companies needing cross ties, and various other lumber markets.[25]

In addition to timber sales, selective cutting, and TSI, early JNF foresters enhanced the young forest with replanting. The Great Depression brought Roosevelt's CCC tree planters onto the JNF and other national forests in a massive, unprecedented—and since unsurpassed—reforestation effort. The CCC planted so many trees that by the late 1930s Region 7 dropped refores-tation as one of its major goals.[26] Even though intensive white pine planting produced a dramatically altered forest ecology compared with the mature or even regenerated hardwood forest, the land certainly improved in several substantial ways. The pine stands created wildlife cover, checked erosion on the steeper slopes, and commercially it artificially initiated a fast-growing crop of trees for later harvest. Even in terms of forest ecology, the pine stands actually created a more diverse (albeit artificial) forest considering the pre-vious predominance of hardwoods. Otherwise, abandoned farmland would run the slower evolution from grasses to cedar trees and eventually to other evergreens—Virginia pines, hemlocks, spruces—before hardwoods established themselves. Artificially planting such areas in the highly adaptable white and scotch pines virtually created a young forest in the course of a spring planting season.

White pines, however, did not grow without problems. Tree inspectors first discovered white pine blister rust in New England forests in 1906 and concluded that the disease had arrived from Europe. As the rust spread, eastern states began a concerted effort to combat it together, which cause eventually contributed to the organization of the Association of State Foresters. By 1924 the rust had spread south to western Maryland and invaded the Appalachians around 1928. During the 1930s, when the chestnut blight reached its most devastating extent, the blister rust also spread throughout the Appalachians. By 1933, 12,000 CCC enrollees combated the blister rust in twenty-two states and utilized a budget of more than $2 million. Richmond, Virginia, became the headquarters for combating the blister rust in the southern Appalachians, and between 1928 and 1937 various relief workers (including CCC enrollees) combated the disease on 338,600 acres of the Shenandoah National Park, George Washington and Jefferson national forests. Their primary efforts lay in destroying all currant and gooseberry plants (the alternate-year hosts for the rust) within 900 feet of white pines. William Campbell performed such work on the Holston District during the late 1930s, and other foresters performed similar duties on every district except the Clinch District (whose mature and regenerated hardwood stands contained no white pines).[27] Eventually the blister rust proved so serious a problem that Congress provided funding to combat it through the 1940 White Pine Blister Rust Control Act.[28] And as bad as the blister rust may have been, the damage of its duration and aftermath in no way approached the magnitude of the devastating chestnut blight, also introduced from the Old World.

Once constituting 40–90 percent of some Appalachian stands, the American chestnut, one of the world's fastest growing, most decay-resistant trees, contributed to an Appalachian ecology only imaginable today. Their tremendous mast production alone provided a source of wildlife sustenance unmatched by any other of the region's tree species. Early mountain settlers found numerable benefits from chestnut trees, both as a food source and building material. Settlers' hogs ran the woods at will during the summer and fattened on abundant chestnuts throughout the forest. The tree's high oil content provided good tannin for leather work, and the relatively soft, highly resilient wood made chestnut the most popular building material for cabins, outbuildings, and fences.

The chestnut blight was caused by a fungus (*Endothia parasitica*) and apparently came to North America during the early twentieth century.[29] The effect upon the American chestnut (*Castanea dendata*) species rapidly grew to catastrophic proportions. As everywhere in the Appalachians, the chestnut blight literally transformed the hardwood forest.[30] On the early Glenwood purchase unit, where chestnut constituted 60–90 percent of the forest. Rangers first detected the blight around 1912, with probably little idea of its implications. In 1914 the Glenwood's silvicultural techniques still favored the chestnut as a slope species, as though the chestnut were still a viable resource. But the blight's rapid progress drastically changed the foresters' perspective. By 1922 Glenwood personnel suddenly favored salvaging "as much chestnut as possible prior to its destruction" and advocated cutting all chestnut trees in heavily populated stands that showed an 80 percent or greater degree of blight infection.[31] By 1926 rangers fully realized the impact of the destruction under way, even though they could not fully anticipate the final result. Nevertheless, they made plans to carefully favor other species in an effort to fill the impending void.

Clinch District ranger John Shipley indicating a tulip poplar illegally peeled for creating a huckleberry bucket. U.S. Forest Service photograph.

By 1936 most of the Glenwood's chestnut had already died, and rangers devoted their efforts toward salvaging what remained. In 1938 JNF Ranger J. N. Van Alstine estimated between 15,000 and 20,000 cords of dead or infected chestnut on his New Castle District alone. He sought its removal both as a stimulus to the local economy and as a way to remove a huge wildfire hazard. The chestnut tree's natural oil now became as much a drawback as it had been an asset; where split-post fences and cabins were famous for endurance, the masses of deadwood now alarmed foresters with their incendiary potential. In 1940 the Virginia mountains contained 758 million board feet of dead standing chestnut, and the accessible timber alone constituted a fifteen-year industrial supply of tannin.[32] Obviously the board feet and cordage figures following the aftermath of the blight indicate what a huge component of the forest the American chestnut had been. The tremendous loss of mast alone dramatically altered the food supply for wildlife species.

After the chestnut tree faded away, a radically altered forest persevered. Local residents lost a source of free food for themselves and their free-ranging hogs and lost one of the best wood species ever known to humanity. But local people continued their traditions of forest foraging for nontimber resources. These practices had prehistoric origins, of course, and Native Americans had sometimes shared their knowledge with the early newcomers to the region. Certainly forest foraging coincided with the earliest historic occupation of the southern Appalachians.[33]

Much of the early JNF's interaction with local people involved permits for gathering "herbs" such as ginseng, galax, elderberry flowers, polk berries, buck vine, lobellia, moss, and cherry bark. The JNF used permits as an attempt to regulate or restrict resource use. In the case of cherry bark, in demand during the 1930s and 1940s as an ingredient for cough medicine, the JNF tried to prohibit its gathering altogether, for cherry trees were not common, and stripping their bark, of course, killed the trees.[34]

Locals also requested what they sometimes called "woodmits" for gathering firewood. During the late 1930s, Ranger Van Alstine issued numerous free firewood permits to area residents. He also advocated typical Forest Service multiple-use-sustained-yield philosophy with the idea that New Castle District timber could supply fifteen of the twenty-two local portable sawmills that were rapidly cutting the area's remaining private woodlands. In this vein, he observed "that the U.S. Forest Service [had] gone a long way in bettering the social and economic problem" on his district.[35]

While the larger operations such as the Hassinger Lumber Company shipped more than 99 percent of its timber to old business associates located no closer than Philadelphia, the JNF encouraged local market, small-scale forest-dependent industries and tried to supply them with a regular source of wood. As outlined in the *National Forest Manual*, 1928–33, the Forest Service gave priority to local economic needs, from the planning stage down to the ranger in the field, ensuring protection of isolated timber markets from "exorbitant or unreasonable prices," even to the point of completely closing the market to outside interests. The 1922 management plans of the JNF's Glenwood Ranger District (then part of the Natural Bridge National Forest) favored "local operators as much as possible" and even encouraged local cutters to harvest their timber during the autumn and winter, when farm work demanded the least of their attention.[36]

The old 202-C permits, named after the Forest Service form designation, dominated JNF timber sales to local woodsmen. Many of these sales amounted to less than $500, so the districts scaled them for price instead of offering them in open-bid. The sale could range from a few select trees to a maximum of sixty acres. The average sale was less than 50,000 board feet. In 1922 the Glenwood working circle reported few sales exceeding $100.[37] As a 202-C customer during the 1930s and 1940s, Hermie Medley described a good working relationship with New Castle rangers Van Alstine and W. W. "Buck" Taylor, from whom he bought much timber through 202-C sales. "I always made money," he remembered, but noted that the rangers did not sell timber merely for the asking. The rangers could not trust certain area loggers to cut only specified trees, clear brush, leave low stumps, and generally practice good forestry, and therefore would not issue them permits. On the other hand, Medley always tried to practice conscientious logging, and as a result always received permission to cut a certain amount of national forest wood every year.[38]

Recreation

The timber resource naturally dominated Forest Service ideology, policy, and budget from its earliest history, and only from within this established context did the recreation resource make its first inroads. Despite several prominent Forest Service employees who recognized the importance of recreation during the 1910s, recreation had always suffered inadequate funding.[39] Agency foresters tended to view recreation as a noncommodity use and resisted deviation from the Forest Service's early utilitarian focus. Also, with the creation of the National Park Service in 1916, some in Congress saw national forest recreation as redundant and refused funding on those grounds.[40] Finally, in 1922 Congress gave the Forest Service its first recreation budget of $10,000, but not until the Great Depression and subsequent New Deal did national forest recreation receive significant expenditure.[41]

Southwestern Virginia's magnificent scenery constantly attracted local hikers and campers and, since the 1930s, progressively more tourists and other outdoor enthusiasts from outside the area. By geographical determination the JNF acquired much of southwestern Virginia's most mountainous land, ill suited for agriculture and more expensive to log than valley land. JNF territories such as High Knob, Mount Rogers, and Mountain Lake became some of the higher elevation areas that attracted visitors to their scenic beauty. During the first decades after the 1916 establishment of National Park Service, the Forest Service grew much more interested in providing outdoor recreation despite a consistent dearth of funds for such endeavors. Some of the Forest Service motivation arose from interagency jealousy, and from fear of having the Park Service acquire national forest land.[42] But national forest recreation traditionally differentiated itself from concentrated national park recreation by providing a dispersed (and less expensively maintained) outdoor experience. The Park Service, as a single-use agency, has always focused on recreation in preserved sites, while the Forest Service, as a multiple-use agency, provided recreation within a forest management system including other demands.

As the 1929 *National Forest Manual* stated, "It is not the purpose of the Forest Service to duplicate within the national forests the functions, methods, or activities of national, State, or municipal park services, nor to compete with such parks for public patronage or support."[43] In 1940 a journalist echoed these sentiments in writing about the JNF's Cave Mountain Lake Recreation Area in Rockbridge County, Virginia, describing it as democratic, "rustically simple," minimally policed, and unregimented.[44] Such a description typified early JNF recreation.

Between the 1910s and 1930s recreational interest in southwestern Virginia began to coincide with the slow, early development of the recreation profession within the Forest Service. A "Division of Recreation and Lands" appeared in the Forest Service directory for the first time in 1935, and Depression-era work programs resulted in tremendous recreational-facility construction. Between 1934 and 1941, the Emergency Conservation Work program spent from $3–$15 million annually for eastern national forest land acquisition, especially geared toward CCC-led recreational development. Among other facilities, the CCC developed the Cave Mountain Lake and High Knob facilities on the JNF.[45]

Tourist boosterism is almost a perennial disease among certain local business people, especially in impoverished areas of scenic beauty. Nowhere on the JNF has this idea appeared more apparent than on the Clinch District near the Virginia-Kentucky coalfields. Since the nineteenth century absentee ownership of minerals and the early years of exploitative coal mining labor has contributed to this area's poverty. To some, tourism seemed like a viable alternative for local prosperity, and even before the NFRC began buying land in that locale, various business and community leaders perceived recreational economic development potential.

In 1937 a Pikeville, Kentucky, group comprised of the Pikeville Chamber of Commerce, Rotary Club, American Legion, and others advocated the creation of a national forest in the vicinity of the Virginia-Kentucky border. Here they perceived an impoverished area of poor quality timber and exhausted mines, where a depressed coal industry had contributed to a farm tenancy rise of 88 percent in five years. Steep hills and poor logging practices had caused severe soil erosion and seemingly exacerbated flooding in the area. An average acre rendered only thirty-five board feet of timber. By 1939 the JNF had bought 15,000 acres of this territory from the Clinchfield Coal Corporation, thus ending "a campaign which business leaders and sportsmen of Kentucky and Virginia" had been advocating for many years.[46] Almost a century later we see that tourism has not solved the area's economic problems.

Wildlife

Linked to outdoor recreation, but even more important to local cultural traditions, was the hunting and fishing that revived in southwestern Virginia in direct connection with the development of the JNF. Overhunting, overtrapping, and habitat destruction decimated much of the native wildlife that once lived on JNF territory. Wolves disappeared during the late nineteenth century, and elk soon followed. Beaver had vanished completely by 1890, and deer, turkey, bear, and mountain lions experienced almost the same fate. Native trout suffered terrible losses, resulting from deforestation that deprived streams of vital shade (and thus temperature control) and lack of watershed retention, which choked streams with sediment.[47] In 1901 H. B. Ayres and W. W. Ashe observed in their study of the southern Appalachian forests, "trout have been dynamited, deer hunted, and turkey, quail, and pheasants slaughtered until game is nearly exterminated."[48]

The Forest Service began working with local game associations directly after the 1911 Weeks Law established the first purchase units in Virginia. Early cooperative work centered on fire protection and restocking. At this point, wildlife management was only at its earliest stages, with rare individuals such as Aldo Leopold advocating game conservation. In fact, most wildlife rehabilitation programs did not gain momentum until the 1920s.[49]

Cooperation between Virginia and the federal government gained an early advantage during the early 1930s when Justus H. Cline, a native Virginian and director of the American Wildlife Federation, met Allen R. Cochran, who later became forest supervisor for the JNF. Men like Cline and Cochran became essential for rehabilitating wildlife on the national forests, for they emphasized wildlife as much as most of their colleagues advocated timber. Cochran recalled his own lack of training in wildlife management as typical of others he met in the Forest Service.[50] He wrote in 1949, "Although a few

JNF rangers release hatchery trout into Stony Fork, 1962. U.S. Forest Service photograph.

foresters may have caught the vision of Aldo Leopold, the rank and file of foresters failed to see the potentials of managing wildlife as a resource until it was thrust upon them."[51]

By the 1930s both the Forest Service and state of Virginia were making joint plans for wildlife revitalization. In September 1935 they established the Virginia Cooperative Wildlife Research Unit.[52] With help from CCC enrollees, the JNF created and improved habitats, then began importing and restocking game. The 1937 Pittman-Robertson Act helped facilitate cooperative efforts between Virginia and the JNF by funding restocking efforts, habitat management, and by bringing the federal Fish and Wildlife Service onto Virginia's national forests.[53] All these efforts culminated in the Virginia Plan of 1938, which heralded one of the most cooperative wildlife management programs between any state and its national forests.[54] The 1938 Virginia Plan's federal-state efforts called for joint wildlife management planning and development involving law enforcement, stocking, predator control, emer-

Dismal Creek elk hunt, Giles County, VA, 1958. These men were among the last to hunt the western elk transplanted onto JNF territory. U.S. Forest Service photograph.

gency game feeding, and habitat improvements. The plan also introduced the national forest stamp, which hunters bought for annual hunting rights. The Virginia Commission of Game and Inland Fisheries director met at least annually with JNF and George Washington National Forest supervisors to design wildlife management budgets.[55]

The procedure worked as follows: a state game technician studied various forest watersheds, then advised the ranger about wildlife in relation to a particular habitat. That ranger prepared a work plan for a state game manager, who implemented the actual wildlife or habitat management. The game manager encouraged forest plant food species by creating forest clearings or releasing fruit trees, restocked game species, and monitored hunting and fishing activities.[56] In 1938 the JNF established the 5,660-acre Hurricane Branch Area and the 5,000-acre Feathercamp Area for cooperative wildlife conservation management. Other such areas followed.[57]

Between 1935 and 1939, the JNF released tens, if not hundreds, of thousands of trout into local streams and rivers.[58] Game managers also introduced western elk in a follysome nod to their extinct eastern forerunners. Before poachers finished them off in the 1970s, what had come to be two small herds represented the only wild elk in the eastern United States.[59] But possibly more than any other species, white tail deer thrived in the renewed habitat. Deer

The leavings of poachers, Wythe Ranger District, 1961. U.S. Forest Service photograph.

stocking began as early as 1927 on the Clinch Ranger District and continued until the late 1930s, when the population became self-sustaining. In 1939 the JNF and CCC began building wire-fence Special Wildlife Refuge areas in which high concentrations of game multiplied, later released on the districts at large.[60] Without the presence of natural predators, and with a limited deer-hunting season, white tail deer soon grew to enormous numbers. These large, unprecedented herds would eventually become their own worst enemy.

Moonshine

The enlightened approach JNF rangers and ranger assistants generally took to local moonshiners operating on the national forest reflected a remarkable harmony with the local culture.[61] In general, those working on the early JNF were savvy to the local moonshine practice. William Campbell found many moonshine stills during his wanderings over the Holston District during the 1930s but only left notes asking that they be moved, as was the district's informal policy. "We didn't bother them, and they didn't bother us," Campbell remembered. He asked the whiskey makers to move their stills mostly for fear of fire danger, since moonshiners cooked their mash with open flames. Campbell recalled one still in particular he found on Tall Mountain on the Clinch District; a beautiful device of solid copper—quite remarkable among the more common, makeshift contraptions that often included such devices as car radiators. He complemented the owner of the still the next time he saw him, but the owner appeared shocked. "You didn't mess with that still, did you?" he asked, alarmed. Campbell replied, "No, but you better get it off of national forest land." The still disappeared within a couple of days.[62]

Brad Clark, who began working as a fire lookout on the Clinch Purchase Unit in 1931, declared that "moonshiners were our best friends." After all, fire control alone demanded a much larger workforce than the Clinch District's three full-time employees, and Clark praised moonshiners as some of their best firefighters; men who would only take an occasional day off to run their mash, then return right back to the fire line.[63] James N. Jefferson, who helped build the High Knob recreation area on the Clinch District during the late 1930s, agreed that moonshiners there tended to be helpful to the Forest Service and also tried to be very careful with their own mash-burning fires.[64] Both Clark and Campbell refused to reveal the locations of stills to inquiring Internal Revenue agents, claiming they "had never seen a single still" in their entire career. The rangers could not afford to alienate themselves or the Forest Service from the community, for certainly a rise in incendiary fires, at the very least, would have followed. Besides, Clark remembered, Clinch District personnel saw no point in doing the Internal Revenue Service's job for them.[65]

New Castle ranger Ed Jenkins poses beside a moonshine still that started a forest fire, 1940. U.S. Forest Service photograph.

Technically, though, moonshiners who operated on national forest land were trespassing. Hence, JNF employees would attach notes to their stills, politely asking them to move.[66]

Moonshiners were highly inclined to locate their stills on national forest land, since Internal Revenue officials would then have to catch them *in the act* of making whiskey; whereas merely possession of whiskey-making equipment on private property were grounds for arrest. Sometimes Internal Revenue agents did catch moonshiners in the act on JNF land, such as the two men, 125 gallons of whiskey, and 1,600 gallons of mash they discovered in April 1949. Found in Giles County, near the border of the Wythe and New Castle Districts, the agents described the operation as the "most complete illegal whiskey manufacturing outfit ever found" in Virginia.[67]

The JNF's cooperation with moonshiners remains best understood within the overall context of Forest Service interaction with the people of southwestern Virginia. Land acquisition, timber and other forest flora management, wildlife habitat and species recovery, fire control, and recreation all reflected the JNF ingratiating itself with the local public. Such an approach would be expected in a scenario where JNF personnel lived in the same community with local people, attended weddings and funerals, hosted boy scout meetings, and waved at everyone they saw while driving down the highway—because they recognized everyone they saw. Local people went to see "the ranger" if they wanted to cut firewood, gather ginseng, or fell a particular walnut tree to make a single piece of furniture. The national forest provided hunting and fishing grounds for all local sportsmen, and the entire family could picnic or relax at a number of JNF recreation areas. The national forest was a public resource, and before interstate highways penetrated southwestern Virginia, local people constituted the most immediate public.[68]

JNF cooperation with moonshiners may seem exceptional, but certainly it was the best alternative for the circumstances and reflected the overall harmony that JNF staff sought between the Forest Service and local people. The local people remained strongly attached to their southwestern Virginia homes, despite the difficult economy. When asked why people remained in post-timber boom Konnarock (then adjoining the Holston Ranger District), Dora Testerman, born in 1916, said, "Because they loved it."[69]

Chapter 7

World War II and Postwar Transitions

Among the many aspects that differentiate the early JNF from the agency at mid and late century, probably no other underwent as tremendous an evolution as fire control. Even by the 1950s local situations had drastically changed from the earlier era of annual hillside burning. Indeed, in 1953 Ranger J. N. Van Alstine, who once recalled seeing "Meadow Creek and Sinking Creek black all the way to New River" from ash runoff, reflected that mowing sedge grass had replaced burning it, and the use of limestone fertilizer was improving the land.[1] He joked about selling two of his New Castle District's fire towers, one forty feet high on Sinking Creek Mountain and another sixty-five feet high on Johns Creek Mountain. He explained how the towers had proved essential during the early years for fire detection, fire reporting, and communications. But, he said, "Today we have few fires and most of them are reported by local residents as soon as a lookout could report them. Improved communication facilities . . . and good citizens have canceled out the value of these towers."[2]

World War II gave tremendous impetus to the campaign against forest fires. The Allied Effort, so highly dependent upon timber resources, helped make woodland arson "tantamount to sabotage," and even the FBI began investigating forest incendiarism. During the 1940s a most remarkable drop took place in JNF incendiary fires, accounting for 46 percent of the Forest's fires in 1941, down to 25 percent the following year, 20 percent the next, and only 3 percent in 1947.[3] The 1940 Virginia state law prohibiting open-air burning before 4 P.M. contributed tremendously toward ending accidental and careless fires,[4] and naturally, all the JNF's educational efforts, public relations, and disincentives for "job fires" contributed to a reduction in fire incidence and damage. Throughout the 1940s, fires caused by smokers, debris burners, and railroads also dropped.[5]

During World War II all national forests channeled their resources into the Allied effort through such programs as the Timber Production War Project. In January 1942, Region 7 organized a Unit of War Activities and by September devoted more than 80 percent of its staff and forest resources toward the war effort. These activities included military designation of critical national forest and private timber supplies, added allocations for firefighting equipment, and a more intensive media campaign against firefighting, mainly warning that forest fires aided the Axis powers. In order to meet wartime demand, timber harvesting far exceeded the previous conservation era's limited cutting. Loggers working on the JNF and other national forests felled much wood considered otherwise unsuitable for cutting. Annually, national wartime timber demands claimed about 17 billion cubic feet of wood, 50 percent more than the forests could regenerate in the same time period.[6]

The JNF supplied oak ship timbers, yellow poplar airplane veneers, saw timber for army trucks, and pulpwood for explosives. The Clinch District particularly supplied chestnut-oak, which the military shipped to Seattle for construction in nonmetallic bottoms for minesweepers. Other war demands on wood products included pulpwood for shell encasements, camouflage nets, and even collapsible fuel bags in airplanes. Counties in southwestern Virginia began pulpwood drives encouraging local landowners to meet the nation's need, advocating a "cord of pulpwood for every serviceman." Where Virginia loggers harvested only 450 million board feet in 1932 (continuing the decline after the 1909 peak of 2.1 billion), the war years saw the annual harvest rise more than a billion board feet.[7] JNF timber contributed to such demands, but as Supervisor Cochran commented in 1943, the JNF war effort's shortage lay not in wood but in available labor to harvest it. Nevertheless, during fiscal 1943 the JNF supplied over 8.6 million board feet of such wood.[8]

Fire Control and Timber Management

The successful campaign against fire of the JNF and other Appalachian national forests began to wind down after World War II. Concurrently, the JNF diversified and broadened its firefighting endeavors by utilizing modern fire-suppression technology (such as aerial detection and suppression and even smoke-jumpers) and participating in a national network of Forest Service firefighting.

Throughout the 1950s and 1960s, JNF firefighting technology followed all the modern developments. The JNF began using power pumps, radio communications, and portable fire simulators that used overhead projectors to portray fire, smoke, and background scenes, imitating the fire conditions of various woodlands for training. Old fire towers disappeared as the JNF started using helicopters and planes for aerial detection and suppression.[9] Helicopters saw their first use in fire suppression during April 1963 with the infamous MacAfee Gap fire on the New Castle District, among the very worst in JNF history. During the mid-1960s Region 7's concept of the JNF as a fire "hot spot" continued, contributing to fire prevention as a high funding and policy priority. After the 1950s JNF firefighters began to fight fires on other forests throughout the nation, and after the termination of Region 7 (in 1966), the JNF became something of a training ground for Region 8 firefighters.[10]

Trucks loading pulpwood cut from New Castle District, 1955. Cutting wood for paper manufacture became a major endeavor on many parts of the JNF during the 1950s and 1960s. U.S. Forest Service photograph.

After the war, timber sales on the JNF dropped dramatically, from 19.1 million board feet in 1946 to a postwar low of 5.7 million in 1952. In 1949 timber surveyors classed more than half of Virginia's mountain forests in the nominal category containing less than 2,000 board feet per acre. Only 2 percent exceeded 10,000 board feet per acre. The JNF's diminishing harvest occurred despite an overall Region 7 increase, whose timber sales broke $1 million for the first time in 1953 with the sale of almost 95 million board feet. However, 1953 did mark the beginning of the JNF's rebound, and sales steadily climbed.[11] Such an increase matched a trend throughout the nation.[12]

Following World War II, the timber industry operations themselves generally stood in great contrast from the frantic logging practices of a half century earlier. Companies restocked their landholdings instead of liquidating cutover acreage as rapidly as possible. They increasingly improved their overall efficiency. They also gained greater public approval and more political influence.[13] But it was within the context of this increased harvest that the clear-cutting phenomenon arose, and the subsequent controversy over clear-cutting on national forests. This eventually became a noted problem with the JNF's neighbor, the Monongahela National Forest, in West Virginia. On the JNF, however, comparatively less intensive cutting continued throughout the late 1940s, during the 1950s, and into the early 1960s. Selectively cut, small sales to portable circular saw operators predominated.

On July 13, 1960, Forester L. S. Gross wrote of the JNF's proposed timber management plan, "I was shocked to read . . . that 50 percent to 70 percent of the annual cut during the planned period is expected to be marketed in sales with a value of less than $2000. We cannot permit Region 7 to continue in this foolishness."[14] Perhaps Gross failed to appreciate the nature of the local timber economy, the area's limited transportation facilities (still lacking interstate highways), and the prevalence of practical small sales to area farmers. The JNF's timber sales policy actually suited the region quite well, considering the state of the recovering forest. During the 1950s, JNF timber sales remained around only 8 million board feet a year for the first half of the decade, and around 11 million for the latter half. JNF foresters continued to practice timber stand improvement as a favored silvicultural approach, giving such attention to 24,034 acres by the mid-1960s.[15] This kind of traditional stewardship tended to characterize all JNF timber management after the war and until 1964, when clear-cutting fatefully became part of the JNF timber management plan.

Recreation

In 1948 Region 7 inspectors C. M. Granger (assistant chief of the Forest Service) and A. R. Spillers praised the JNF's recreation value as one of its most valuable resources. In fact, with the exception of municipal watershed protection, they felt that recreational use of land "strategically located would seem to take top priority over other resource use."[16] Granger and Spillers's 1948

High Knob Recreation Area, Clinch Ranger District, 1955. Outdoor recreation also became a major postwar focus of the JNF and all other national forests. U.S. Forest Service photograph.

Green Pasture Recreation Area, early 1950s. This was the JNF's only segregated recreation area until the Civil Rights Movement contributed to the integration of all Forest Service recreation areas. U.S. Forest Service photograph.

observation followed on the heels of the JNF's first stage in recreational development, born amidst the social work programs of Roosevelt's New Deal, typically the dispersed type of recreation commonly found on national forests throughout the United States. Post–World War II prosperity brought a renewed era of recreation use and development in the United States with a surge in population, automobile manufacture, highway construction, and added leisure time for many Americans. Recreation visits to national forests nationwide surged after World War II, from 10 million in fiscal 1945 to almost 134 million in fiscal 1964.[17] The JNF clearly reflected this national trend with a quadrupling of recreation usage between 1952 and 1957 alone.[18] Unfortunately, inadequate funding again characterized Forest Service recreation, just as it had during the pre-Depression years. Despite the sharp rise in outdoor recreational use and demand after World War II, Forest Service and congressional efforts to win greater funding for recreation failed. Sponsors could not win approval for either the 1949 Tackett Bill or 1955 Metcalf Bill, both of which aimed at bolstering funding for national forest recreation. Instead of actual facilities implementation, recreation staff ended up emphasizing planning.[19] Greater congressional funding and Forest Service recognition of the recreation resource, however, began to gain momentum during the 1960s. For the JNF this phenomenon most saliently came to light with the creation of the Mount Rogers National Recreation Area, discussed in the following chapter. Ironically, the Mount Rogers venture partially backfired in JNF's face, in large part due to local opposition to eminent domain and national opposition to concentrated recreation at the cost of environmental impact.

Watershed and Soil Reclamation

By 1948 JNF rangers had gained an appreciation of scientific watershed protection and management, largely through the findings of the Coweeta Hydrological Experiment Station in North Carolina. After the 1954 Watershed Protection and Flood Prevention Act,[20] the JNF and other Region 7 national forests conducted watershed surveys and made plans and impact reports for specific watersheds. The JNF focused on areas under its complete management, such as Johns Creek, the North Fork of the Pond River, and the John Flannagan Reservoir.[21] But other areas, where mineral companies owned mineral rights, left the JNF virtually powerless in preventing environmental and watershed damage. In certain locations, forty and fifty years after the Weeks Act, erosion caused the JNF more trouble than ever before.

The most severe early mining damage on land that became part of the JNF occurred in Smyth County around Glade Mountain and Georges Branch, where manganese mining companies stripped hundreds of acres during the first half of the twentieth century. The damage to the land here was indeed much more detrimental than logging operations, since mining not only entailed cutting the forest but also removing all the topsoil and subsoil in the endeavor to retrieve minerals.

In 1934 the Belmont Corporation reserved mineral rights on Glade Mountain in what became the JNF's Holston Ranger District two years later. The Belmont Corporation held these rights until 1966, and during these thirty-two years various mining operators, such as Union Manganese Com-

"Waste land" near Blacksburg, Virginia, spring 1931. Foresters recommended that this overgrazed land be planted in pine and locust trees. Virginia Department of Forestry photograph.

pany and Manganese Mining and Contracting Company, extracted many tons of ore from the area.[22] JNF personnel constantly struggled with the Glade Mountain manganese miners in an effort to keep their operation within even the most basic environmental standards, such as grading their roads at no more than 7 percent. Actual corporate rehabilitation of exploited mining sites remained beyond the law. The JNF's Glade Mountain Rehabilitation Plan of the late 1950s revealed that the Forest Service could not "require the operator to completely rehabilitate the entire area," even though the JNF made every effort to persuade them.[23] As Region 7 Forester Hamilton Pyles wrote to Murray Stevens of the Appalachian Trail Conference in 1960, "Where minerals are outstanding or reserved, the Forest Service has very limited authority to control the location and method of operation and in only a small number of cases can we require restoration and rehabilitation of the area."[24]

When Glade Mountain streams ran red and brown with sediments, the Holston Ranger District—not the mining company—built settling ponds to check erosion. In the long run, the Glade Mountain area became, in terms of acreage, the JNF's largest strip mining site and soil restoration project.[25] The JNF inherited this problem and, without actually owning the specific territory's mineral rights, had to cope with the environmental damage without

Land damaged by Union Manganese Company, 1955, near Sugar Grove, Virginia. The JNF eventually rehabilitated this property. U.S. Forest Service photograph.

the benefit of any significant legal authority. Where the JNF owned a tract's mineral rights, however, they had much more control over its extraction, even during this era preceding stricter environmental laws.

Throughout the 1950s and 1960s, the JNF approved, rejected, or required further measures on applications to lease federally owned iron and manganese. For instance, in January 1958 the American Manganese Company of Athens, Ohio, received permission to lease 157 acres in the Bland-Giles area on the new Blacksburg District. The JNF required the company to agree to "special terms and conditions" entailing restoration, repair, and reclamation. These efforts would include such practices as limiting the height of spoil banks, building sedimentation ponds, and reseeding disturbed areas. Other applications required modifications. Joseph B. Hyman's September 1957 prospecting permit for 575 acres on the Clinch District's Pine Mountain met with a certain amount of resistance from JNF Supervisor W. C. Curnutt. He felt that the land in question presented a "difficult and critical" operation and that "necessary safeguards and rehabilitation measures will be complicated and expensive." Supervisor Curnutt withheld permission to begin until Hyman submitted a more detailed plan that would include soil and water protection, rehabilitation measures, and proof of financial capability in performing these measures. In another instance, the JNF approved an iron and manganese mineral permit in Giles County but reduced the proposed territory from 800 to 610 acres in order to protect wildlife habitat along the crest of Flat Top Mountain.[26]

One of the most conspicuous mining controversies on the JNF erupted in 1960, when E. L. Keesling from Bramwell, Virginia, proposed to lease 10,500 acres in the Bland-Giles area for manganese mining. Some of the public, especially hunters, voiced their opposition to this proposal. They expressed particular concern regarding the proposed mining's effects upon the 8,000-acre Dismal Wildlife Management Area, where one of the JNF's only elk populations struggled for survival. During July 1960 Supervisor Curnutt recommended complete rejection for the entire lease. Curnutt explained that this proposal resembled earlier manganese operations on the JNF, adding that "we have not been able to cope with large scale operations of this type. Until it can be demonstrated on areas already disturbed that these can be handled to adequately safeguard surface values, we cannot recommend these operations on undisturbed areas."[27]

The Izaak Walton League soon joined the opposition, as did R. H. Cross of the Virginia Commission of Game and Inland Fisheries.[28] Among the proponents, Virginia Polytechnic Institute geology professor Byron N. Cooper stressed economic benefits, and by November 1960 the Pulaski Town Council and Virginia Governor J. Lindsay Almond also advocated Forest Service approval. Other town governments, including Narrows and Bluefield, and other state politicians, including West Virginia Senator Robert C. Byrd and Virginia Governor-elect W. W. Barron, began to voice their endorsement, mostly on the basis of economic development. In December, E. L. Keesling submitted a detailed mining plan to the JNF, and finally (in January 1961) Forest Supervisor Curnutt gave his approval. The Bureau of Land Management granted Mr. Keesling his permits the following month, and core drilling began in August. Supervisor Curnutt based his reluctance to approve the project on painful experience. Much more than logging operations, old mining sites required tremendous soil reclamation and watershed protection, involving regrading slag heaps, building check dams, fertilizing and liming soil, reseeding, and reforestation. The eventual costs of a single restoration project would run into hundreds of thousands of dollars.[29]

In retrospect, and in light of later federally mandated strip-mining laws, the Forest Service's original watershed protection ideology actually served the JNF quite well. Leaders such as Supervisor W. C. Curnutt had been indoctrinated in the strong advocacy of curtailing erosion, and the mineral extraction practices of the time certainly damaged the environment in a manner unacceptable to a later generation of lawmakers and the American public. Thus, while the Forest Service may have overextended its original watershed protection crusade, an element of wise environmental practice certainly resulted, remained central to the agency's mission, and was positively borne out by many following instances.

Wildlife Management

Earlier wildlife regeneration efforts began to show results after the war. The JNF opened its first deer season in 1945, and over the next ten years deer hunting on the national forest increased 400 percent, coinciding with a similar rise in trout fishing. The wild turkey population on the George Washington National Forest and JNF rose from an estimated 2,600 in 1938 to 3,400 nine years later. Black bear showed an even greater rise, from 500 in 1938 to 1,200 in 1947. By January 1953 the New Castle District allowed trappers to capture beaver, which had rapidly reproduced after the introduction of New Hampshire beaver fifteen years earlier. By the 1950s the first evidence of overpopulation of deer in certain areas was already evident.[30]

The U.S. Fish and Wildlife Service began sponsoring Cooperative Wildlife Research Units around the nation, and Virginia became one of the earlier participants. Among all of the JNF's wildlife regeneration efforts, the 10,000-acre Broad Run Management Area on the New Castle District became the most significant. This highly cooperative project involved five agencies: Virginia Polytechnic Institute (VPI), JNF, Virginia Commission of Game and Inland Fisheries, the Wildlife Management Institute, and the U.S. Fish and Wildlife Service. The university provided office space, utilities, and the basic necessities for a research environment. The JNF provided 10,000 acres of habitat, and the Virginia Commission and Wildlife Management Institute provided technical and scientific advice. The U.S. Fish and Wildlife Service's Burd McGinnes, leader of the Research Unit, became an adjunct professor at VPI.[31]

On the Broad Run area, the JNF used timber stand improvement and logging to encourage the proliferation of deer, turkey, and grouse, especially with hunters in mind. VPI students helped staff check stations that measured the game harvest, primarily deer and turkey. Some VPI Fisheries and Wildlife graduate students wrote their master's theses or doctoral dissertations based on research they conducted on the Broad Run Area. About eighteen written studies resulted, ranging from assessing deer and turkey population distributions, to climatic influences on hunting, to impacts of timber cutting on various animal populations and their habitat.[32] Other, less intensive habitat improvement consisted of maintaining the "edge environments" of forest openings. By the mid-1960s Clinch rangers had worked with more than a hundred such habitats, seeding clearings with clover and various grasses, and encouraging the proliferation of food-bearing trees and plants. By this time small game populations had risen over most of the JNF, with the Blacksburg District experiencing its highest numbers of turkey to date, and with the New Castle especially high in squirrel.[33]

Since the U.S. Fish and Wildlife Service began keeping statistics (1965–present), Virginia residents have consistently represented about 90 percent or more of registered hunters,[34] which would support the idea that the JNF as a hunting ground has overwhelmingly appealed to local residents. Game species still dominated public concern, but quietly a few scientists conducted studies of nongame animal species and other aspects of the forest ecosystem. These were among the first efforts that would soon gain tremendous momentum, beginning in the 1970s with new federal laws and ecological awareness of rare and endangered flora and fauna species and the importance of biological diversity.

Science, Pure and Applied

In 1962 Arthur Shields completed one of the most intense botanical studies on the JNF when he focused upon the "Isolated Spruce and Spruce-Fir Forests" around the Mount Rogers and Whitetop Mountain area. Prior to Shields's dissertation, many earlier botanists, scientists, and other observers had made limited notes concerning various flora and fauna species of this unique region, including salamanders, bryophytes, *Picea,* and "herbaceous species." Many other earlier studies had focused upon regions near the Mount Rogers / Whitetop area, such as the Great Smoky Mountains or Allegheny Mountains. But Shields took the opportunity to characterize the forest itself, study the ecosystem behind its endurance, compare it with northern forests (more characteristically spruce-fir) and the other isolated southern Appalachian spruce-fir stands in the Smoky Mountains. Shields documented north slope stands on Mount Rogers in 1954 and 1959 prior to JNF timber sales, before the Forest Service ensured further study when they set aside 1,300 acres as a scenic area in 1961.[35]

Some forestry scientists used the JNF as their living laboratory for white pine and yellow poplar studies.[36] But perhaps most intriguing were ongoing efforts to revive the American chestnut. In 1939, shortly after the chestnut blight ran its course in the southern Appalachians, scientists began a succession study on Beanfield Mountain, near the Mountain Lake Biological Station in Giles County, on what was then the JNF's New Castle Ranger District. Where an oak-chestnut forest had predominated prior to the blight, various oak species (red, white, and chestnut oak) initially filled the American chestnut's void, as did hickory in later years.[37] As part of the original American chestnut ecosystem, the JNF provided a constant habitat for various blight-resistant strains that scientists developed and released over the years, such as the special Chinese chestnut strain that the Forest Service experiment station out of Beltview, Maryland, planted on all the JNF districts in 1964.[38]

Multiple Use and Sustained Yield

Many of the conflicts over national forest management during and following the 1960s often stemmed from concepts of multiple use. Eventually this morphed into a more dichotomous debate over the forest for its commodity versus ecological values. But in the meantime, how various people defined "multiple use, " its propensity toward contradiction, its specific instances of outright impossibility, and its feasibility on a generally limited budget remained the crux of Forest Service controversy from the 1960s to the 1990s. The multiple-use idea preceded creation of the Forest Service, going back to the 1897 Organic Administration Act.[39] The concept, if not its practice, continued with the 1911 Weeks Act that established eastern national forest lands. In fact, throughout its history, the Forest Service laid claim to a consistent but informal approach to multiple use. In any case, Congress deemed it appropriate to underline this tradition with passage of the 1960 Multiple-Use and Sustained Yield Act (MUSYA).[40] Congress saw MUSYA as supplementing the 1897 Organic Administration Act. Multiple uses included recreation, range management, timber management, watershed protection, and management and protection of wildlife and fish. Prophetically, MUSYA also recognized wilderness as among the multiple "uses,"[41] a value that the 1964 Wilderness Act would significantly elevate.

Many in the Forest Service felt that the agency's multiple-use approach was a proven and effective way to address the needs of diverse public and industrial interests without favoring any one group too much and thus becoming a "captured" agency. According to this approach, engendering a certain amount of discontent from all quarters could ensure the survival of multiple-use philosophy. As Forest Service Chief Richard McArdle wrote in 1953, "Since we must consider the interests of all the people, so also we usually find ourselves in the middle."[42]

Philosophy aside, however, informal concept did not necessarily indicate formal implementation, and the fact that "timber was king" became a cliché in descriptions of the pre-1960s Forest Service and a tradition of criticism that persevered into the 1990s. Some foresters and other observers claimed that MUSYA merely officially sanctioned what the Forest Service had always informally practiced. The old competition with the Park Service over recreation management likely played into the Forest Service's initiative behind MUSYA as well. If the Forest Service recognized recreation and wilderness more overtly, they might keep their management initiatives.[43]

Despite various proponents and opponents of MUSYA, a fair amount of confusion surrounded the 1960 legislation itself.[44] No clear definition of "multiple use" emerged. The legislation stated that multiple use was the "management of all the various renewable surface resources of the national forests so that they are utilized in the combination that will best meet the needs of the American people . . . with consideration being given to the relative values of the various resources."[45] Congress defined "sustained yield" as "the achievement and maintenance in perpetuity of a high-level annual or regular periodic output of the various renewable resources of the national forests without impairment of the productivity of the land."[46]

Little public comment accompanied MUSYA's passage. The period following the 1960s, however, saw an increasing number of scholars and agency personnel alike criticize the Forest Service and the congressional funding behind it as traditionally emphasizing timber production at the expense of other resources.[47] Various sectors of the public, such as wilderness advocates and recreation users, increasingly challenged Forest Service resource management. Whatever "true" multiple use entails has remained debatable ever since.[48]

Evolution from the Bottom Up

During the 1960s and 1970s, the JNF rapidly grew from an integral and intimate part of local communities into a complex and sometimes inaccessible bureaucracy. As Wythe District's Ray Walters recalled, the "whole crew" used to go out on field projects, marking timber, constructing roads, building picnic areas, and the like. By the early 1990s some Wythe District personnel never left the office, and no one project ever involved the whole crew again.[49] This small example typified the broader scenario. While the district ranger of the 1930s spent most of his time in the actual forest, the 1960s saw him (and eventually "her" on some ranger districts) faced with the loss of the old field diaries, discontinued employment of local fire wardens, and increasing paperwork and administrative duties. The district ranger became an office manager.[50] The public appeals process, once a relative rarity, became much more common and frustrated many rangers, some of whom were already disillusioned with their indoor desk jobs. Forest rangers had to learn to play a very political game involving often conflicting environmental, industrial, and social interests.[51] Given this broader context, the JNF also underwent an administrative shift that catalyzed this process.

JNF administrative history falls into two main categories: its diminutive place on the southern flank of the national forest system's old Eastern Region (Region 7) and its dynamic place at the northern edge of the current Southern Region (Region 8). The first phase lasted from the JNF's inception until 1966, when the Forest Service dissolved Region 7 and transferred the JNF into Region 8. At this time the Forest Service actually considered merging the two Virginia national forests and administering them similarly to the single supervisor office national forests in Florida and North Carolina. This idea, as it turned out, would have to wait nearly thirty years before it was realized. These thirty years, however, became quite dynamic ones for the JNF, and in quite marked distinction from the neighboring George Washington National Forest. Much more than a shift in territorial redistricting, the JNF's transition between regions coincided with dramatic changes in the greater agency (not to mention the nation) that ushered out an entire era, and made way for the highly visible, multifaceted, and controversial national forest of the postwar twentieth century.

Within the Region 7 scenario, the early JNF was an isolated, inconspicuous forest far from Pennsylvania headquarters and the demographic expedience that focused regional office attention on the White Mountain and Green Mountain national forests of New England. Most of the early JNF foresters came out of Pennsylvania or North Carolina colleges and universities, and some spent their entire careers on the JNF. Most districts employed only two full-time men: the district ranger and his general district assistant. In the small employee pool, workers did not see much influx of new people or even rotation of personnel, leading some of the first newcomers out of Region 8 to characterize the old JNF as provincial and "inbred."[52]

The transfer into Region 8 revolutionized the JNF. Early JNF employees invariably described Region 7 as very conservative in both policy and funding, forcing them to handle operations on a very limited budget. To a degree, tight finances characterized Region 7 for much of its early history. Franklin Reed, Region 7 forester during the 1920s, felt that the greatest value of the eastern national forests generally lay in demonstrating scientific forestry for private woodland owners—not in any major timber-production capacity. Before and during World War II, Region 7 forest administrators simply tried to maintain the silvicultural status quo and relied mostly on natural timber regeneration. They practiced only a nominal amount of disease and erosion control. After World War II regional administration began to shift more toward resource management, but the JNF, a timber-poor, underfunded, and isolated young forest, essentially remained in the former mode.[53]

Beginning in 1966 a major infusion of Region 8 personnel came to the JNF, many of whom represented an entire new breed of professional specialists entering the Forest Service nationwide. Employees from the earlier era tended to be diverse, hands-on personnel highly skilled in fire prevention, heavy equipment operation, mechanical maintenance, telephone and radio installation and maintenance, and general labor. Around the late 1950s the JNF began to emphasize particular specialized skills to some degree, but these skills remained tied to the old era of fire control and timber management. The new personnel that poured into the JNF after 1966 tended to be specialists in fields such as business administration, recreational planning, civil engineering, and personnel management. Along with the retirement of many of the older (and some of the original) JNF employees came the end of their way of conducting Forest Service operations.[54] Internally, the Forest Service underwent a bureaucratic restructuring that centralized authority and policy making as never before. After half a century of relative autonomy, the agency faced a dramatic transition into an era of markedly greater congressional regulation, public participation in its management decisions, and increased and conflicting demands upon its resources, especially in regard to timber and recreation. Traditional decentralization, so highly valued by those in the agency, suffered in the new age of environmental politics and an increased use of the public appeals process.[55]

The advent of new specialized fields necessitated more supervisor office control in order to launch new programs.[56] Wildlife biologists, landscape architects, recreation planners—and eventually archaeologists, botanists, air quality specialists, and other specialists—entered an agency hitherto almost exclusively dominated by foresters.[57] The JNF's proximity to Washington, D.C. (with the new highway, Interstate 81, providing easier access), combined with an especially extroverted approach among certain key personnel, contributed to the JNF's leadership within the Forest Service and national policy. For instance, the cooperative wildlife program between the JNF and the state of Virginia became a model for Region 8 and then the nation. Or, in terms of recreational planning, the JNF personnel involved with developing the Mount Rogers National Recreation Area helped foster a new era of public involvement in Forest Service policy decisions and their implementation four years before the National Environmental Policy Act required such procedures of all national forests. While public involvement with Mount Rogers became at times highly controversial (discussed at length, below), the entire issue was so conspicuous partly because of the JNF's progressive attitude. In fact, such a leadership orientation contributed to individual JNF personnel, such as former Supervisor Michael Penfold (son of prominent conservationist Joseph Penfold), to participate in the national debate surrounding the 1973 Monongahela Decision and the policy formation that culminated in the historic 1976 National Forest Management Act.[58] In later years, in something of an ironic twist, the JNF, as one of the worst offenders, became one of the central contributors to congressional testimony concerning the below-cost timber sale phenomenon.[59]

Evolution from the Top Down

The Forest Service and JNF of the 1960s and 1970s coincided with a nation of greater mobility and a mushrooming population that demanded more recreation from the Forest Service than ever before, which helped spur such programs as Operation Outdoors.[60] Other resources also came under increasing demand. In 1963 JNF timber management staff anticipated selling around 26 million board feet annually during the latter half of the 1960s but ended up selling far more, such as the almost 39 million board feet in 1969.[61] In response to greater timber harvesting demands, and typical of the other national forests, the JNF greatly expanded its forest road mileage.[62] At the same time, wilderness enthusiasts began to advocate preserving the JNF's remaining "undisturbed" lands (a controversial concept in itself, especially in the eastern United States), especially following the 1975 Eastern Wilderness Act. And while the Forest Service had remained relatively autonomous during its first fifty years, the 1960s and 1970s witnessed increasing congressional regulation, resulting in legislation such as the 1964 Wilderness Act, the 1969–1970 National Environmental Policy Act, and the 1976 National Forest Management Act. The timber resource of the eastern national forests themselves had finally reached a marketable age, and timber and pulp companies sought more wood. More hunters and fishers combed the woods and waded the streams, unprecedented numbers of hikers discovered the Appalachian Trail, and drivers crowded the Blue Ridge Parkway (which traverses the JNF's Glenwood District). More than any other time in its history, the JNF began attempting to answer the needs of a great variety of interest groups, all claiming priority on the public lands. The old caretaker job had thus grown much more complex, both in terms of national forest resources and the public demands upon them.[63]

Changing Public Expectation

During the civil activism of the 1960s and 1970s, a large sector of the American populace awoke to a number of environmental and natural resource issues, including national forest timber policy. Given this public awakening, the controversies surrounding Forest Service policy during this period involved several coinciding situations. Many Americans had gained a first-time interest or new assessment of their public lands. Economic prosperity placed greater demands on Forest Service resources, particularly (and contradictorily) on timber for housing demands and older growth forests for recreation demands. The Forest Service did not or could not satisfy these interest groups with balanced utilization of national forest resources and as a result began a reconfiguration of national forest policies and procedures.[64] An unprecedented redefinition of the Forest Service mission followed.

The 1969–70 National Environmental Policy Act (NEPA) required public involvement in national forest management activities for the first time. NEPA required the Forest Service to draw up "unit plans," which allowed public involvement with Forest Service decisions. For instance, when the JNF proposed its Poverty Creek Unit Plan on the Blacksburg District in 1971, the environmental impact statement reflected consultation with seven national, state, and community entities, including the New River Valley Planning Commission, Council on Environmental Quality, and Virginia Tech University.[65] The JNF and other national forests began publishing Draft Environmental Impact Statements (DEIS) and Environmental Impact Statements, which addressed specific Forest Service management plans in accordance with NEPA. Among the issues the JNF's 1985 DEIS specifically addressed included land acquisition, clear-cutting, off-road vehicle use, the then-lingering possibility of Devil's Fork Wilderness Area designation, and other controversial topics. With all these issues, the JNF tried to measure possible social and economic consequences and effects on natural resources and the physical environment.

The 1974 Forest and Rangeland Renewable Resources Planning Act (RPA) represented a response to previous years of public outcry over Forest Service resource management and addressed a new age of multiple-use planning. Congress also designed the act to help measure the federal funding for the Forest Service's new responsibilities. This act required the Forest Service to assess its resources and prepare a management program every ten years but rejected the idea of public involvement in the traditionally autonomous organization.[66] Two years later, the National Forest Management Act (NFMA) went much further, reemphasizing multiple use and now requiring public involvement at every level of timber management decisions. A tremendous amount of public involvement indeed followed.

Ideas surrounding multiple use became especially intriguing when played out on a national forest such as the JNF, with its comparatively small acreage, irregular holdings, close proximity to high population concentrations, hardwood timber resource (as opposed to softwood tree farm resources), and complex forest ecology. The JNF encompasses several major forest ecosystems, including cove hardwoods, other hardwoods, planted pines, and, perhaps most notably, the now-diseased or deceased unique spruce-fir habitat surrounding Virginia's highest mountains (killed through poor air quality, described below). Scientists had identified many endangered or threatened animal or aquatic species throughout the JNF, and in compliance with federal law such life forms merited protection.

1970s Conflict

The civil unrest of the 1960s involved several important fronts, including the fragmentation of liberal politics in America, the black civil rights movement, a blossoming antiwar movement, and a general disillusionment among white middle-class youth with the "Establishment" that expressed itself through the counterculture.[67] The momentum of these trends during the 1960s continued and expanded into the next decade with the rise of the American Indian Movement, Women's Liberation, and a growing environmental movement. The very doctrine of "progress," long the impetus behind Western civilization's industrialization, came under extensive attack and expressed itself, in part, through a "back to nature" movement among urban middle class.[68] The JNF in general, and the Mount Rogers National Recreation Area (MRNRA) in particular, would directly feel the impact of this demographic shift.

The creation of the MRNRA in 1966 thrust the JNF to the forefront of a whole new Forest Service recreation era. The MRNRA also presented the JNF with its greatest modern conflict involving a series of complex concerns, such as environmental preservation, forced relocation of area residents, and recreation-oriented economics. In 1973 Harold Calhoun, a JNF forester from the traditional period, spoke of this controversy. His perspective not only captured the difficulty the JNF had come to face but also indicated the end of an earlier era.

> Not many years ago, we in the Forest Service often heard people expressing their admiration for the way we were managing the public's land. We hear a lot of voices now [1973] expressing opposite views. It seems that the opposition's voice is a lot louder—or maybe we only imagine this. A lot of this criticism is unjustified. On the other hand, some of it is, and I'm probably not qualified to distinguish the difference. However, it looks like a caution light is before us and we need to take a long, hard look at any proposed increase in land acquisition. Even though the NRA is a law and we're supposed to make it work, we'll need a lot of public backing to cause that to happen. Acquiring land we don't actually need, or our own failure to utilize to the fullest our present ownership, is a sure way to lose any support we now have.[69]

Lost upon most outside observers (especially the critics), the era of the late 1960s to mid-1970s saw a generational and philosophical rift develop within the JNF itself. Older, traditional foresters and rangers like Harold Calhoun, who had spent the majority of their careers fostering close community relations, found themselves suddenly immersed in some of the fiercest environmental political wars. They did not tend to find allies in a younger generation of Forest Service employees who had different educational backgrounds and different professional ambitions, such as concentrated recreational development.

Chapter 8

The Mount Rogers National
Recreation Area

Rising to 5,729 feet on the Smyth and Grayson county line, Mount Rogers and neighboring Whitetop Mountain (5,520 feet) have attracted the attention of explorers, scientists, sightseers, and other visitors for centuries. Wilburn Waters, as legendary as Daniel Boone in some parts of southwestern Virginia, lived the life of a hermit, hunter, trapper, and backwoodsman in the Mount Rogers and Whitetop Mountain area during the nineteenth century.[1] In 1835 University of Virginia professor of natural science William Barton Rogers, who later founded MIT, made a geological survey of various areas in southwestern Virginia, including Whitetop Mountain, the coalfields around Pocahontas, and what came to be named after him—Mount Rogers. Dozens of other scientists studied the Mount Rogers area throughout the nineteenth century and up to the present, attracted to various unique aspects of the environment's spruce-fir forest, an isolated remnant from the last glacial age.[2] Eleanor Roosevelt visited the area in 1933 for the music festival that Whitetop Mountain hosted annually throughout the 1930s.[3] Following World War II, as car travel became increasingly popular, a growing number of tourists discovered the Mount Rogers area's remarkable landscape.

During the late 1950s Congressman Pat Jennings, a Smyth County native, visited the Wenatchee National Forest in Washington State to help establish a national wildlife area. At this point, Jennings realized the unique beauty of the Mount Rogers area and began working toward having Congress and the Forest Service give the area some kind of special designation to preserve its scenic qualities.[4] In 1961 the Forest Service established the 1,300-acre Mount Rogers Scenic Area, with further studies directed toward a possible Mount Rogers Whitetop Recreation Complex.[5]

In 1957 and 1958 the Forest Service and Congress launched various planning efforts to address the growing public demands for outdoor recreation, including the creation of the Outdoor Recreation Resources Review Commission and the Recreation Advisory Council.[6] In 1962 the Outdoor Recreation Resources Review Commission released its massive twenty-seven-volume series, which studied recreation in the United States and covered such diverse topics as economics, wilderness, land acquisition, user satisfaction, fishing, and hunting. The bulk of America's NRAs arose out of this context, specifically designed to serve a growing and more diversified recreation clientele.[7]

In 1963 the Recreation Advisory Council, a cabinet-level committee, envisioned NRAs as fulfilling a function that fit between those of national, state, and local parks.[8] The Forest Service would manage NRAs with more flexibility than the Park Service managed national parks in order to accommodate a diversified and, in places, intensive recreational use. NRAs had to be accessible to large urban areas and interstate tourism.[9] The choice of the Mount Rogers area, midway between the already overcrowded Shenandoah and Smoky Mountain national parks, and near the future crossroads of interstates 81 and 77—within a day's drive from Washington, D.C., and other eastern urban areas—became a natural choice. In fact, for some years, a parallel recreational interest in the area developed at state level. In early 1962, State Delegates Lox and Hodges of Grayson County introduced a bill into the Virginia House of Delegates asking for $5,000 to survey the Mount Rogers and Whitetop Mountain area for a possible state park. The House of Delegates approved funding, and by August 1964 the state had begun acquiring land for a proposed state park southeast of Mount Rogers on Haw Orchard Mountain.[10] In the following years, the state acquired 4,754 acres for this park, later named the Grayson Highlands State Park. Meanwhile, Congressman Jennings continued to promote Mount Rogers on a federal level.

On March 13, 1963, Jennings introduced House Resolution 4824, or the Whitetop Wonderland bill, to the Congressional Committee on Agriculture. This represented the next significant step toward designating Mount Rogers as a special recreational area. Jennings later modified his proposal to fit the Recreation Advisory Council's NRA criteria.[11] Finally, on May 31, 1966, President Johnson signed Public Law 89-438 creating the Mount Rogers NRA in Carroll, Grayson, Smyth, Washington, and Wythe counties, Virginia. At this point the NRA's proclamation boundaries encompassed 154,000 acres, 84,000 of which the JNF already managed as part of the Holston and Wythe ranger districts. Initial NRA plans called for further land acquisition and recreational development, including 900 family camping units, 100 day-use units, and a "winter sports site," later designated a ski area atop Pine Mountain in Grayson County. Other proposed developments included three water impoundments and a scenic highway to link the town of Damascus with the eastern edge of the NRA near the New River. While only 170,400 people visited the Mount Rogers area in 1965, congressional supporters anticipated as many as 1 million annual visitors by 1972 and 5 million by 2000.[12]

From the outset, a concerted solicitation of local public input characterized the Forest Service's implementation of the NRA legislation. Charles Blankenship and other planners, such as Lionel Melancon and Reginald

A 1964 "Show Me" trip to the New Hale Lake construction area on the Wythe Ranger District, which was incorporated into the Mount Rogers NRA two years later. Such trips were designed for both entertainment and education. U.S. Forest Service photograph.

Kinman, continually visited area community centers, schools, churches, and civic groups, gave radio talks, and contributed NRA information to local newspapers. They encouraged local people to influence NRA development. As Blankenship remembered, the JNF did not intend to "sell" an idea; rather, they wanted local people to help shape a legislative decision.[13] In this manner, they anticipated future legislation that would require such public participation in management decisions.

Zoning of private land within or neighboring the NRA's proclamation boundaries immediately became centrally important to NRA planners in their effort to avoid uncontrolled development. The earliest NRA plan envisioned leaving 30,000 privately owned acres within the NRA boundary, and thus inholding became an early consideration. NRA planners worked with county boards of supervisors and local citizens to encourage development rights, property easements, and zoning through a regional planning commission.[14] A notable result arose in August 1972, when the Washington County Planning Commission unanimously voted to zone Konnarock as a "village," a designation that NRA planners hoped would control growth and keep this large inholding harmonious with the NRA's Rural America landscape theme.[15] This single incidence, however, became the exception, and in cooperative private land controls the NRA encountered one of its earliest disappointments.

Despite a few signs of progressive zoning in Vermont, Oregon, and Hawaii during the late 1960s and early 1970s, rural people throughout the United States vigorously resisted rural zoning.[16] Landowners in proximity to the Mount Rogers NRA proved no exception. Historically, rural areas without zoning were the most susceptible to uncontrolled recreational growth, and the real estate development boom of the mid-1960s and early 1970s exploited such vulnerability. An absence of zoning eventually became the very basis for the various negative aspects that accompanied tasteless recreational development elsewhere in the United States, and was precisely what NRA planners sought to avoid.[17] As NRA plans moved into the 1970s the absence of land controls on private property had far-reaching consequences for projected recreational development plans. In the meantime, between 1972 and 1979 a conspicuous debate overshadowed the amicable public involvement that had characterized Forest Service interaction with community leaders during previous years. Some local people remained supportive of the NRA throughout the following struggle,[18] but NRA opponents vociferously denounced recreational development and further land acquisition, particularly when such acquisition involved condemnation. Controversy over condemnation became the most emotional issue surrounding NRA development.

Under the Mount Rogers NRA legislation, the Forest Service could acquire acreage by condemnation, donation, exchange, or transfer, as needed.[19] The NRA's legal right to condemn land marked a significant exception to the Forest Service chief's standard policy. To gain such an exception, the NRA had to assemble an interdisciplinary team of lands staff and recreational planners to identify and justify priority tracts, which NRA personnel based on topography and projected developments. The 84,000 acres under JNF management in 1966 comprised almost only mountainside and ridgetop land and was unsuitable for developing the proposed concentrated recreation facilities. The chief of the Forest Service approved the condemnation exception in September 1969, and the NRA based its subsequent land acquisition strategy on the 1968 Master Plan for recreational development.[20]

Unfortunately, the 1965–67 rapid acquisitions of the neighboring Grayson Highlands State Park cast the NRA in a bad light before its land acquisition program even began. The Virginia Department of Conservation and Economic Development, Division of Parks, began rapidly acquiring the 4,754-acre park in 1965. Park officials originally anticipated complete acquisition in 1966, but had to extend their plans into 1967 because of lingering condemnation cases. By November 1967 the park had or was negotiating for only thirteen parcels of land compared to condemning thirty-five, sixteen of which were still under appeal in the state supreme court.[21] The fact that the Grayson Highlands State Park was originally named the Mount Rogers State Park, and did not change its name until after 1967 to avoid confusion with the NRA, further muddled matters.

JNF personnel who shaped their careers in an earlier era of harmonious community interaction grew disturbed over forced relocation, in itself not part of the Forest Service tradition. NRA Lands Staff personnel like Harold Calhoun, George Wolfel, and Don Martindale regularly worked in direct contact with area residents, were often sympathetic with their plight, and sometimes advocated their interests.[22] In June 1972 Lands Staff Officer George A. Wolfel expressed a desire to see either lifetime reservation rights or special use permits for owners willing to sell their land in the Fairwood Valley, rather than condemnation.[23] In September 1971 Harold Calhoun advocated dropping priority acquisition of the Paul Atwood Estate altogether, based on the fragile psychological condition of the elderly widow Marie Atwood following the deaths of her husband and one of her sons.[24] After 1971 all NRA staff had the advantage of offering area residents greater federal relocation assistance and funds, based on improved legislation.

In January 1971 Congress passed the Uniform Relocation Assistance and Real Property Acquisition Policies Act (Public Law 91-646), thereby providing owners of condemned property an unprecedented level of financial compensation and relocation aid. Public Law 91-646 marked a culmination of a decade of congressional studies and bills aimed at mitigating adverse relocation procedures, so that displaced persons should "not suffer disproportionate injuries as a result of programs designed for the benefit of the public as a whole." Public Law 91-646 provided landowners with as much as $15,000 in tax-free relocation funding, help in locating replacement housing, and generally "to assure that such displaced persons receive the maximum assistance available to them."[25]

While not a panacea, Public Law 91-646 at least enabled a dimension of humanitarian accommodation not previously possible. A noteworthy instance involved the NRA's fight to win relocation funding and replacement housing for Sessie Pruitt. Instead of relocating, Pruitt sold her fourteen acres in Grayson County to the Forest Service in 1970 and retained a "life estate," or lifetime rights, to live in her house and occupy a surrounding acre of land. During following years, however, the NRA gradually acquired the tracts surrounding Pruitt's property, and by 1976 Pruitt, a widowed, elderly lady in poor health who had relied heavily upon her neighbors for sheer existence, found herself alone and isolated. The JNF recognized her plight and asked the Regional Office in Atlanta to consider allocating Pruitt relocation funds. After the JNF's first request, in September 1976, the Regional Office refused, especially since the NRA had bought Pruitt's land before enactment of Public

Law 91-646 (which Congress did not pass until January 1971). But JNF personnel like Harold Calhoun and George Wolfel, who advocated interpreting this situation in the "spirit of the [Public Law 91-646] act," continued to advocate Pruitt's cause and successfully won her replacement housing and relocation funds in 1977.[26]

NRA area residents also employed the legal system with greater effectiveness than property owners from earlier eras. Where displaced people of the 1920s and 1930s ineffectively resisted or even influenced Park Service or TVA acquisition of their land, Mount Rogers area citizens successfully employed the legal system in their struggle against the Forest Service. Norris Basin residents contested only 3 percent of the land condemned by the TVA and did not receive higher compensation for their trouble. NRA area landowners who entered condemnation proceedings all received higher land prices, with only one exception.[27]

In 1972, for instance, the JNF appraised Belle Johnson's six acres in Grayson County at $2,680. A second Forest Service appraisal, in 1973, valued the property at $3,800. Finally, on August 12, 1974, the court awarded Johnson $7,000. Similar procedures won Oscar Parsons an additional $3,400 for his twenty-five acres in Grayson County, Paul Plummer an additional $9,366 for his 117 acres in Grayson County, and Wythe M. Hull Jr. an additional $8,813 for 280 Grayson County acres. The single exception involved the case against W. S. McCarter et al., in which land originally appraised (in 1967) at $27 per acre eventually received $25 per acre in 1971.[28] Of the 25,000 acres the JNF added to the Mount Rogers NRA between 1966 and 1980, they condemned, through forty-eight lawsuits, fifty-three tracts totaling 6,581 acres between 1968 and 1975.[29] Of the 6,581 total condemned acres, 4,238 involved land with no dwellings. Twenty-four of the forty-eight lawsuits, however, did involve relocating residents, whose holdings ranged from a third of an acre to 143 acres.[30]

The Reverend William Gable, who moved to Konnarock from Chicago in 1971, first learned of the NRA condemnations through members of his congregation. Gable was part of a massive demographic reversal in Appalachia during the 1970s, in which a million more people moved to the mountains than left them. This reversal followed thirty years of heavy out-migration among natives, and especially centered upon the rural, undeveloped areas of the region.[31] Since many of the new migrants were escaping from the city and urban problems, they tended to bring distinctly antidevelopment ideas into the mountains with them, and Gable's particular sentiments along these lines had significant effects on the Mount Rogers NRA. He began leading the first organized opposition by mailing newsletters to local residents and attending board of supervisors meetings in Smyth, Grayson, and Washington counties to protest condemnation and question NRA development plans.[32] In his August 17, 1973, newsletter, which he mailed to Mount Rogers area box holders, Gable warned that NRA development would bring about an onset of a "city, with city-type problems," unless land was kept in current local hands. He wrote, "We should each begin thinking about refusing to sell to [businessmen, real estate developers, outsiders] for even up to $10,000 per acre."[33]

The people Gable represented, primarily from the Helton Creek area in Grayson County, and some in the Fairwood Valley of Smyth and Grayson counties, were a silent minority before Gable arrived. Though it maintained an open, voluntary membership, the Mount Rogers Citizens Development Corporation (MRCDC) certainly did not represent a cross-section of the region's people. In February 1967, with encouragement from NRA Planner Blankenship, the MRCDC met for the first time, under the auspices of the Council of Southern Mountains.[34] Both the Council of Southern Mountains and the MRCDC wanted to include local residents in the economic growth that promised to accompany federal recreational development. The citizens formed the MRCDC "for the purpose of assisting in planning for and executing the orderly development of the economic and human resources within the Mount Rogers area," and specifically to avoid a glitzy tourist trap.[35] But the MRCDC was comprised of local community leaders, including merchants and county government officials, characterized by some as a local "elite" who "actively and intensively lobbied for a high density, capital intensive NRA development [in the hope of] increased economic gains for the area and themselves." Additionally, these people tended to live in area towns, such as Marion and Abingdon, rather than within the NRA boundaries itself, and did not face relinquishing their own property for the NRA.[36] In contrast, Gable described the people he represented as not wanting to stand on a stage, use a microphone, or speak for allotted times in the formal meetings that were taking place.[37] And, of course, the people Gable described also had property at risk of condemnation.

Part of the ensuing unrest involved an emotionally charged struggle, which Gable portrayed as rural farm people, deeply attached to their land, against a more powerful adversary, the federal government. As Gable said in August 1973, "relocation is an emotionally very upsetting thing for mountaineers . . . the land has been lived on for whole lives and generations even . . . it has been made livable by sweat and blood."[38] Gable and the Council of the Southern Mountains, who eventually turned against NRA development plans, helped enlist Charlottesville Attorney John C. Lowe to represent some of the local people in condemnation proceedings. Both Gable and Lowe saw part of their historic roles as an attempt to counterbalance the Forest Service in the land condemnation battle.[39]

Aside from outright condemnation, one grievance Gable, Lowe, and others levied against the Forest Service involved land prices.[40] When appraising land, the Forest Service produced several "comparable sales" of similar property in the region, designed to help establish a fair price. The process that preceded condemnation usually involved seven steps: (1) The Forest Service made its initial appraisal. (2) Subsequent Forest Service attempts to negotiate a willing-seller deal with the landowner proved unsuccessful. (3) The Forest Service filed a Declaration of Taking, which transferred the land title to the United States and represented the first step in actual condemnation proceedings. At this point the Forest Service could negotiate no further, and the case became the responsibility of the U.S. Department of Justice. (4) The U.S. attorney scheduled a court date for the hearing, often a year or two in the future. (5) The U.S. attorney contacted the Forest Service and required an updated appraisal based on the property value at the time the Declaration of Taking was filed. This second appraisal inevitably produced a higher price based on rising land values in general and/or any improvements the landowner had made, such as well-drilling or building construction. (6) The Forest Service filed its updated appraisal to the U.S. attorney. (7) The U.S. attorney then could negotiate further with the landowner, indicating the amount of legal fees that would be incurred if the suit went to court.[41]

Under Public Law 91-646, the federal government could not alter their appraisal to compensate for property values that escalated solely because of federal development activities, sometimes called "project-generated" values.[42] Delineating between property values that rose directly because of federal government development, anticipated or actual, and those that rose for other reasons, such as inflation, greater access, or uneven land price fluctuations in other parts of rural America, obviously became practically impossible but did not diminish the impression among some that the Forest Service was cheating landowners. The court eventually based its award on the results of testimony from the Forest Service, its appraisers, and landowners and any private appraisers they may have hired. The result was usually an award between the government and private appraisals. Some property owners let paperwork go into the last stages of condemnation, then settled out of court. The word got out among some landowners that the initial condemnation proceedings would inevitably result in a higher offer through the U.S. attorney, and by avoiding lawyers some landowners directly collected this higher price without relinquishing a percentage of the overall award for legal fees.[43]

Gable and Lowe did not pretend to represent the entire demographic of those losing property. However, anyone focusing strictly on their advocacy would likely develop an additional misconception that potentially distorted overall NRA acquisition—particularly if they did not distinguish between NRA acquisitions and the earlier, aggressive Grayson Highlands State Park acquisitions. By August 1973 the NRA had acquired approximately 17,000 acres through 149 tracts from willing sellers, and only 18 through condemnation.[44] Among the condemnation cases, the four largest landowners—respectively holding 1,441, 886, 705, and 541 acres—lost only pasture and woodland and certainly did not experience relocation.[45] In these four cases, and in twelve others, condemnation involved no full-time or seasonal dwellings and, as a result, no displacement. Other condemned property that *did* involve residences could hardly be considered as "made livable by sweat and blood." For example, Sinclair Preston's summer lodge situated on 483 acres, and the 105 acres and recreational cabin of John Kabler, a Florida savings and loan executive, were clearly recreational properties. Three other lawsuits involved recreational cabins only. In fact, the size of all these larger holdings suggested that the owners were among the region's more prosperous people. But the emotionalism over condemnation overshadowed these facts and clearly involved a strong element of cultural self-determination that Gable and others feared the NRA threatened. And like land condemnation, cultural issues in Appalachia during the 1960s and 1970s followed a long and troubled legacy that did not encourage NRA development.

The origins of the hillbilly stereotype arose following the Civil War when "local color writers" visited the region and reported back to periodicals like *Harper's Monthly* to feed the minds of a reading public with what came to be a convention depicting Appalachian mountain people as ignorant, inbred, ill-clothed, and constantly feuding amongst themselves. The legacy continued and, indeed, still lingers, as the comic strip Snuffy Smith demonstrates.[46] But the consequences could prove far more severe than insulting and inaccurate literature. Promoters of the Shenandoah Park employed the hillbilly stereotype as part their rationale for removing area residents, particularly citing the 1933 "study" undertaken by psychologists Mandel Sherman and Thomas R. Henry entitled *Hollow Folk*.[47] And after the Great Smoky Mountain National Park displaced hundreds of residents of Cades Cove, Tennessee, during the 1930s, the Park Service created "an authentic living museum of Southern Appalachian culture" from the remnants of Cades Cove abodes, carefully removing more modern, frame-type structures and retaining the stereotypical log cabin.[48] Obviously, by the 1960s and 1970s many people living in the mountains had grown tired of such patronizing exploitation, inadvertent or otherwise.

Opposition leader Gable was among the first to challenge the Great Society's impoverished-Appalachia cultural depiction.[49] In September 1973 he commented, "the question is, will this area remain a rural Appalachian underprivileged area, which I believe is what most people want, or will it be allowed to be commercially developed and destroy our mountain culture?" JNF supervisor Michael J. Penfold questioned "whether such a mountain culture still exists and whether people still want to preserve it." Gable maintained that the Forest Service did not recognize the "present culture."[50] And while

Blankenship had envisioned the rural America landscape theme as a way to preserve the rural setting and promote authentic native crafts, others perceived it as only a patronizing sideshow of stereotypical hillbillies.[51]

An ironic aspect of the debates surrounding local culture lay in a predominance of nonnative people engaging in them. In 1974, in his contracted study for the JNF, William and Mary Sociology Professor Edwin H. Rhyne inadvertently captured this paradox by writing, "no matter how much of the NRA impact area is backwoods in comparison with the rest of the nation, it is still very much a part of the nation's mainstream."[52] The exceptionally rapid changes that accompanied the NRA's first decade probably exacerbated the cultural clash. But the threat to local culture was only one of many such scenarios taking place in other areas of the United States where a rural agricultural economy and way of life gave way to recreational development.[53] Central to this transition lay questions concerning recreation-led or recreation-driven economics.

Economic concerns surrounding the NRA arose on two levels: Forest Service recreational funding at a federal level and local debates focusing on the economic pros and cons of recreational development. Inadequate funding at a federal level prevented the Forest Service from proceeding with NRA developments and thereby eroded some of their former local support.

The boom in postwar outdoor public recreational demand that coincided with a maturation of national forest timber and subsequently more intensive management and production aggravated an intra-agency tension between timber and recreational management that continued into the early 1990s, especially in regard to budgetary competition.[54] The financial struggles of the Mount Rogers NRA epitomized this intra-agency tension. With the advent of the Nixon and Ford administrations, 1969–76, the massive financing necessary to develop a visionary project the scope of the Mount Rogers NRA did not arrive. As early as 1970 and 1971, with the first indications against condemnation and development just arising, some local supporters of the NRA grew frustrated at what they perceived as a stalled proposal. Every year following the Johnson administration the NRA suffered inadequate congressional appropriations and by fiscal 1973 endured a 40 percent budget freeze.[55] Between fiscal 1977 and 1980, Congress gave Forest Service recreation much better funding, but by then the time had already passed for the NRA to take full advantage.[56]

A key aspect of this issue lay in a lack of monies to implement *development,* as opposed to acquisition. Acquisition, after all, engendered the most significant opposition to the NRA, while NRA proponents hinged their support on anticipated development. The accomplishment of one without the other doomed the NRA to much criticism and loss of support. Land acquisition funding for the NRA came from the 1964 Land and Water Conservation

Fund (LWCF), but the Forest Service, as a federal agency, and unlike state and local agencies who received LWCF monies, could not use them for development. The Grayson Highlands State Park abutting the NRA, already quickly acquired and developed by 1970, cast the NRA in a poor if misunderstood light in this way as well.[57] As inflation worsened during the 1970s LWCF monies themselves lost their originally anticipated purchasing power, and in addition to the JNF's inability to use LWCF monies for recreation development, congressional line-item funds for such development remained unavailable or inadequate.[58] Finally, property owners legitimately complained that federal acquisition took lands off the county property tax base, and without generation of recreational dollars to compensate, such a loss of income hurt county budgets.[59]

At a more immediate local level, and interrelated with cultural values, were questions surrounding recreation-driven or recreation-led economics. Beginning with John F. Kennedy's campaign stop in West Virginia in the spring of 1960, rescuing Appalachia from poverty became one of the most important federal goals of the 1960s and a central focus of President Johnson's Great Society, particularly with passage of the 1965 Appalachia Regional Development Act.[60] Recreation and tourism seemed to present a natural economic solution and were central considerations as early as 1964 in regard to the Mount Rogers NRA.[61] NRA planners operated with these same optimistic assumptions and emphasized economic benefits within a scenario of strictly controlled growth and retention of rural atmosphere. Their economic emphasis clearly reflected the Great Society's War on Poverty perspective, which generally viewed Appalachia in terms of deprivation. As Supervisor Penfold read to the Mount Rogers Planning District Commission in September 1973, "You clearly have a choice. You can preserve the so-called 'mountain culture' with its attendant poverty, hopelessness, and lack of opportunity or, in this case, because of the National Recreation Area you can accept the challenges, seek ways of making the necessary changes as humanly [sic] and with as little impact as possible and carefully mold an environment that offers hope for the future, not desperation and poverty."[62]

The Economic Development Administration and Appalachian Regional Commission represented some of the higher profile agencies working with the Forest Service in their attempt to empower local business people. During the early planning process, before they sided with Gable against the Forest Service, the Council of the Southern Mountains sponsored meetings in Damascus focusing on economic issues. Local landowner Albert C. Mock became very active in these meetings and worked to initiate new local industries such as the Iron Mountain Stoneware pottery manufacturers.[63] The MRCDC and the Forest Service emphasized spreading commercial ownership throughout the community through a broad-based, vertically integrated approach to maximize economic benefits. Local empowerment would be achieved through selling business shares.[64]

Gable himself remembered that sometimes he had to "tread a thin line" between his personal attitude against development and the obvious need for local economic stimulation. He based his opposition to such development on the perception that most of the NRA-generated jobs would involve seasonal, service-oriented, minimum-wage employment.[65] Early published economic studies of recreational development tended to stress positive aspects, but the field of recreation economics in itself was just emerging. Interestingly, economists began to reassess the situation by the late 1970s, and their conclusions tended to echo Gable's contemporary doubts.[66]

In general, the Great Society's dreams of a tourist-boosted economy in Appalachia faded amidst the recessionary 1970s. The entire southern Appalachian region did not experience substantial economic growth through recreational development during the 1960s and 1970s, and the JNF experienced no dramatic visitor growth between 1972 and 1980, the crucial years of the Mount Rogers controversy.[67] Already backed against the wall financially, the NRA continued to face William Gable, who would not compromise on the condemnation issue, and increasing attention from state and national groups who challenged NRA development on environmental grounds.

The National Environmental Policy Act (NEPA), enacted 1969–70, caused NRA managers to rework and expand their initial plans in order to comply with the new legislation. Over the next eight years, the NRA staff worked through the environmental impact statement process that entailed specialized studies of air quality, rare and endangered plant and animal species, historic and prehistoric cultural resources, a sociological profile and impact assessment of the NRA on area people, and environmental studies concerning the effects of potential road, impoundment, and ski area construction. But the environmental movement began to clash with tourist development projects throughout the nation, and the proposed NRA developments proved no exception.[68]

After the first Earth Day in 1970, the nation's environmental movement acquired considerable momentum and began to show its influence in specific locales throughout the United States. The southern Appalachian region gained early focus in regard to coal strip mines and the hydraulic engineering projects of the Tennessee Valley Authority and Army Corps of Engineers.[69] The Izaak Walton League led a lawsuit against the Monongahela National Forest in West Virginia over improper clear-cutting procedures that were damaging the environment. All these events indirectly affected the NRA with a growing public consciousness over the potential detrimental effects that development could have on the environment. Of direct significance for the NRA, area residents, the state of North Carolina, and national environmental groups such as the Sierra Club and Audubon Society all banded together to defeat the Blue Ridge Reservoir proposal in North Carolina and Virginia, abutting the southern border of the Mount Rogers NRA.[70]

Since the early 1960s the Appalachian Power Company had planned the huge Blue Ridge Reservoir that would inundate parts of Grayson County, Virginia, and neighboring Ashe County, North Carolina. Water-based activities had proven a central draw in other developed recreation areas, and besides their own planned water impoundments NRA planners maintained a keen interest in the Blue Ridge Reservoir proposal. In 1970 and 1971 the Appalachian Regional Commission and URS Research Company both recognized the importance of the reservoir to NRA development plans.[71] Besides the Blue Ridge Reservoir's additional tourist value, the projected impoundment would flood much of the private land just south of Pine Mountain, in Grayson County, beside the NRA's proposed ski area, and thus protect this unzoned private property from land speculation and uncontrolled development—the only possible protection, given the county's unwillingness to implement zoning. But by 1977 opponents defeated the Blue Ridge Reservoir proposal, and thus helped influence NRA planners to drop the proposed Pine Mountain ski area, already under attack from NRA development opponents.[72]

A hiker atop Pine Mountain in the Mount Rogers NRA. This area became the most controversial in JNF history, with eminent domain used to acquire property to serve the demands for outdoor recreation. U.S. Forest Service photograph.

Opposition to acquisition and development challenged the original congressional vision of the NRA and the six years of planning and implementing this vision. NRA designers instinctively defended their work. As Charles Blankenship wrote in 1973, "clearly, Congress did not intend a small, quiet NRA. In fact, by definition I don't think there can be such a thing as a small, quiet NRA."[73] As the Forest Service came under increasingly severe attack, NRA planners began casting back to the original NRA legislation, which clearly specified concentrated development in response to the nation's recreational needs. Nevertheless, 1973–74 became a time of reassessment for the JNF.

In an atmosphere already charged with suspicion, the failure of the JNF to produce a Unit Plan outlining the exact tracts slated for acquisition probably engendered further distrust among acquisition and development opponents. As NRA area ranger Keith Argow said at the Atlanta regional staff meeting in August 1973, "our Unit Planning package will show for the first time the 30,000 acres that will remain in private ownership in the five counties. The fear of the unknown by the local people is a serious and difficult thing for us to work with."[74] Instead of a unit plan, a bureaucratic quagmire ensued, reflecting the predominant reassessment of the NRA at the regional Forest Service level.

In August 1973 the NRA continued to operate on a land plan stemming from 1967 that included future acquisition of 340 additional tracts, involving 116 yearlong and 65 seasonal residences.[75] Some of these acquisitions would inevitably involve condemnation, which assured continued controversy. In September 1973 the NRA lands staff met to reconsider their situation. They recognized that land acquisition should proceed more along ownership boundaries rather than topography, in order to avoid condemnation if possible. Inholdings remained desirable.[76] In January 1974, four years after NEPA initiated such studies, NRA staff determined a need for further examination, specifically regarding the area's sociology, economics, archaeology, and history. In early March 1974 NRA planners met for two weeks to analyze each aspect of the proposed NRA development, such as the scenic highway, the possible impacts of such aspects, and alternatives to the plan.[77]

Amid their own uncertainties and in dealing with a public still divided over NRA plans, the Forest Service feared that public debate could go on endlessly. As JNF Supervisor Penfold wrote in March 1974, "We frankly fear that review of these two documents [the Unit Plan and EIS] has the frustrating potential to drag out indefinitely. Opportunities for argument and discussion on both philosophy and substance of the plan have no limits."[78] Amidst the uncertainty and reassessment, a consensus gradually developed at all levels in the Forest Service to relinquish condemnation, and the lawsuits quickly diminished, with sixteen filed in 1974 and the final four in 1975. By now the

NRA had clearly changed directions, and additional, external events and influences reinforced this new course.

The NRA's 1977 Draft Environmental Impact Statement elicited input from a large and varied number of local, state, and national groups. The local Citizens for Southwestern Virginia group opposed condemnation and concentrated development, and the growing environmental movement made its voice heard through groups like the Virginia Wildlife Federation and Sierra Club, who opposed the construction of recreation features on environmental grounds.[79] The Mount Rogers Final Management Plan and Environmental Impact Statement, published in January 1978, documented the new order, based on public response: the Forest Service faced significant opposition to both the ski area and the scenic highway, many locals resisted land condemnation and displacement, and many feared an intensive commercial development that would raise their property taxes without necessarily benefiting them economically. "The social and cultural changes envisioned by respondents range[d] from nightmares of overcommercialization and second home development to visions of progress, modern conveniences, and higher standards of living."[80]

To seal their opposition stance, the Citizens for Southwestern Virginia began gathering signatures to oppose further NRA development, and by May 1978 they had over 11,000 signatures protesting Forest Service land condemnation, the proposed ski resort, and the scenic highway. They advocated maintaining dispersed development on the NRA and continued to circulate petitions. They reportedly collected more than 23,000 signatures by March 1979.[81] The Citizens of Southwestern Virginia based many of their objections on what they perceived as contradictions or lack of careful study within the NRA plan itself, involving such topics as sewage treatment, garbage disposal, and transportation logistics. They also recognized the local lack of zoning and the unwillingness of county governments to implement such zoning, and thus the danger of the NRA proceeding along the lines of concentrated development.[82] At this point, however, the trend of the NRA had already turned away from concentrated development. Henceforth, the NRA featured "camping, picnicking and interpretive sites rather than the flashier developments."[83] The NRA maintained a relatively low profile during the 1980s and into early 1990s. The 1985 Forest Plan elicited very little public response concerning the NRA.[84]

That so much contention arose over the NRA acreage indicated, on one hand, the ambitious ideas that professional recreation planners had for territory they appreciated for its unique public appeal. For the residents, this land was also unique, but it was home. A wide spectrum of feelings about the territory had only one thing in common, and that was some sense of its exquisite beauty. Even the language of one of the federal condemnation suits would have evoked the envy of a nineteenth-century romantic. It is worth quoting in its entirety:

> To reach "White Top," as the land to be evaluated is called, from the west you motor over a well-engineered smooth paved two-lane very crooked state highway located along the bank of Straight Creek, a clear swift stream, at the base of a deep canyon, walled by bold and almost perpendicular cliffs that rise in naked majesty hundreds of feet above, and down the sides of which in gulch and glen mad waters rush to the ravine below, and there chafe and madden at their base in the stream filled and choked by the shattered fragments riven from the summit. Here and there the hand of nature, the great restorer, has softened the asperities of the scene and clothed in verdure the beauty both of beetling [sic] cliff and precipitous gulch. Here the spruce towers in somber majesty, there the laurel, ivy and rhododendron throw their dark mantle over the spectral rocks scattering their fragrance and, anon, the bramble and muscadine mingle foliage and fruit in wild and graceful beauty, caressing with tender tendrils and purple berries some tempest riven or lightning blasted trunk. The summit is crowned with delicious huckleberries and the ground covered with azaleas of every color of the rainbow. For several miles, this wild and savage scene stretches before us in unbroken continuity, except the occasional glimpse of the graceful fawn, leaping along as if to show its skill. Too soon we pass from this wondrous dream, and the majestic outline of the colossal "White Top" arrests our gaze as it lifts its savage head hundreds of feet into the blue sky crowning as it were some gigantic "Stupa" above the unmeasurable rocks, the fallen cliffs and shattered arches of ante deluvian architecture. You then pass for some miles where the rich pastures gently slope upward from the highway along which you drive, admiring the grandeur of the lofty mountain—a glory of Creation.[85]

Perhaps such landscape has the power to evoke such an effusive description from even sober-minded people more familiar with dry legal text. But clearly there were deep feelings from every sector of society who participated in the NRA controversy, from local residents to national environmental organizations, to federal entities such as judges and Forest Service personnel.

As originally conceived, the Mount Rogers NRA was perhaps too ambitious in relation to its size. An NRA of twice the acreage might have harmonized all of the recreational development aspects, but an NRA of Mount Rogers's 154,000 acres faced an almost impossible task. The JNF's eventual abandonment of the ski area, impoundments, and scenic highway reflected a scaled-down, relatively low-density recreational development scheme more adequately suited to the limited acreage.[86]

The early 1990s debate surrounding possible widening of State Highway 58 through the Mount Rogers area reflected a considerably different initial response among area residents compared to twenty years earlier. Quite unlike the creation of the NRA, when only a few people realized its scope, local opponents and proponents of Highway 58 expansion began advocating their viewpoints at the very beginning of the initial planning process.[87] The timely response to widening Highway 58 is in itself part of the 1970s legacy of public participation in the Mount Rogers NRA.[88]

For all the strife associated with the Mount Rogers NRA condemnations of the 1970s, by the 1990s some people had come to favor JNF management of the area, particularly as it manifested itself in protection of the returned forest and as a buffer against second-home development.[89] Since intensive recreation never developed, neither did its positive or negative components (such as its ambiguous economic impact, increased automobile traffic, and almost certainly higher property taxes to finance increased infrastructural demands on highways, sewers, and water and electrical supply). However, with Damascus one of the most popular places along the Appalachian Trail, proximity to the NRA has offered some modest economic stimulus and perhaps the lower impact tourism many foresaw as potentially more harmonious with the area's rural traditions.

Chapter 9

From Commodity Interests to Ecological Forestry

National forests are now amidst perhaps the most dramatic transformation in their history. What began as a utilitarian venture appears to be evolving into a more holistic consideration of the forest environment. A generational divide within the JNF characterized the Mount Rogers NRA policy controversy of the 1970s, which ended with ambiguous results. Concentrated recreation did not result, but neither could the JNF return to the old Depression-era community-based integration of district rangers. The current and recent transition also involves a generational divide, but it is far more sweeping and involves all national forests.[1] Trajectories of the contemporary evolution are impossible to ascertain, but an abandonment of king timber primacy for noncommodity forest features appears to be under way. As might be expected, a great deal of controversy throughout the last half century has accompanied this dramatic transformation. Not surprisingly, these controversies centered around timber harvesting itself.

JNF timber management has weathered its share of timber controversies since the 1960s. As elsewhere in the nation, protests over clear-cutting and below-cost timber sales (or deficit sales) reflected broad objections to greater Forest Service policy. Added to these aspects were local and regional contentions that recreation and ecological preservation were more important on an eastern national forest such as the JNF, with its limited acreage and more intense human populations in close proximity to these public lands. In recent decades, critics have charged that commodity timber values remained a constant characteristic from which all other forest uses were gauged, despite the rising importance of concentrated national forest recreation, wilderness designation, emphasis on nongame wildlife species, cultural resources, and more holistic forest ecology.[2] At least rhetorically, the JNF and other national forests have finally begun coming to grips with noncommodity timber values and ecological forestry rather than traditional utilitarian forestry.

During the 1960s the small 202-C timber sales, designed for Timber Stand Improvement and for providing local resources and revenue, all but disappeared. Industrial pulpwood corporations began to replace local farmers as the JNF's main wood-buying customer. The traditionally accommodating type of stewardship among local people, as far as timber was concerned, began to fade.[3] For illustration, in 1948 the JNF administered about 700 timber sales for that year's cut of 13,419,000 board feet. In the late 1980s the JNF administered an average of 30 timber sales totaling 21,000,000 to 30,000,000 board feet.[4] The 1960s also witnessed a transition from old TSI-oriented forestry to large pulpwood sales and clear-cutting. Pulpwood companies were beginning to finance logging operations on some of the larger JNF sales and even directly bid on a few contracts themselves. As commercial pulpwood buyers began to take over the market, they brought with them a decided political and economic interest in national forest timber policy.[5] The Appalachian Forest Management Group grew out of this context around 1986 and became quite vocal in advocating timber harvesting on Virginia's national forests.[6]

Clear-cutting

The advent of commercial pulpwood buyers of JNF timber coincided with the most controversial aspect of JNF timber policy from the 1960s to the 1980s: even-age management, or clear-cutting. Clear-cutting came to epitomize a clash in values and reassessment of Forest Service management practices, but it also represented a communications breakdown between the Forest Service and environmental groups (plus other members of the interested public). The Forest Service found itself fighting a public relations battle involving ecology and aesthetics and thus ventured into environmental politics almost completely unprepared, and certainly with little previous experience with such challenges.[7]

Around the end of the nineteenth century, when the idea of national forests first began to capture the consciousness of conservationists, selective cutting and "all-age" or "uneven age" management carried the day among scientifically trained foresters such as Bernhard Fernow and Gifford Pinchot.[8] Their approach reflected their German training in forestry and, in America, their reaction against previous forest stripping. Fernow, Pinchot, and others thus influenced the early Forest Service to practice all-age, diameter-limit timber management. This tradition continued during the first half of the twentieth century. It was during the post–World War II housing construction boom that forest scientists and timber managers first decided that clear-cutting best suited regeneration in western Douglas fir stands. These foresters, ever concerned with efficient and scientific woodland production, began to adopt clear-cutting as a management technique, at least under certain conditions. By the early 1960s most eastern national forests practiced clear-cutting, and the JNF's timber management plan officially included clear-cutting by 1964.[9]

Clear-cutting first began gaining approval on the JNF during the early 1960s. A forest experiment station located at Berea, Kentucky, had been conducting demonstrations of various timber management techniques over the previous decades when traditional selective cutting still prevailed. The clear-cut experimental plot was touted as the "worst possible method" in a scenario of different techniques. Silviculturists grew surprised, however, when strong, straight trees grew out of these plots. Many forest professionals from Virginia began to visit the experiment station, including state foresters, timber staff from the George Washington National Forest, and all divisions of the JNF. The days of eastern woodland even-age management had begun. Public controversy followed close behind.[10]

A definite distinction should be made between clear-cutting and the stripping of trees from entire mountainsides during the late-nineteenth-, early-twentieth-century industrial timbering of southwestern Virginia and other parts of the southern highlands.[11] Early timber industrialists generally cut all trees within their technical capability or economic feasibility. The JNF and other Appalachian national forests tended to limit the size of clear-cuts to as small as forty- or sixty-acre tracts. The JNF deliberately staggered the boundaries of a clear-cut to create more numerous, irregular edge areas that favored the habitat preferences of some wildlife species, then a common favored practice among forest ecologists. For example, two roughly twenty-acre areas might constitute an actual forty-acre cut, with a hundred-foot-wide band of trees left bisecting the entire area.[12] Obviously timber industrialists of the earlier era did not design their cuts in this manner at all. The fundamental distinction arose through silviculture and economic considerations. Through the kind of research described above, foresters had come to a scientific conclusion that clear-cutting presented the best timber management technique for the commodity focus that had always characterized the agency.

By cutting 500-acre tracts in close proximity to one another, the Monongahela National Forest demonstrated that clear-cutting had a potential for abuse, and in such cases clear-cutting more closely resembled the old method of stripping.[13] Timber cutting abuses on the Monongahela led directly to a national reassessment of clear-cutting. In 1969 the House of Representatives defeated the National Timber Supply Act, which would have channeled timber sale receipts back into Forest Service timber management.[14] This action reflected, in part, a growing public mood against increased national forest timber harvesting—and particularly clear-cutting—that culminated during the early and mid-1970s. Probably more than any other event, the noted 1973 Monongahela Decision affected the JNF's timber management policy of this period.

The Monongahela Decision, handed down by the federal court in Elkins, West Virginia, found clear-cutting in conflict with the 1897 Organic Act. Public challenge of timber policy on the Monongahela became something of a test case for all eastern national forests. In direct response to these West

Virginia events, the JNF drastically curtailed its timber sales and harvests. After reaching an unprecedented sale of 38.8 million board feet in 1969, the JNF's timber sales immediately and dramatically dropped, from 21.9 million in 1973 to a new low of 5.2 million in 1976.[15]

The Monongahela Decision's legal precedent directly led to an overall congressional and public reassessment of national forest timber management policy. The 1974 Forest and Rangeland Renewable Resources Planning Act (RPA) required examination of all national forest resources, with documentation on various resource demands or potential demands within the overall management context. The Monongahela situation and the RPA contributed to passage of the 1976 National Forest Management Act (NFMA),[16] which sought to prevent future abuses of clear-cutting by requiring public participation at every level of Forest Service management decisions.[17]

NFMA redefined the Forest Service's mission as no legislation had since the agency's inception and largely reflected congressional, public, and Forest Service response to clear-cutting. Although the Forest Service had long been in the land management planning business, NFMA required new levels of planning that would coordinate the perspectives of their own now-diverse personnel with the interested public. The JNF responded with land management plans, which sought to coordinate forest resources, management functions, and public input. As JNF Supervisor Thomas Hoots wrote, "the major process for determining the feasibility of a timber sale is the Environmental Analysis and Assessment," which involved consultation with cultural resource managers, wildlife biologists, silviculturists, landscape architects, and the public before a timber sale took place. Objections from any of these sources necessitated reevaluation and possible modification.[18] The controversy generated by clear-cutting obviously affected the course of national forest policy, even if controversy surrounding national forest timber cutting continued in different ways.[19]

Even forest professionals often took a mixed view on clear-cutting. As historian Harold Steen noted, both clear-cutting and selective cutting had potential for abuse.[20] Clear-cutting could take place on too large a scale and with disregard for detrimental environmental impact, such as erosion, difficult tree regeneration, or destruction of habitat for certain animal species. Selective cutting, however, certainly changes the forest composition and thus can alter its ecology in more subtle ways. The most commonly cited potential abuse, of course, focuses on destruction of the forest's overall commercial value through "creaming" or "high grading"—that is, harvesting of only the most marketable trees.[21] During the early 1990s JNF employees sometimes said "clear-cutting has its place" or "there's room for both selective cutting and clear-cutting." Some felt clear-cutting did not suit Appalachian hardwood species in particular, but saw a place for such even-age management in Douglas fir forests.[22]

The JNF responded to the clear-cutting controversy by immediately reducing its timber harvest, then reassessing timber management in general. Before 1970, bulldozer-type steel track machines constituted the most common type of logging equipment used in the eastern United States. Obviously, such equipment damaged the surface of the ground to at least some extent. Loggers began replacing such equipment with skidders, which swiveled in the middle and rolled along much higher off the ground on huge rubber tires. At the same time, selective cutting and more conscientious skid road construction began to gain favor. Finally, some loggers began to use cable systems that damaged the environment even less.[23]

The JNF reflected this trend, and new approaches included minimizing the visual impact of clear-cuts through a coordinated effort between timber staff and landscape architects. The JNF, too, began experimenting with Western-style cable-logging systems and "group selection" involving timber cutting on no more than two to three acres, all in an effort to continue harvesting clear-cuts while lessening road construction-oriented damage.[24] But the long term trend indicated a dramatic evolution away from clear-cutting altogether. Controversy over below-cost timber sales probably hastened this change.

Below-Cost Sales

Simply put, below-cost national forest timber sales result when the agency spends more money designing and administering a sale than it actually receives in direct returned revenue from the purchaser.[25] In the mid-1970s forest economist Marion Clawson cited the following characteristics as contributing to problems in such economic efficiency: excessive timber inventory, excessive management costs, difficult accessibility (which contributes to loss of timber through rot, disease, windfall, etc.), inadequate congressional funding for the Forest Service, and too many divergent "owners" ("the" public) who dictate national forest usage, thus leading inevitably to economic inefficiency.[26] Clawson wrote, "the Forest Service emerges from this analysis as a land-poor landlord, one who owns vast areas of valuable property but lacks cash to manage it well, and whose management produces low monetary returns."[27] Objection to below-cost timber sales grew more widespread within the Sagebrush Rebellion during the Reagan presidency, when privatization of various public lands gained advocacy and attention, if not implementation.[28] In 1985 various parties discussed this topic at some length, including the House Agricultural Committee, which heard testimony from a number of experts and other interested parties directly involved with specific national forests, including the JNF.[29] The JNF's proximity to Washington, D.C., relative accessibility to large numbers of eastern urban recreation users, and its unfortunate status as one of the most severe examples of below-cost national forest timber sales contributed its center-stage position.[30]

Typical of most Appalachian national forest acreage, the JNF has never had a particularly rich timber resource. As previously illustrated, the JNF tended to acquire steep, often rocky, difficult to access land ill-suited for agriculture. Naturally, such land generally produced the region's lowest quality timber. Also, as previously discussed, many formerly soil-rich areas suffered tremendously from hillside farming, when centuries of accumulated topsoil washed away.[31] But it became impossible to distinguish between critics of below-cost timber sales who were genuinely concerned over fiscal matters and those who used the issue as another guise in their opposition to national forest timber cutting in general. As William Shands observed earlier, clearcutting had become an environmentalist "shorthand term to express dissatisfaction with a number of aspects of congressional policy and forest management, particularly the funding and management priority given to timber over other uses."[32] Below-cost sales served a similar dynamic.

Of course, if timber sales ended altogether on the JNF, a hefty agency budget would remain. This may reveal a hidden no-cut agenda behind at least part of the below-cost controversy. In the larger picture, as historian Samuel P. Hays has revealed, difficult topography and industry demands has never rendered the national forest timber resource profitable or even self-sustaining to begin with. In the modern era this has led to inventive bookkeeping involving nonmonetary benefits of national forests.[33] Hays's discovery underlines enormous philosophical and political questions about who the national forests should serve and for what purposes.

JNF timber has never had an economic impact comparable to some western national forests. In 1985 all national forest wood accounted for only 3 percent of Virginia's wood resource. Westvaco's paper plant in Covington, Virginia, relied on national forest wood for only 6 percent of its raw materials.[34] On the other hand, smaller, local logging enterprises rely on the JNF for most of their wood. B. A. Mullican logging company has historically cut most of their timber from the JNF's Clinch Ranger District, and the Bennett Logging and Lumber Company of Covington has relied on national forest timber for as much as 80 percent of its supply. The Marley Moldings company moved to Smyth County, Virginia, with specific considerations given to the wood supply of the neighboring JNF.[35]

Wood-using industries that had come to rely on JNF timber supplies defended deficit sales for a number of reasons, including local economic stimulation, improvement of the forest through salvaging trees that would otherwise die from disease, and the usual argument for the long-term benefits of road construction.[36] Below-cost opponents interpreted JNF financing of logging access roads as an aspect of corporate welfare, wherein the JNF and other national forests were subsidizing the timber industry at the expense of the tax-paying public. In general, timber cutting opponents argued that the JNF's true value lay in scenic beauty, recreational usage, and preservation of a forest ecosystem not interrupted by logging.[37] In the end, as part of a direct reaction to below-cost sales, the JNF moved away from clear-cutting, reduced its overall "suitable timber base," reduced logging road construction, and gave non–timber oriented expenditures a greater portion of the budget.

In fiscal 1988, the Forest Service clear-cut a combined 4,703 acres on the Jefferson and George Washington national forest acres. In fiscal 1996 the cut was reduced to a mere 405 acres. Overall volume of timber harvested also dropped, from 69.2 million board feet on the combined national forest acreage in fiscal 1988 to only 37.6 million in fiscal 1996. In fiscal 1990 the JNF allocated $369,600 (3 percent of its total budget) for timber road construction. In fiscal 1994 that amount dropped to $136,365 (or .8 percent). During these same years, recreation funding either matched or exceeded timber management funding. In fiscal 1990, 11 percent of the budget went for recreation and only 8.6 percent to timber management. Fiscal 1995 saw recreation funded at 9.8 percent and timber at 9.1 percent.[38] Obviously, JNF timber management has undergone some revolutionary changes.

The gradual development of ecological forestry concepts coincided with this reduction in commodity timber harvest. In general, ecological forestry approached the woodlands environment with an eye toward valuing and understanding everything and how each forest component invariably influenced and interacted with other components and within the ecosystem as a whole. Nothing could be more radically different than a commodity approach to single forest components, such as marketable trees or game animals. The role of forest fire illustrated how an earlier heresy slowly became part of forest ecology orthodoxy.

Prescribed Fire

An early, highly successful campaign counteracted serious conflagration abuses but did not yet appreciate the ecological aspect that fire played in woodlands that have adapted to this natural force.[39] It was this growing appreciation for fire's natural ecological role that changed the Forest Service's approach to their old enemy. The eastern United States in general remains too crowded to accommodate the "let burn" policy practiced on western wilderness areas. Instead, the place of fire in the forest ecosystem must be imitated with prescribed burning.[40]

Many traditional foresters, who had fostered their careers during the early twentieth century largely on reversing the common human-caused fires throughout the South, resisted the notion of prescribed burning. Opponents feared soil and seed destruction, as well as the danger of losing control of fires. Proponents, on the other hand, claimed fire actually stimulated certain new seed germinations and rid the forest of dangerous underbrush, a potential incendiary or wildfire hazard. In any case, by the late 1950s the Forest Service had largely come to accept at least the concept of prescribed burning.[41] Actual implementation on the JNF, however, would have to wait some years. Most early foresters and even forest scientists did not seem to know how central fire had been for the establishment of the forest they protected.

In the mixed hardwood forest of the JNF, fire prevention has encouraged shade-tolerant hardwood species such as beech and sugar maple.[42] In 1988 David Van Lear and Thomas Waldrop observed that fire actually encouraged certain tree species in the Appalachian forests, such as some varieties of oak, pitch, and table mountain pines. Fire discouraged species with thinner bark, such as maple and poplar. Since maple and poplar tend to dominate former oak sites, prescribed fire could possibly function as a tool for regenerating an oak forest.[43] The effects of fire or its absence presents a science in a nascent stage, to say the least. Many decades of studies would be necessary before a more detailed picture emerges.

For example, the Peters Mountain mallow (*Iliamna corei*), an endangered herbaceous perennial plant, apparently neared extinction largely through absence of fire. Like other flora species, the mallow's hard seeds need an extreme agent in order to germinate. Overbrowsing by deer presents a subsequent problem for young plants, but the recent efforts and attention of the Nature Conservancy, JNF, U.S. Fish and Wildlife Service, and Virginia Tech promise to ensure this plant's survival.[44] The Peters Mountain mallow, named for its location in Giles County, presents only one fire-dependent flora species among many that ecologists are beginning to understand.

Certainly fire played an integral part in Appalachian hardwood ecology in the past, and subsequent Forest Service fire suppression influenced this ecosystem. As JNF Ranger Bob Boardwine observed in the early 1990s, his New Castle District's southwestern slopes' hard pine species were fading—in part, reflecting the successful fifty-year-old fire suppression era.[45] The diminishing numbers of those trees changed the ecological makeup, where any single tree species had numerous interrelations with other flora and fauna species dependent upon or associated with it. Interestingly, in this case, human intervention dramatically influenced the forest environment through protective rather than outright destructive or radically disruptive means. The role of fire in the Appalachian hardwood ecosystem reveals how once seemingly beneficial practices have had unexpected results. Certainly fire's positive possibilities only became conceivable partly through the long-term success of the great campaign against excessive, human-induced fire.

Soil Reclamation

As traditional as its timber stewardship, watershed protection continued on the JNF during the postwar decades with increasing effectiveness and what should have been widespread approval. Unfortunately, ambitious soil reclamation never drew as much public attention as controversial timber management techniques. But while JNF silviculturists were attempting to improve the applications of their science, so too were hydrologists and minerals specialists. Their work related to logging, logging road construction, mineral extraction, and, most dramatically, reclamation of old mineral extraction sites. While the 1977 Surface Mining Control and Reclamation Act most specifically addressed coal strip-mining, clearly the federal legal and public environmental mood had shifted toward protecting lands and waters from any detrimental mining effects.[46]

The reclamation successes of old mineral extraction sites on the JNF present its most spectacular ecological improvements in recent history. Among the more severely damaged acreage the JNF faced reclaiming involved the Glade Mountain area in Smyth County where the Union Manganese and Manganese Mining and Contracting company stripped dozens of acres during the first half of the twentieth century, especially during both world wars and the Korean War, when the federal government subsidized their operations for the combat effort. Subsequent abandonment left the JNF with a large denuded area stripped of any vestige of topsoil or covered with raw spoil banks, all of which actively polluted neighboring streams such as Cripple Creek, Blue Spring Creek, and Killinger Creek. Even initial JNF soil restoration immediately benefited the area, and by the summer of 1967 Blue Spring Creek, draining reclaimed land, showed marked improvement over other creeks still choked with silt.[47]

By far the most expensive JNF watershed project concerned restoring Georges Branch, the 986-acre abandoned mining operation on the Wythe District.[48] This project not only drew the most expenditure but also represented one of the most cooperative projects undertaken on the JNF during the modern era. While active restoration did not get under way until 1985, parties involved in the earliest planning processes of 1980 included all divisions of JNF staff, Virginia Division of Mined Land Reclamation, Northeast Forest Science Lab (Princeton, West Virginia), the Appalachian Trail Club, the Soil Conservation Service, the Army Corps of Engineers, and others.[49] By May 1983 JNF staff had completed the Georges Branch prescription, which Region 8 Staff Director J. Lee Bardwell praised, identifying it as "the largest single project planned with the Region to date." Bardwell also appreciated the project's complexity, its tremendous expense, and the JNF's designation of restoration priorities. These priorities would facilitate the most efficient funding, since "multi-year financing [was] about the only way that the necessary money [could] be made available."[50] Bill Morrison, JNF Watershed Forester at the time, remembered initial predictions foreseeing a project that would take twenty to forty years.[51] A pleasant surprise, however, awaited the JNF.

Tennessee Valley Authority (TVA) matching funds and their lime-spreading helicopter helped turn the George's Branch project into a very short-term success story. Active restoration work began in June 1985 when the JNF and TVA began liming, fertilizing, and planting grass, with plans for initial reforestation the following spring. By late August their efforts had already begun to approach the later stages of the restoration project.[52] The JNF and TVA continued their cooperative efforts into 1989, by which time each party had spent almost $150,000.[53] By the early 1990s the Georges Branch area still required environmental monitoring for several decades. It takes land many years to recover from strip-mining, and a particularly dry year can reverse earlier restoration efforts. As Don Blackburn reflected, only by the early 1990s had the JNF passed the critical stage with the Glade Mountain area, which they restored during the mid and late 1960s.[54] All current JNF mining, logging, road construction, or any other activity disturbing or potentially disturbing the soil requires much careful execution and monitoring. Gas and oil drilling, as well as deep coal mining, now prove relatively safe. In 1992, according to JNF minerals specialist Don Blackburn, even though mining companies of the 1980s and 1990s kept their profit margin in mind, they generally proved reasonable in complying with these laws when operating on JNF territory.[55]

Minerals and Fossil Fuels Management

Though not as significant a factor as in some western or southern national forests, or a coal-rich Appalachian national forest like the Daniel Boone in Kentucky, minerals and their extraction have nevertheless been a constant concern for the JNF since its inception, especially in the coal-bearing Clinch Ranger District area. Iron, coal, manganese, natural gas, various types of surface rock (and possibly oil in the future) have and will play a significant role in the JNF's mineral management. Mineral extraction now takes place under modern legislation designed to prevent soil erosion and protect stream ecology. But mineral ownership and extraction rights continue to present the JNF with a complex situation. On many eastern national forests, private landowners have retained mineral ownership or extraction rights after Forest Service purchase. In these cases, the national forest manages only the land's surface rights.

Originally, the 1920 Mineral Leasing Act gave the Department of the Interior jurisdiction over Forest Service mineral leases, exclusive of the southern Appalachian national forests.[56] The 1947 Acquired Lands Act extended the Department of the Interior's jurisdiction to include the JNF and other regional national forests.[57] Region 7's director of recreation and land summarized these legal circumstances for the JNF this way in 1958:

> On national-forest lands where mineral rights are owned by the United States, the management of mineral resources is the joint responsibility of the Bureau of Land Management and the Forest Service. The BLM has primary responsibility for disposal of minerals and receives applications for permits and leases on such lands, but must obtain the consent of the Forest Service before issuing a permit or lease. It is the responsibility of the Forest Service to determine whether the disposal of minerals would be commensurate with the management objectives of the national forests, and to require such land protective stipulations as are in the public interest.[58]

By the early 1990s minerals involved with the JNF posed three possibilities: outright federal ownership of minerals, reserved mineral rights on federally owned land, or outstanding mineral rights on federally owned land. In 1992 the JNF owned about 87 percent of its minerals, with 13 percent either reserved or outstanding. This 13 percent lay mostly on the Clinch Ranger District and mostly involved coal. In 1992 Don Blackburn described the JNF as a moderate Region 8 national forest in terms of mineral activities—light compared to national forests in Texas and Louisiana and heavy compared to a forest like the George Washington, which contained relatively few mineral resources.[59] Outright federal mineral ownership presented the JNF with its simplest circumstances, in which prospective mining companies secured a lease through the BLM and proceeded to extract minerals according to federal environmental guidelines. Such a situation gave the JNF direct control of mining, just as it exercised direct control over timber harvests.[60]

Reserved and outstanding mineral rights present a far different situation, and one that can grow quite complex when anyone, including the Forest Service, tries to sort out actual mineral ownership. The legalities involved in such cases stem from stipulations that corporations or private individuals included in various transfers of land, and they often precede Forest Service ownership by several transactions.[61] Numerous heirs, disputed land titles and deeds, the notorious "broad deed,"[62] and other problematic mineral leases—sometimes extending back to the turn of the nineteenth century and before—continue to characterize the entire southern Appalachian coalfields. For instance, a thirty-six-year-old Wise County lawsuit involving an 1874 mineral lease for thirteen coal seams came to involve about 500 heirs.[63] Mineral rights or ownership remain completely unknown on some tracts in the coalfields, on both federal and nonfederal property. In far greater proportions than many other places, lawyers heavily populate the predominantly rural coalfields, and much of their work centers around sorting out minerals rights, leases, or ownership.[64]

In any case, reserved mineral rights on JNF property means that the Forest Service owns only the surface of the land. The mining company who uses reserved rights must obtain a permit from the Forest Service to mine, but such a procedure has now become almost a formality, and the private party pays some token fee, maybe one dollar per acre per year, to exercise their reserved rights. Outstanding rights, however, give the holder almost free rein and allow mineral extraction to proceed much as it would on private land. About the only Forest Service prerogative lies in observing the operation and threatening or enacting a lawsuit if the mining company violates state environmental standards, which are essentially on par with federal environmental standards.[65] As in the rest of the nation, environmental protection has made significant progress surrounding JNF mineral extraction, no matter who owns the mineral rights.

Before 1977 no policy for strip-mining on national forests existed, and surface mining for coal posed a serious environmental threat on national forest lands. For the JNF, this pertained to the Clinch Ranger District. Coal companies regularly stripped for coal and mountains-turned-mesas appeared throughout the tri-state area of Kentucky, West Virginia, and Virginia. Much uncertainty lingered during this period, and the JNF, the BLM, or a private interest might control strip-mining of coal on a given situation on the Clinch District. The 1977 Surface Mining Control and Reclamation Act effectively prevented strip-mining coal on all eastern national forests.[66] Some holders of reserved or outstanding mineral rights in the Clinch District area argued their privilege to surface mine thereby, but the JNF successfully countered this strategy in court by indicating that strip-mining, as a coal extraction procedure, was not commonly practiced or did not exist at all during the late nineteenth and early twentieth centuries, when the private individuals in question secured such mineral rights. As such, strip-mining does not threaten JNF territory. By the early 1990s, the Clinch Ranger District featured only two deep coal mines, one tapping federally owned minerals and the other tapping reserved minerals. For the most part, these mines promised to remain environmentally safe.[67]

During the 1970s energy crisis, geologists and mining companies once again grew excited over the oil and gas potential of the southern Appalachian eastern overthrust belt, included in JNF territory. Companies like Gulf and Chevron began acquiring gas leases on the Clinch District during the late 1970s, and by April 1981 the JNF had received over three hundred oil and gas lease bids involving forest-owned mineral rights, most of which were on the Clinch. After 1987 various companies, including the American Natural Resources Production Company, paid the Clinch District $52,000 for gas and oil leases on 13,000 acres.[68] By the early 1990s no company had found oil on the JNF, though the right market circumstances could encourage exploration once again, when deep wells, hard rock, and the very uncertainty of finding oil at all would justify drilling expenses. Gas extraction, on the other hand, has proven quite successful. Geologists predict many more years of gas production from the Coeburn and eastern Pine Mountain fields; at between 1,200 to 1,500 feet below the surface drillers have found methane or natural gas associated with coal beds. Like deep coal mines, gas operations present relatively little environmental threat. The most disruptive phase accompanies only the actual drilling, when mining companies cut roads and bore their wells. But after the few weeks involved in this initial stage, with the well lined and tapped, extraction generally continues quite safely.[69]

Among the more substantial postwar legal changes, the 1987 Federal Onshore Oil and Gas Leasing Reform Act required environmental assessments of mining activities on all national forests, placed bids for federal minerals on a competitive basis, and raised the minimum bid from one to two dollars an acre.[70] The 1987 act especially gave eastern national forests considerably more control over their mineral resources, by requiring the BLM to accept (previously optional) Forest Service recommendations surrounding mineral extraction.[71]

Modern Hydrology

During the late 1960s and 1970s, professional hydrology and soil science began to come into their own on the JNF and other national forests. Former JNF employee Bill Morrison recalled himself as one of the last "watershed foresters" who worked through this transitional period with professional hydrologists and soil scientists before retiring in 1986. During this time the JNF relied quite heavily upon the most current scientific data gathered at the Coweeta Experiment Station located in the highlands of North Carolina, in a forest environment somewhat similar to that of the JNF. The JNF built stream buffer strips, logging truck stream crossings, and water bars—all designed according to the latest Coweeta techniques. The Coweeta Station became not only a source of specific scientific data and hydrological techniques but also a channel through which JNF hydrologists communicated with other hydrologists nationwide.[72]

Early hydrologists focused on hydroelectric power potential, protecting municipal water supplies, and erosion mitigation. From this basis, later interests grew to include hydrology as part of a crucial botanical and ecological imperative. Within the specialty itself, Forest Service hydrology began with an emphasis on water *quantities* as affected by different land uses and evolved into greater emphasis on *quality* of water as related to its cycles and its nutrients' cycles within various forest ecosystems. Finally, in the modern era, the Forest Service recognized hydrology as one ecological component or medium among several that included soil biota, insects, and biotic flora and fauna.[73] The advent of ecological considerations have especially marked the second half of the Forest Service's century-long history. An organization once peopled by forestry pragmatists came to contain a significant number of employees geared more toward the forest's aesthetic or scientific qualities, or both. Along with the rise of ecological forestry, the JNF has entered an entirely new era in which ecological considerations maintain a prominent role.

Developments in Modern Wildlife Management

The 1960 Multiple Use and Sustained Yield Act helped bring the JNF its first full-time and professionally trained wildlife biologist when Paul Shrauder moved from South Carolina, where he had worked on the Sumter National Forest as a state employee. Shrauder spent his first year examining wildlife habitat and staking out vegetative plots over the JNF. At that time, during the early 1960s, three wildlife biologists and sixteen game managers worked for the state of Virginia, and Shrauder helped coordinate federal-state wildlife management efforts.[74] The personnel shuffle that followed the JNF's 1966 transfer into Region 8 resulted in Shrauder moving to the Atlanta regional office that year. There he built upon his JNF experiences and embarked upon the Southeast Wildlife Habitat Survey. By 1969 he and two other wildlife biologists, Herman Holbrook and Bill Zeedyk, had produced the region's first *Wildlife Habitat Management Handbook*,[75] which they and other wildlife biologists updated periodically. This handbook addressed such topics as "Management of Understory Plants for Forage and Fruit Production," "Management for Mast," and "Forest Type-Groups and Associated Wildlife Species." It also addressed the positive effects of clear-cutting for game habitat. Clear-cutting encouraged mast-producing trees, and the edge environments created by clear-cuts attracted deer; thus, Shrauder and his coauthors became early proponents of this aspect of clear-cutting. They also introduced the "featured species" concept and management, which included analyzing habitat, native species, the featured species' effects on other resources, as well as public interests, needs, and involvement.[76]

The arrival of Shrauder in some ways marked the first sign of the new, more ecologically aware era that the JNF was about to enter. Timber people might have been suspicious or confused by Shrauder or the wildlife biology endeavor in general, but they took some assurance in his emphasis on game animals. After all, hunting and fishing were old American traditions, and those oriented toward pragmatic timber commodities could relate to some degree to game animals as commodities. The rise of wildlife biology and game management helped facilitate the transition from the earlier primacy of timber utilitarianism and the later development of what came to be called "ecosystem management."[77] Shrauder's early focus differed from other JNF foresters in a fundamental way. Rather than concentrate on trees as the forest's crop, he focused on the forest as animal habitat.

After the 1960s, wildlife management underwent some tremendous conceptual changes that affected the JNF and every other national forest. The 1964 Land and Water Conservation Fund supported threatened species protection,[78] and afterward Congress reinforced such protection with a series of laws that culminated in the 1973 Endangered Species Act (ESA).[79] The ESA in particular heralded a shift toward ecosystem management on national forests.[80] The ESA encouraged biologists to conduct endangered species inventories on the JNF that eventually contributed to intensive studies of faunae such as the log perch, bog turtle, and Yonahlosse salamander. Professors and students from area colleges and universities, along with the U.S. Fish and Wildlife Service, sometimes coordinated this fieldwork with timber sales, and the JNF integrated these findings into their land management plans. Even the state, whose traditional wildlife management focused almost exclusively on game species, began to take an interest in nongame species. With Shrauder's help, Virginia initiated their peregrine falcon program, and eventually Virginia residents could contribute state income tax to nongame wildlife programs.[81]

During the 1960s and early 1970s, the old rationalization of clear-cutting providing wildlife habitat began to wear thin with some members of the public and scientific community. White tail deer were already overpopulated in some areas and actually needed no more of the temporary edge environments created by clear-cutting.[82] Critics still saw national forest wildlife management as favoring game species, with little consideration for holistic ecology or biodiversity.[83] Forest researchers compared clear-cuts with selective cuts on the JNF and concluded that, in some cases, selective cuts actually helped deer more than clear-cuts.[84] At the very least, their conclusions countered what had begun to seem like an overly simplistic argument.

Congress modified the 1976 National Forest Management Act in 1982 to include protection for threatened and endangered species, as well as directing attention toward maintaining diverse plant and animal communities.[85] As a result of such legislation, JNF mineral leases and timber sales had to provide protection for threatened or endangered fish species, such as the spot-fin chub, yellow-fin madtom, and Roanoke log perch, and endangered mollusk species such as the fan shell, Appalachian monkeyface, and the dromedary mussel.[86] Similarly, forest uses could not interfere with endangered plants, such as the interrupted royal fern (Craig County), spreading marginal wood fern (Mountain Lake), and the Peters Mountain mallow (Giles County), or threatened birds, such as the Bewick's wren, found in the Mount Rogers / White Top Mountain area.[87]

Clearly flora and fauna management considerations have grown tremendously more complex and challenging on the JNF. Wildlife biologists now appreciate how edge areas favor only some species (red fox, white tail deer, ruffed grouse, and many species of songbirds) but not others (such as warblers, wood thrushes, mountain lions, and black bear), which prefer large, dense areas of forest. Reviving fish populations by importing brown and rainbow trout is not the same as restoring the native brook trout, a fragile, char species indigenous to the southern highlands.[88] Ultimately, southwestern Virginia's forests and wildlife have experienced at least three dramatic phases during the past century: destruction, rehabilitation, and now a more holistic ecological management in an era of diverse human pressures. The advent of areas set aside for a minimum of human disturbance (poor air quality being a major exception) represented a major milestone in the latter phase.

Wilderness and Air Quality

The establishment of wilderness areas within the JNF created interesting possibilities for continuing a higher degree of biodiversity in southwestern Virginia.[89] With the 1964 Wilderness Act, Congress and the Forest Service had only considered wilderness as a western United States possibility.[90] At this point they did not ignore eastern national forests; rather, by its very definition, the prevailing concept of "true wilderness" could not include eastern lands, already significantly altered through past agricultural and industrial endeavors. In the 1960s, wilderness purists saw advocates of eastern wilderness as diluting an otherwise strict concept, and some preservationists felt designating eastern wildernesses would endanger the entire concept of wilderness. They feared a "grow your own wilderness" mentality would threaten undisturbed areas since, if wilderness could be grown, then it could also be (temporarily) destroyed. Nevertheless, in 1975 Congress passed the Eastern Wilderness Act, and the JNF immediately recommended the 8,850-acre James River Face area for designation.[91] The James River Face became Virginia's first wilderness area, and one of the very first in the east.[92] The Eastern Wilderness Act stated management of these eastern areas in consistency with the 1964 Wilderness Act and recognized them as also having "specific values of solitude, physical and mental challenge, scientific study, inspiration, and primitive recreation for the benefit of all."[93]

The Eastern Wilderness Act also allowed for "wilderness study area" classification, which indicated possible future wilderness designation. The JNF put the Mountain Lake, Peters Mountain, and Mill Creek areas under this consideration. The JNF's wilderness areas arose through processes involving public input, and by 1990 numbered eleven places with a 59,834 combined acreage.[94] Until 2009 the Clinch had been the JNF's only ranger district not featuring a wilderness area, mainly because of mineral concerns. But in 2009 the JNF designated the 3,270-acre Stone Mountain Wilderness in Lee County, as well as five additional wilderness areas on the Eastern Divide Ranger District and the Mount Rogers NRA, altogether constituting over 27,000 acres.[95]

Unfortunately, setting land aside from timber harvesting, road construction, or use of motorized vehicles does not alter the fact that the JNF and the southern highlands in general are downwind from major sources of air pollution, thus making this region the recipient of some of the most toxic accumulations of acid raid and ground level ozone. The destructive force of polluted air and its associated precipitation have especially impacted the highest elevation areas, which disproportionately lie within the JNF.

Creation of the James Face Wilderness in 1975 brought air quality considerations to the JNF.[96] The 1984 Virginia Wilderness Act required air quality studies for potential wilderness areas. Possible Class-I wilderness designation for the James River Face Wilderness Area, as well as possible wilderness designation for four Virginia wilderness study areas on the Jefferson and George Washington national forests (Shawvers Run, Barbours Creek, Rich Hole, and Rough Mountain) initiated air quality monitoring to the JNF for the first time. However, continuing operation of upwind midwestern coal-fired power plants meant no mitigation of poor air quality.

By late 1986 scientists determined that Whitetop Mountain area rain proved five to seven times more acidic than in neighboring lower altitudes,[97] and early 1992 brought some momentous news: for the first time, scientists claimed an unquestionable connection between acid rain and the incidence of high elevation forest decline in the eastern United States (not to be confused with so-called natural balds found in the southern Appalachians).[98] Professor Donald DeHayes, leader of the study, and other researchers determined that acid rain weakened trees on Whitetop Mountain and made them more vulnerable to disease and cold temperatures. Subsequent study seemed to contradict their findings, but continuing investigation seems to verify DeHayes initial conclusions.[99] It will likely remain difficult to reach any objective conclusion, since nonscientific factors invariably influence such experiments. But at the very least, the JNF and other public lands remain important for providing researchers with a living lab. This was especially important in the Whitetop air study case, for as scientist Frank Thornton wrote, most previous experiments had taken place in an indoor laboratory or other controlled situations.[100]

Our entire energy consuming society has caused the compromise of air quality and resultant effects on the JNF and other regional forests. Unless some radical improvement, such as carbon sequestration, becomes technologically and economically viable, the problem is likely only to get worse. In this unfortunate scenario, the JNF and other public lands in southern

Appalachia actually stand out most in their roles as barometers for decline. But at least contemporary environmental scientists enjoy traditional national forest access, and so the JNF may function in part as a control group in contrast to surrounding private forest lands. In this way we may continue to get an idea of the state of our air quality and overall forest health.

Forest Science and Ecology

Timber surveys represented some of the earliest scientific work on the JNF and reflected the prevalent commodity interests of an earlier era. The first Forest Survey (1933–40) involved a basic inventory; the second (1946–57) attempted to measure timber resource trends; and the third (1957–66) incorporated more sophisticated timber analysis, including human removal of trees and natural mortality as agents of forest change.[101] But the 1974 Forest and Rangeland Renewable Resources Planning Act (RPA) encouraged the most comprehensive forest survey of all. The 1974 RPA prompted three researchers to conduct a multiresource inventory study, which included the JNF in an overall reassessment of southern national forest resources. This study integrated the three previous forest surveys, and one of its main efforts lay in providing the Forest Service and other land managers with a modern analysis of nontimber resources. The researchers also voiced a new appreciation of the forest as an immense and complex ecosystem.[102]

By the 1980s oak species on the JNF had begun an apparent decline. Though nothing approaching the devastating chestnut blight, the gypsy moth, so-called overmature oak stands, and possibly acid rain contributed to increasing instances of disease and timber quality deterioration on some sites. Other Appalachian national forests had begun to show more severe decline than private hardwood forests, possibly because national forest trees were generally older and located on poorer sites.[103] Ironically, the chestnut blight was largely (though indirectly) responsible for the heavy concentration of oaks, which, in turn, have experienced gypsy moth infestations. Like the chestnut blight, gypsy moths were an exotic "pest" from abroad and spread, similarly, into the southern Appalachians from the northeastern United States. Agriculturalists appreciate how monocultures invite devastating infestations, and the oak forest had done just that. In the longer-term biological picture, of course, such infestations ultimately consume themselves, as does any parasite killing its host. The Forest Service recognized this and decided against spraying insecticides in an effort to combat the moth.[104]

Perhaps the most remarkable aspect of gypsy moths, in terms of JNF management, is what did not happen. Timber cutting proponents excitedly advocated salvaging as many trees as possible through harvest before moths could defoliate and kill them. Such alarms had arisen over the decades all along the eastern corridor of the moth's migration. The JNF did not harvest the trees, and dire predictions of defoliation and forest destruction simply did not materialize. Instead, Jim Sitton, timber management staff officer, chose to incorporate oak harvests related to gypsy moth defoliation into the regular "allowable sales quantity" rather than pursue a traditional tactic of authorizing additional salvage sales to the regular cut. Jim Loesel, a close monitor of JNF activity for many years, recognized this as a fundamental philosophical shift in timber management. "It is not our moral duty to grab every tree that dies," Loesel said in 1997, reiterating the biological value of decomposing trees.[105] Loesel's comment also reflected the agency trend toward greater attention to noncommodity forest values.[106] Certain JNF employees appreciate such biological value in great detail.

The JNF hired its first botanist in January 1992. Fred Huber and other Forest Service botanists stood at the very forefront of a fledgling new field involving biological emphasis on national forests. Huber's responsibilities included identifying threatened and endangered plant species, monitoring old-growth timber, making field inventories prior to various JNF projects (such as timber sales, road construction, or recreational development), and coordinating all these efforts with other JNF specialists as well as various federal and state agencies. With cooperation from the Virginia Natural Heritage organization, one of Huber's early projects focused on the possibility of establishing the very first Research Natural Areas on the JNF.[107] Huber's hiring represented an early indication of a continuing shift away from traditional commodity timber emphasis. Like archaeologists and other "oligists," Huber's duties actually often obstructed traditional timber harvesting and road building, for the discovery of rare or endangered flora would mean the end or modification of a planned timber sale or road access construction. In areas set aside from commodity considerations altogether, forest science could proceed with less conflict.

One of the most dynamic aspects in scientific research involving the JNF lies in relation to the University of Virginia's Mountain Lake Biological Station. University of Virginia biology professors Bruce Reynolds and Ivey Lewis established this facility in 1929 and immediately attracted scientists from all over the nation. From the first National Forest Reservation Commission purchases to the present day, adjoining national forest land has increasingly augmented the station's environment. In 1959 station director Horton Hobbs persuaded the JNF to protect 1,500 acres near the station as an undisturbed natural area. The Virginia Wilderness Acts of 1984 and 1988 that created the Mountain Lake Wilderness, of course, tremendously improved upon this earlier Forest Service protection.[108]

Where previously experiments like the white pine provenance study and poplar tree studies were applied science, the Mountain Lake Biological Station's researchers most often worked in the realm of pure science, with studies on such topics as land snails, scarlet oak sawfly, and macrophyte ecology. Their location on Salt Pond Mountain (popularly known as Mountain Lake) has come to play an even more important role during the past couple of decades, as actual field study and ecological systems have gained more prominence in the traditionally lab-oriented sciences of botany and biology. The unique qualities of the Mountain Lake and southern Appalachians area—at over 400 million years old, among the world's very oldest uplands—now attracts scientists literally from all over the world.[109] By the early 1990s, around fifty scientists and their assistants, some supported by National Science Foundation grants, conducted research out of the station. Researchers and science professors offered eight to ten lab and field-oriented courses during the summer on such subjects as ornithology, insect biology, and behavioral ecology.[110]

Previous reluctance on the part of scientists to conduct long-term experiments on JNF land was all changed by the designation of the huge (by eastern standards) Mountain Lake Wilderness. Some of the station's experiments involve designated plots that scientists study over many decades and, indeed, their entire careers. Understandably, earlier scientists set up their permanent plots on the station's leased land, or the adjoining Mountain Lake Hotel land, both of which promised more long-term stability than JNF land, then subject to possible timber cutting at any time.[111] But even before wilderness protection, the national forest land surrounding the Mountain Lake station enhanced their own study areas by involving them in a larger forest ecosystem. Also, Wilbur and other professors often used the JNF for a field laboratory and for conducting various scientific surveys. But these scientists grew especially pleased with the prospect of conducting major studies within the Mountain Lake Wilderness itself, an additional forest habitat constituting more than 10,000 acres.[112]

Along with Mountain Lake, the Mount Rogers / Whitetop Mountain area constitutes an area of noteworthy botanical value.[113] The area's red spruce (the southern extent of a northern species) and Fraser fir (the northern extent of a southern species) evolved into unique hybrids. By the late 1980s the balsam woolly adelgid (*Adelges piceae*) had already decimated more than 91 percent of the Fraser fir population in the other isolated areas in the southern Appalachians supporting this species.[114] For a while scientists hoped the Mount Rogers fir would continue to resist the adelgid through some kind of natural, genetic makeup. For a few years it did. By the mid-1990s, however, the Mount Rogers Fraser fir succumbed to the adelgid after all. Poor air quality almost definitely contributed to the extinction of this subspecies.

Ecological Forestry and Ecosystem Management

Many decades before the Forest Service officially embarked upon ecosystem management,[115] forest ecologists were quietly building a corpus of scientific evidence that rendered commodity oriented silviculture oversimplified and obsolete. Around the time of the founding of the JNF, forest ecologists were split into two major schools. The mechanistic school pursued an inductive approach that focused on field experiments. The holistic school deduced concepts like the climax forest and forest succession in the now-discredited balance-of-nature worldview. By the 1970s the holistic school fell out of favor, even if the more perceptive ecological forestry worldview replaced their false one.[116] Also beginning in the 1970s, this new scientific work began merging with public and congressional sentiment against clear-cutting. These two great, partially overlapping forces had much to do with the trend toward ecological forestry and at least a significant modification of previous utilitarian forestry.

In 1992 U.S. Forest Service Chief F. Dale Robertson directed all national forests to embark upon ecosystem management. A year later, for the first time in agency history, a nonforester became Forest Service chief. The rise of Jack Ward Thomas, a prominent biologist, to the top of the agency signaled a change in national forest management approach and atmosphere. In 1996 fisheries biologist Michael P. Dombeck succeeded Thomas. Despite overall budget cuts, the Forest Service doggedly stood by old multiple-use aspects of national forest administration, but ecosystem management sought to incorporate such varied uses within a comprehensive measurement and protection of the biological whole. Ecosystem management includes a bureaucratic aspect of couching old multiple-use policy in new environmentally correct jargon,[117] but decades of public and congressional pressure and an increasingly profound shift in forest science circles themselves indicate substance behind the rhetoric. Furthermore, specific evidence indicates a shift on the JNF.

After the beginning of the administrative merger with the George Washington National Forest in 1995, the GW-JNF created a new budget category called Ecosystem Planning, Inventory, and Monitoring. In both fiscal 1996 and 1997, they allocated more than $1.6 million to this effort, representing about 8.5 percent of the total respective budgets, and on a par with timber management budget allocations.[118] Furthermore, in 1994 Region 8 forester Bob Joslin incorporated the JNF into environmental management considerations involving all southern Appalachian national forests. With many commonalities among ecosystems, the idea was an efficient, coordinated, regional approach to common policy.[119]

Ecosystem management has quite a history of evolution within the agency, with early focus on game management growing into greater wildlife and habitat protection, with special focus on rare and endangered species. The interrelation among the many flora and fauna had long preoccupied ecologists, and gradually such an approach to forest management worked its way into policy. The JNF now uses fire as part of ecosystem management. Air quality and possible effects of acid rain remain monitored. But the gradualness of the development of ecosystem management, with a series of federal laws relating directly or indirectly to one aspect or another, would seem to indicate the new approach's solid grounding. Habitat is addressed from a number of angles, including wilderness designation and specific species protection. Specific flora and fauna receive direct protection. NEPA and NFMA promise continued scrutiny over the entire plans. Much has changed since Paul Shrauder arrived as the first and sole JNF wildlife biologist, who even then had to divide his attentions between timber management and forest habitat enhancement.

Ecosystem management raises larger philosophical questions regarding the natural environment and the role of humanity. Green Calvinists reject the role of humanity altogether, in consistency with their general misanthropy. Forest Service employees, on the other hand, predictably embrace the human agency, even if they are markedly swinging away from commodity emphasis toward ecological forestry, at least in their personal opinions.[120] Ecologists now recognize that constant change and periodic instability characterize forest ecosystems and that humans are but one agent effecting or mitigating these changes.[121]

In any case, the Forest Service's approach to ecosystem management recognizes current ecological thinking but also takes into account human economic and cultural values. This becomes all the more important for an eastern national forest like the JNF, both historically and in a contemporary and future sense. The contemporary and future scenario for the JNF is one of comparatively intense human visitation, whether by neighboring resident populations, Appalachian Trail through-hikers, or vacationers from the urban and suburban East Coast. Historically, the human component becomes even more profound. The National Forest Reservation Commission purchased the JNF's entire acreage from private holders. Like other eastern national forests, this has rendered the JNF an enormous wealth in both historic and prehistoric cultural resources.

Chapter 10

Cultural Resources

In all, the scope of JNF cultural resources spans the entire period of human prehistory and history of southwestern Virginia. Documentation and preservation of these resources has contributed, in its modest way, to a more detailed understanding of the greater region, Appalachian history, and environmental history. The cost-sharing, or matching-funds policy, has allowed the JNF to coordinate its own work with nonagency archaeologists, anthropologists, historians, and geographers. But, technically, JNF cultural resource work has fundamentally differed from that undertaken in academia. In the past, perhaps especially during the difficult early years of launching a new and very different endeavor within the agency, some Forest Service archaeologists themselves complained that cultural resources in general received the very least attention of any division. These critics charged that the Forest Service did not treat cultural resources as a true resource at all, but rather only as a support resource for timber management and road construction, or as a necessary and unavoidable evil, given the federal laws requiring protection of historic and prehistoric sites and artifacts. Early wildlife biologists and later botanists might have levied the same charge.

Increasingly since its inception, the Forest Service has attempted to some degree or another to protect or at least recognize prehistoric and historic cultural materials on all national forests.[1] The once-private and corporate landholdings now constituting the JNF have rendered the public a great variety of historic cultural resources. The evidence and interpretations of this past human activity not only holds significant inherent value but also contains keys to understanding the actual physical state of the land and the woodlands themselves. The current forest environment, of course, reflects substantial previous human impact.

In 1935 Congress passed the Historic Sites Act in an attempt to "provide for the preservation of historic American sites, buildings, objects, and antiquities of national significance."[2] Thirty-one years later, as part of the Great Society's urban renewal goals, Congress enacted the 1966 National Historic Preservation Act.[3] The 1969 National Environment Policy Act required Environmental Impact Assessments, which initiated a cultural resource survey (particularly for archaeological sites) prior to timber sales. But it was the 1974

Archaeological and Historic Preservation Act that put archaeologists permanently on national forest staffs.[4] This law required all federal agencies, including the Forest Service, to survey its territories for archaeological artifacts and evidence of historic significance before building roads, conducting timber sales, or otherwise disturbing the ground in a manner that could potentially damage such material.[5]

The 1979 Archaeological Resources Protection Act gave substantial punishments for crimes involving the looting of archaeological sites.[6] Only a small fraction of southwestern Virginia's prehistoric sites have been excavated or studied, and many more remain to be discovered. Unfortunately, much of the fragile archaeological record has already been permanently destroyed or damaged to the point of seriously compromising its scientific value. Careless or crude excavations and indiscriminate artifact gathering in the past have all seriously marred the prehistoric record. Even more ominous, late-twentieth-century looting of prehistoric sites, which has become something of a cottage industry in some parts of the southern Appalachians, threatens to obliterate precious information still awaiting discovery and analysis.[7]

With archaeological sites now legally protected from destruction by building development or highway construction, the misguided (if predictable) assignment of monetary value to projectile points and other artifacts perpetuates the human threat to southwestern Virginia's distant past. Despite the destruction wrought by a long history of private collectors and pot hunters, JNF archaeologists and other professional and amateur prehistorians have managed to piece together at least the beginnings of southwestern Virginia prehistory. By the 1990s JNF cultural resource management began enjoying a shift in emphasis that gave somewhat more favor to nontimber resources.

Southwestern Virginia features even more historical evidence than archaeological artifacts on present JNF holdings. Since private individuals or corporate entities once owned all of the JNF's 700,000-plus acres, the historic cultural resources are especially rich. Since 1911 federal acquisitions have rendered the JNF sites of old logging and mining operations, railroad beds, iron furnaces, farmsteads, and other homesteads. Notable features include two Civil War sites (though many Civil War expeditions crossed present JNF territory), an incline railroad operation, and several nineteenth-century turnpikes. Significant architectural features and their surrounding histories include the Green Cove Train Station, the Konnarock Lutheran Girls School, and part of the Hagan Estate, once the home of a wealthy landowner near the Virginia coalfields. Some of these sites have received research and documentation, others preliminary investigation. Still others remain promising potential projects. For management purposes, the JNF has incorporated cultural resource information into visitor centers, published articles or booklets interpreting such sites, and given lectures and slideshows for area school-children and civic groups. Particular buildings or sites may be nominated for listing on national or state historic registries.

Following is a prominent example of a JNF historic cultural resource and an indication of its historic context. Green Cove Train Station (located on the Mount Rogers National Recreation Area) involves the logging history of the surrounding area, mostly in Washington County, Virginia. It also reveals something of the community history of this small valley before out-migration, discontinued train service, and other factors ended an era of local history. Green Cove State reflects the kind of public history pertaining to the JNF and other federal and state lands.

Green Cove Station

On October 6, 1991, Green Cove residents, steam railroad enthusiasts, politicians, JNF personnel, and others gathered at Green Cove to dedicate the former station as an official Forest Service Information Center.[8] The JNF thereby inherited the structure with the most social significance of the early twentieth century in the Green Cove Valley. The history of the Green Cove Station represents the sort of public history possible on the JNF and other eastern and midwestern national forests.

The building that once housed Green Cove depot is located in the far southeastern corner of Washington County, Virginia, in close proximity to Grayson and Smyth counties. The Virginia-Carolina Railroad Company (VCRC) completed the structure in 1914 at a cost of $2,600.[9] The single-story frame building measures sixty-one by twenty-one feet, consists of board and batten siding, and is capped by a standing-seam tin roof. The station agent heated the depot with two pot-bellied stoves, using some wood, but mostly coal, initially brought in by train and later by truck.[10] The structure has two large freight doors, one opening to the rear, the other opening to a track-side loading dock. There are also two pedestrian doors, one situated track-side, next to the freight door, the other on the building's end, facing the old stationmaster's residence. Green Cove Creek flows nearby, through the beautiful Green Cove Valley amidst some of Virginia's highest mountains. White-top Mountain is behind the station to the south, and Chestnut Mountain is on the valley's far side, to the west. Mount Rogers is nearby, about six miles to the east.

Traditionally, early writers described the Green Cove Valley and greater Holston headwaters area as exceptionally rich hunting and fishing grounds, void of permanent Native American settlements and possibly a buffer zone between the territory of the powerful Cherokee and Shawnee tribes.[11] After various frontiersmen (including Daniel Boone and Wilburn Waters) traversed and lived in the region, subsistence farmers gradually arrived in permanent settlements. Sawyers and those working in wood products trades operated on a relatively small scale, cutting and processing only the highest quality and most accessible trees. But even before the late nineteenth- and early-twentieth-century timbering frenzy, Washington County figured prominently in sawed lumber and wood products production.

In 1859 Washington County produced the most wooden carriages in the entire Upper Tennessee Valley Region,[12] and along with the Jefferson and Washington counties of Tennessee, processed 53 percent of the area's lumber. Along with the Jefferson and Sevier counties of Tennessee, Virginia's Washington County produced 60 percent of the region's furniture. Washington County's sawed lumber production continued steadily for the next twenty years, ranking second in the Upper Tennessee Valley Region in 1879.[13]

Some of Washington County's least accessible timber lay in its rugged southeastern region, around Green Cove. The topography between Abingdon and Damascus still lies within the Great Appalachian Valley region. But the area to the southeast, toward Green Cove, changes dramatically, from fairly wide vistas to steep mountains, with little flat land between their bases and winding creeks flowing among them. In 1876 the Washington County Court created its Holston Magisterial District in this mountainous territory.[14] The sparsely populated district constituted about one-fifth of the county and consisted of thousands of acres of virgin forest. In 1880, before the railroad arrived, mostly Virginia-born white farmers or farm laborers populated the Holston District. A few blacksmiths, millers, cobblers, and teachers also lived there, as well as two physicians and one lawyer. Other occupations, represented by a single individual, included merchant, minister, teamster, brick mason, and saddle harness maker. At this point, workers in the wood trades included four carpenters, three wagon makers, and a single sawmill worker. Indicatively, only one man listed his occupation as railroad worker.[15]

Twenty years later, occupations had changed little. Farmers and farm laborers still predominated in 1900, and only six men worked as sawyers or sawmill workers, and still only one man worked for the railroad. However, a significant population increase began to indicate the first stirrings of change. Outside the Holston District, Washington County grew only 7.5 percent between 1890 and 1900, while the Holston District leapt 45 percent.[16] Many more carpenters and salesmen began operating in the region, and the population growth and diversification began to reflect the impending logging boom.[17] By 1910 the socioeconomic picture had changed considerably.

Between 1900 and 1910, when the rest of Washington County grew only 6.5 percent, the Holston District population grew 60 percent. In 1910 almost two hundred men worked in the lumber/railroad industries in the Holston District alone. Forty-nine worked in various capacities at lumber mills, seventy in logging camps, forty-two in lumber yards, and thirty-eight for the railroad, either the main line or the "log train." Many more merchants worked in this area, as well as bark extract contractors, boardinghouse employees, the first constable, and the first mailman. A tremendous number of the new workers came from outside the state, especially from North Carolina, but also from West Virginia, Tennessee, Pennsylvania, Ohio, Kentucky, and as far away as Sweden.[18] Pennsylvania native Martin Hassinger, whose father owned the massive Hassinger lumber operation in Konnarock, estimated that they employed a total of 400 people in the 1910s during maximum production.[19] Industrialization came to this rugged corner of Washington County like a storm.

On July 20, 1912, Calvin Toliver sold a narrow strip of land beside Green Cove Creek to the Virginia-Carolina Railroad Company for a thousand dollars.[20] The VCRC was in the process of extending its track from Creek Junction through Green Cove on the way to Elkland, North Carolina. The railroad followed trails originally traversed by animals, Indians, and early explorers.[21] Beyond Damascus, the railroad snaked through the mountains over dozens of trestles. But the landscape opened back up a little bit when it reached the Green Cove Valley, and this geography helped make the Green Cove Station a central commercial and social gathering place.

When the train arrived at Green Cove in 1914, area farmers raised corn, oats, wheat, beans, and other garden vegetables, as well as cattle and hogs. Suddenly they gained access to outside markets. In this way, the train helped transform Green Cove Valley agriculture from subsistence farming to partial commercial production. The train also introduced an overall way of life that included more cash exchange, a wider variety of consumer products, and greater contact and communication with neighboring and distant regions. For example, Green Cove train station agents sold commercial cloth and provided telegraph and mail service.

The train itself also carried mountain people outside the region, as well as brought tourists and entrepreneurs into the area. During its peak operation, from 1914 into the mid-1920s, as many as two passenger and six freight trains ran daily along the main line, with logging spurs connecting at various points. The original railway builders chartered their organization in 1887 as the Abingdon Coal and Iron Railroad Company. These men anticipated tapping iron ore deposits they thought to be considerable around the Damascus area. The absence of this resource, along with the general economic crisis of the 1890s, quelled the company's efforts. Other capitalists, however, discovered that this region's true wealth lay in timber. Backed by northern lumber interests, Wilton E. Mingea took charge of the railway project and in 1900 renamed the company the Virginia-Carolina Railroad. Five years later he extended the line over thirty miles into the heavy forest around Konnarock, where the Hassinger Lumber Company had begun operation in 1902.[22]

The Hassinger Lumber Company, responsible for cutting most of the timber in the Konnarock, Green Cove, and surrounding areas between 1902 and 1928, participated in the intense timber harvesting described in chapter 4. Mostly local sawyers had hitherto only selectively cut in the Appalachian Mountain forests, usually only near waterways, and left vast stands of uncut timber on the mountain sides, some measuring as much as eight feet in diameter.[23] During twenty-six years, the Hassinger Company cut some 30,000 acres in Washington County, with an average of "fifteen to eighteen million board feet of lumber annually," as well as thousands of chestnut poles and mine props and thousands of cords of tanbark, extract, and pulp wood.[24] By 1912 Washington County alone produced more lumber than the entire state of Pennsylvania and, at peak production, possibly became the most productive lumber-producing county in the nation.[25] By the time the timber boom ended in the late 1920s, at least 3.5 million board feet had left the Washington County mountains, almost all of it shipped north.[26]

Though not a company logging town like Konnarock, Green Cove became a social and commercial center for this dynamic logging activity. The Hassinger Lumber Company soon exhausted its timber supply around Konnarock and began expanding into the Green Cove area.[27] Logging expansion meant railroad extension, which necessitated new stations like Green Cove. Adele Edwards, surviving daughter of lifelong stationmaster William M. Buchanan, recalled hearing of loggers who, awaiting the train, stacked mounds of pulp wood next to the depot siding. A logging spur ran up the Star Hill Branch near Green Cove on the west side of Whitetop Mountain, a lumber yard lay across the tracks from station itself, and logging camps dotted the surrounding area, making Green Cove a true "lumbering village."[28] Naturally, the logging industry helped foster other economic activities, and these activities centered around the Green Cove Station.

In 1912 W. E. Mingea began selling VCRC stock to the Norfolk & Western Railway Company. The new stockholders extended the railroad south into North Carolina, utilizing free or cheap laborers. White foremen oversaw white laborers and black convicts building the railroad. Community people called some of the black laborers "doggets," presumably after a certain George Emmett Dogget who supervised their building activities. The African Americans lived in shacks about a mile up the tracks from Green Cove, near Horseshoe Bend, in an area called Dog Town to this day.[29] Blanch Clark, retired station employee and longtime Green Cove resident, recalled that her father, who came from Tennessee with other white laborers to help build the railroad, worked for little money.[30] Other railroad workers included a group of Italians from Pennsylvania, and at least a dozen Swedish families.[31] In the tradition of early-twentieth-century capitalists, labor was inexpensive, if not free.

The VCRC asked W. M. Buchanan, a schoolteacher in Green Cove, to be stationmaster for the new depot. He agreed and performed his station agent duties in the early morning and late afternoon so he could continue teaching.[32] In this sense he became a railroad employee. However, he soon found himself operating far more than just a train station. In addition to his station agent and storekeeper duties, Buchanan began operating the post office on August 8, 1914, and a Western Union telegraph near the same period.[33] On December 9, 1919, he registered as notary public and began performing marriages and notarizing deeds in a capacity that lasted for more than forty-four years.[34] Surrounding farmers sold him chickens and homemade cheese, local hunters and trappers brought in animal pelts for shipment to the north, and various other industrious souls came to the depot to trade in chestnuts, chinquapins, ginseng, elderberries, mayapple root, burdock root, beedwood leaves, and mountain laurel. Buchanan soon relinquished his teaching post and devoted all his energy to the station.[35]

Though lumber continued to dominate the market, by 1915 a variety of enterprises had sprung up along the Virginia-Carolina Branch. There were manufacturers of various lumber products, including barrels, bark extract, wagons, and "coops and cases," as well as producers of lime, blanket makers, a soft drink bottling operation, and a soapstone quarry. The railroad anticipated other commerce in livestock, talc, mica, and possibly copper ore.[36] Outside capitalists relied upon the train for their various operations in the mountains, and people in the mountain communities sought the train for contact and commerce outside the region.

Buchanan's wife and daughters oversaw the family home, not a hundred yards from the depot, which functioned as something of a restaurant and hotel for certain area residents and for the transient populace at large. The Buchanans and their helpers sometimes produced fifty to seventy-five bag lunches a day for travelers, either receiving a telegram in advance indicating the number of passengers or simply anticipating a full train, since "you could be sure that there were plenty of travelers on that train" during the boom years.[37] Providing such meals proved a considerable endeavor, because the Buchanans and their workers produced all the food locally, from growing vegetables to plucking the chickens. Electricity did not arrive in Green Cove until after World War II, so kerosene-powered compressors provided refrigeration. The family house also provided long-term lodging for area schoolteachers or short-term accommodations for traveling salesmen who passed through selling groceries, dry goods, or other wares. And on at least one occasion, a snowstorm blocked the train's passage, causing an entire trainload of passengers to spend the night in the Buchanan house, filling all the beds and much of the floor.[38]

Green Cove Station's most dynamic years coincided with the logging boom. By the late 1920s that boom had begun to fade. The Hassingers cut their last trees in 1928 and shortly thereafter salvaged the tracks of the Konnarock spur that had led up to their old mill. The economic thrust that had inspired the railroad was over. Many timber workers left for the nearby coalfields. The Great Depression followed, and the Abingdon Branch never saw glory days again. By 1931 the first mixed trains (passenger and freight) replaced first-class passenger service, and by 1932 freight-only trains had ceased completely. Another casualty came in 1932 at the North Carolina terminus, where the line was shortened to West Jefferson, discontinuing service to Elkland.[39]

At Green Cove, the train continued to pick up a little remaining pulpwood and extract, as well as mountain laurel, chestnuts, various herbs, and cattle. Cash was rare. And if the Great Depression and the logging boom's demise were not enough, the Chestnut blight hit during the 1930s, ending a highly significant source of mast for humans and animals, as well as the most important source of bark extract. Economic boom years had clearly left Green Cove. The community, however, in its commercial and social capacities, still centered around the train station. Postal and telegraph activities and any remaining commercial exchange still took place in Buchanan's store.

Green Cove Station witnessed some of its most poignant years during World War II. Edwards remembered, "there was always the feeling of the boys being called to go to war [because] the boys that were called into service had to board the train here. And it wasn't a very pleasant sight, to see their families; see their boys board the train, maybe to never see them again." And it was through the train station's telegraph that tragic war news arrived. Edwards recalled:

> At the station . . . this message came in about James Hurley, one of our neighbors, that he was lost on the beaches of Normandy. My father said to me, after I came in from school that afternoon, "I think you'll have to take this message to the family." I said, "But I can't." He said, "I think you'll have to." So I did, I delivered that message. His name was James Hurley, and he was lost on the beaches of Normandy. And that is still very fresh in my memory. So that's part of what I remember about the war years. They were rough.[40]

During this somber time, however, Buchanan did not lose his sense of humor. He tried to alleviate some of the tension in his role as telegraph operator. Blanch Clark, lifelong resident of Green Cove, recalled: "My brother called [from] Germany, for Mother's Day; and [Mr. Buchanan] would joke with you when you come in. He'd say, 'I've got a telegram for you.' He wouldn't smile or anything. You'd just feel like, 'Why, he's dead.' But when he'd give you the telegram, why, there it'd be: Happy Mother's Day, or something like that."[41]

Economics and nature began to discourage rail operation. The wealth of timber that had initially financed the railroad was long gone, and now the topography began to wreak havoc. Floods in the high mountains roared down steep ravines, destroying bridges and trestles. By the 1940s Norfolk & Western sought to discontinue the now-costly line, but the Interstate Commerce Commission forced them, as a common carrier, to repair and continue the route.[42] Buchanan's overall livelihood was so ingrained in the station building itself that he continued to operate out of it even when Norfolk & Western retired the depot from its line in 1957. Norfolk & Western sold Buchanan the building outright and struck up a land lease agreement for the quarter-acre lot beneath and surrounding the structure.[43] Mixed-train service continued into the early 1960s, but by 1962 Norfolk & Western officially ceased its passenger service and began reducing its property taxes by removing various features along the line, such as water towers and sidings.[44] By then, only small vestiges of the former era remained.

The train continued to make occasional, informal stops at Green Cove. Blanch Clark remembered, "They'd just stop and talk to you, even when they had no more passengers or mail to drop."[45] By the late 1960s, however, freight trains passed Green Cove only once a week,[46] and many of Buchanan's non-station endeavors in the former depot had ceased as well. The 1958 discontinuation of mail and telegraph service marked a major turning point.[47] After July 1958, Green Cove residents began to receive their mail through rural carriers who delivered letters marked "Route 2, Damascus."[48] Around 1964, after a six-year telecommunications absence, telephones finally replaced the telegraph.[49]

Three or four other stores sprang up in the Green Cove area between the late-1940s and the 1970s and competed with the station's remaining retail business.[50] With declining health, Buchanan discontinued his notary public services sometime between 1963 and 1967,[51] and his daughter Eleanor took up more and more of the storekeeper duties. Eleanor Buchanan eventually became the sole and final proprietor, and the Green Cove Station lingered into the mid-1970s, after which it closed forever.[52] On March 31, 1978, the train itself made its final run. Midwest Steel and Chicago Contracting bought the steel rails for $402,000.[53]

The kind of personal affinity that had contributed to operating the train like a family business began to decline. Railroad presidents who had once worked their way up through the ranks now hailed from law offices, with no previous railroad experience. As early as the 1930s, trucks had begun to pose serious competition for Green Cove's train, but trucks remained isolated, individual carriers of goods and people, while the steam train continued to function as something of a corporate citizen and part of the community. People living along the line knew the crew by name, and trains made unscheduled, informal stops just to socialize, give candy to children, or trade refreshments.[54] Adele Edwards recalled that when the railroad left Green Cove, the community was devastated. "Just wasn't much left. When you lose a way of making a living, [it changes] one's way of life." Less than seven decades after the industrial timber boom began, Green Cove Valley once again grew quiet.

☞❧

Given the previously private ownership of all JNF lands, the potential for historical study of topics such as the Green Cove station is almost endless. Unfortunately, historians for government agencies generally enjoy none of the academic freedom of a university scholar and thus tend to avoid controversial topics. There is tendency toward antiquarianism rather than local history integrated into national themes or larger philosophical frameworks. Agency reviewed publications for the Government Publications Office too often become a sad parody of the peer reviewed scholarly piece when topics and language are checked for political correctness rather than intellectual content. In some ways it is like high school history compared to college history. But, if the Forest Service focus continues to shift from commodity to noncommodity values, perhaps there is some hope that its cultural resources endeavors can move closer to academic scholarship and away from what is too often the whitewashed public relations endeavor. There is a reason why I did not endure as an agency historian. The JNF would prefer the Green Cove station stories, not the Mount Rogers National Recreation Area story.

It has been about a century since the National Forest Reservation Commission began buying the first lands that became the Jefferson National Forest, making this federal agency the caretaker of some of the highest elevation lands in southwestern Virginia. As mentioned in the introduction, the JNF's particular pattern of landownership distinguishes it from the overall region because it concentrates management along the ridges and mountain plateaus rather than in the valleys. The valleys have increasingly featured population concentrations and improved transportation corridors; the rougher terrain, by its very nature, discourages such changes (once overwhelmingly considered "improvements"). With the exception of logging and access roads, and inherited cultural resources such as Green Cove Station, JNF land is excluded from such development altogether. Therefore, these publicly owned mountains, now something of a territorial commons, offer some connection with the primeval past. This may sound absurd amidst discussions of silvicultural methodology, ecosystem science, cultural resources, and tourism management—and yet in some important ways the former supersedes all of the latter. If we have any commonalities in our perspectives, and even if those commonalities are consistent or shifting, our very conceptions of the place have everything to do with what we seek from the land.

Chapter 11

Sacred Land

I tell you, I know places in these mountains
back off the big roads,
up the coves and hollers
old home places
with barns and apple orchards, cattle gaps,
haunted by spirits
spirits of people who left there
taking everything with them but their spirits
for their spirits wouldn't leave that place
and their spirits are there yet

—Jim Wayne Miller, "Brier Sermon—'You Must
Be Born Again'"

Throughout the world, and throughout history and prehistory, mountains and trees have held religious and spiritual significance for any number of peoples. Scholars and writers like Mircea Eliade and Edwin Bernbaum have demonstrated how peoples everywhere have embraced and continue to embrace concepts of the cosmic mountain and sacred tree.[1] In a philosophical sense the forested mountains of southwestern Virginia present but a variation upon this broader theme. As the famous mythologist Joseph Campbell observed, Mount Taylor in New Mexico and Harney Peak in South Dakota were specific, religious land formations for aborigines—but the "holy land is everywhere," for individual places signify but one instance in a universal theme.[2] Still, only a few people are going to adopt the broad anthropological views of a Campbell or an Eliade, so a specific people's concept of home in relation to a particular landscape remains vitally important for understanding that territory in its social and historical context.

Mountains in relation to lowlands represent places closer to the sky, which many peoples have associated with a spiritual dimension. But also very common is the concept of an underworld, which has included mountain caves. Whether representing benevolent, malicious, ambivalent, or ambiguous religious forces, mountains have captured peoples' spiritualistic imagination from time immemorial.[3] We have already glimpsed something about the aboriginal regard for the southwestern Virginia landscape; now we might consider a more recent historical as well as contemporary view of these highlands.

It is impossible to understand how the native-born people of southwestern Virginia have regarded their home without first understanding something about their worldview, and to understand their worldview what is needed more than anything else is an understanding of their religion. As scholar Deborah McCauley wrote of Appalachia, "mountain religion is the truest mirror reflecting who *are* mountain people, apart from which they cannot be understood."[4] This religious tradition is overwhelmingly Calvinistic Christianity, and yet historically it is distinct from mainstream American Protestantism. Furthermore—and most important for understanding the highlands—there are naturalistic elements reminiscent of ancient, pre-Christian, animistic, or numinous religions.[5] Planting gardens according to the Zodiac signs, a practice found throughout southern Appalachia, evokes ancient Babylonia astrology that precedes the Christian era by many centuries.[6] Some have apparently rationalized this paganism by quoting Genesis 1:14 or Ecclesiastes 3:1–2, the "signs and seasons" or "a time for everything."[7] Of course, Genesis and Ecclesiastes are pre-Christian, Old Testament books and likely reflect any number of ancient religious influences possibly dating back to the ancient Sumerians.

Perhaps an epitome of ancient numenism was reflected in the twentieth-century preacher Charlie Bry Phillips and the "praying rock" he used in the Appalachian forest of Rabun County, Georgia.[8] The microcosmic landscape of water-stone-trees represents among the most ancient of numinous associations.[9] While Phillips relied on the Bible's guidance to "go out in the woods and pray,"[10] his use of a forest boulder and smaller rocks reflects a primeval fascination with the religious qualities of stones.[11] Any person of a certain sensitivity may have intuited this phenomenon in the forests along the Appalachian Trail when encountering glacial "erratics" or other boulders, most strikingly in places like Dragon's Tooth and the enormous boulder maze hikers have dubbed the "Devil's Kitchen" or the "Devil's Marbleyard" on the trail to McAfee's Knob.

This is not to say that all parts of the mountains evoke sublime awe. As Lee Smith wrote in her 1983 novel *Oral History,* Hoot Owl Mountain was an eerie place of fog, shade, and death that evoked fear. Very appropriately, its summit featured a cemetery. Those walking on or near Hoot Owl Mountain quickened their pace and resorted to carrying a buckeye in their pocket as a talisman against its dark force.[12] While Smith's Hoot Owl Mountain is fictional, any person of the right sensitivity who has spent time traversing the forest has encountered such places. There is likely no scientific explanation for this sense of foreboding.

The "Devil's Marbleyard," just south of McAfee's Knob on the Appalachian Trail. This sort of landscape evokes primeval ideas of numinism. Photograph by Teresa Chen Shaw.

Mountain imagery figures prominently in the hymn "Palms of Victory," which many in the Grayson-Carroll counties area (adjoining the Mount Rogers NRA) claim as their own, despite its apparent origin elsewhere.[13] In 1982 folklorist Charles Wolfe wrote that nowhere else had he heard this hymn sung so frequently, aware that in previous generations the hymn was sung throughout the South.[14] The opening verse—"I saw a wayworn traveler . . . struggling up the mountain"—indicates spiritual endeavor. The metaphor continues wherein the traveler realizes deliverance or enlightenment upon reaching the valley on the other side.[15] The mountains here may serve only as a metaphor, but it is interesting to note the hymn's enduring popularity near some of the most rugged landscape of the southwestern Virginia highlands.

Even the "lonesome pine" of John Fox Jr.'s old novel has drawn the attention and curiosity of natives and visitors alike. *Trail of the Lonesome Pine* (1908), one of Fox's most popular works, remained in print as late as the 1970s and continues to be performed as an annual outdoor drama in Big Stone Gap, Virginia.[16] No doubt for some the solitary pine tree amidst the deciduous trees has been a mere curiosity, but at the very least it took on metaphorical weight for some. Fox perhaps saw himself as a semi-outcast in Big Stone Gap, and thus the pine tree represented some autobiographical symbolism.[17] The tree's location likely would have become a pilgrimage destination if it had not remained obscure, despite repeated inquiries.[18] It seems unlikely that many would have recognized their own numinous associations with the tree, and yet these implications seem plausible given the enormous amount of tree-oriented religions throughout the historically forested world.[19] As we shall see regarding the Appalachian Trail, even seemingly secular phenomena such as hiking and adventure have a tendency to acquire sacred attributes.

Given the overarching influence of Calvinism, no one would mistake the naturalistic elements of this religion for that of the preceding indigenous peoples; and yet there are elements that resemble Indian religion, which may have been acquired independently, from historic proximity to Native Americans many, many generations ago, or perhaps some admixture of both. In any case, as Catherine Albanese has pointed out, some of the famous "Jack Tales" of the southern Appalachians have "an almost Indian sense of kinship with the animals."[20] The redbird grants its viewer a wish in southwestern Virginia.[21] More profoundly, a bird personified God's messenger for a Mrs. Allen of Swain County, North Carolina, who had a mystical experience in the forest when she was a small girl. Also tellingly, instead of going to the afterlife as she expected, she spotted a large boulder and climbed onto it and went to sleep, where her father later found her. In stark contrast to Germanic tales of the ominous woodlands, Mrs. Allen's parents assured her that she was safe in the forest and never to fear it.[22]

Quarter-blood Cherokee Harrison Caudle, native of Wilkes County, North Carolina (about twenty miles southeast of Mount Rogers), possessed an intimate knowledge of every detail of his immediate homeland and thought of himself as a "forest child."[23] Not unusual in the southern Appalachians, Caudle saw far more value in his own mystical experience (albeit biblically based) than in formal, institutionalized religion.[24] Another example of this arises in the Konnarock area, where, at least in years past, residents recognized the only proper baptism as occurring in natural stream water—never with water from a spigot.[25] Natural water baptism was and is common throughout the South, but unlike their flatland counterparts, mountaineers enjoyed fresh highland water, highly oxygen-laden and free of sediment. Water in nature, the primeval religious element, could not be corrupted by artificial plumbing, even if for a Christian ceremony supposedly far removed from ancient worldview concepts.

This mystical approach to religion apparently lasted the longest in the more remote areas of the mountains.[26] However, as late as 1989, man of religion Coy Miner of Pennington Gap (Lee County, Virginia) felt that people who lived in the remote areas of the highlands were more spiritual. In the valleys, Miner felt, all the distractions of mainstream America had contributed to a superficial religiosity instead of the genuine article.[27] Another man of religion,

William F. Bryant, definitely expressed his personal attachment to the Appalachians in spiritual terms. In 1912 he remembered his boyhood home in the highlands of eastern Tennessee, in stark contrast to his subsequent urban surroundings. The mountains, he felt, "lift up their heads as it were, in praise to the Almighty; and I, in imagination, once more feel and breathe the clear cool air as it comes bounding down the hills . . . through the dark forests . . . and my heart seems to ache with an inexpressible hunger to once more be there."[28] Sam Johnson, of Kentucky's Knott County, recognized a distinct strength of "God-consciousness" in the mountains directly because of the landscape. He asked rhetorically, "When you have concrete buildings and concrete streets, and all your food out of cans, how can you have a strong sense of your identity with nature?" For Johnson, nature and the Creator were one in the same.[29]

There seems no denying that the land has profoundly influenced the people who live there, and how these people see that land. Writing during the early twentieth century, Emma Bell Miles (who lived in the mountains of eastern Tennessee) expressed an appreciation for the mountaineers' attachment to place. They cherished the wilderness removed from concentrations of people. Miles herself felt the lingering sensation of the land's history.[30] Miles also saw the land as giving rise to one of the mountaineers' depictions, now well-established—that of fatalism.

> A man born and bred in a vast wild land nearly always becomes a fatalist. He learns to see Nature not as a thing of fields and brooks, friendly to man and docile beneath his hand, but as a world of depths and heights and distances illimitable, of which he is but a tiny part. He feels himself carried in the sweep of forces too vast for comprehension, forces variously at war, out of which are the issues of life and death, but in which the Order, the Right, must certainly prevail. This is the beginning of his faith as he had it from his fathers.[31]

Many decades later Loyal Jones reiterated this argument. The Appalachian wilderness of Miles's time had diminished and the population had swelled, but the then-obvious marginal agricultural productivity was a conspicuous contributing factor to the region's poverty. "Hard work did not always bring a sure reward," Jones wrote. "Therefore, the religion became fatalistic and stressed rewards in another life."[32] Thus it seems that the Appalachian religious worldview has shaped the peoples' conception of land and vice versa; the two influences remain inextricably intertwined. There was and remains a commonly found general belief that God made the beautiful land for the people to appreciate.

During the late 1960s a certain Mrs. Workman of Wolfe County, Kentucky, said that God have given her family the beloved hollow where they lived. The Workmans and other families wanted to know how their interviewer, Robert Coles, felt being "so high up and so near to God" in their homeland.[33] Mrs. Allen of Swain County, North Carolina, was equally convinced that her particular mountains were "the most beautiful things God ever made."[34] Knott County, Kentucky, resident Verna Mae Slone remembered her father speaking of Jesus as the creator of all nature, which for him represented the world's greatest art.[35]

Edward O. Guerrant, the sometimes-notorious home missionary of the late nineteenth and early twentieth centuries, regularly appreciated such imagery. While traveling throughout the southern Appalachians he saw God's presence in nature, particularly when that nature was free of people.[36] In this way Guerrant was almost like a latter-day Romantic, but definitely a biblical one. Guerrant saw a direct correlation between the holiness of the southern Appalachians and the mountains associated with Christ's crucifixion as well as Christ's first sermon.[37] A more recent expression of biblical correlation with the southern Appalachians lies in the Jesuit priest Al Fritsch's 1986 book, *Appalachia: A Meditation*.[38] Here we find biblical verses interspersed with excellent black-and-white photographs and contemplative observations. For example, relating to Habakkuk 3:10, "The mountains shiver when they see you," Fritsch writes, "God's power affects the hills, the rivers, the valleys. The land and all that is in it also reacts to human activity—that sharing in divine power through technology."[39]

But strictly speaking, numenism has its limits. Whatever religious connection that visitors and longtime residents alike may feel to the land, most of this lies within an overarching belief that the mountains and all the earth are but a transitory illusion in light of eternity.[40] This is the concept behind the title of James Still's 1940 novel, *River of Earth,* wherein the mountains are but mere passing waves "controlled by an Almighty."[41] Many Buddhists might agree, since Buddhism holds the transitory nature of life central to its teachings.[42]

Intimacy of Place

Whether expressed in religious terms or not, there is a markedly strong attachment to home and land in the southern Appalachians. In traditional times, parents would sometimes place their infant children on the earth itself, as if they were "ritualistically given over to the land."[43] When American Electric Power wanted to build a high-voltage electrical power line across portions of the JNF, a predictable resistance developed among all residents of the area. One of the proposed routes was across Peters Mountain in Giles County. Area residents expressed their objections, saying that the mountain was like a living relative to them.[44]

Perhaps this is but a variation upon an American theme, heavily derived from a Lockean worldview actuated on the westward sliding line of the

Anglo-American frontier.[45] Rural people all over the nation are convinced they live in "God's country."[46] Still, the holy land may be everywhere, but it exists in very particular places for specific people, and one person's sacred ground may inspire indifference or even alienation in a visitor. As McCauley observed, in the southern mountains home was not conceived of in formal property lines, but rather in terms of natural topography, such as hollows and ridgelines.[47] There is something primeval in this attachment to home. Ironically, in some ways native-born mountaineers of the twentieth century came to resemble seventeenth- and eighteenth-century Indians coping with an outside, dominant, and in some cases overwhelming culture. On the other hand, this should present no surprise considering the commonality of both groups making their living directly from the land (precontact Indians, of course, exclusively so).

Twentieth-century southwestern Virginians were typical of the southern highlands regarding traditional hunting, fishing, and woodlands gathering practices. With this came an intimate knowledge of the land and its flora and fauna. Folk medicine reflected a heavy orientation toward and knowledge of local plants.[48] Individual tree species knowledge, especially regarding practical uses, was common.[49] Naturalistic "superstitions" and beliefs (especially related to agriculture) abounded but also reflected a great awareness of weather, seasons, plants, and insects.[50] Detailed awareness of the land cropped up in Old World folktales adapted to southwestern Virginia and other parts of the southern highlands. Konnarock resident Bill Haga, a teenager during the late 1940s, told folklorist Richard Chase the tale of "Whitebear Whittington," which included his custom additions of a chinquapin, hickory nut, and walnut (all local items).[51] The story involves bewitchery and an ordeal, common elements in many folk tales, but also includes "Piney Mountain," possibly alluding to Pine Mountain neighboring Mount Rogers.[52]

In another tale gathered by Chase, residents of Bristol (Virginia/Tennessee) and Wise County, Virginia, narrated the story of the "Tall Cornstalk," replete with the southwestern Virginia landmarks of Cumberland Gap, Tazewell, and the Clinch and Powell rivers.[53] In any case, in all these ways the traditional mountaineers had far more in common with the precontact aborigines than they did with postindustrial mainstream America. And, without a doubt, intricately linked with awareness of this particular land came a fierce sense of home.

Attachment to Place

During the late 1970s anthropologist Allen W. Batteau spoke with Caucasians in the southern Appalachians who remained strongly attached to the mountains despite working in northern cities for years at a time.[54] Asked if they had ever "lived" up north, they initially replied "no," paused, then went on to describe working in Chicago or Detroit for years at a time.[55] But they insisted they had never "lived" in these places. They had never bought property nor married into a "foreign" family. Home had always remained back in the mountains. As one old resident of Green Cove told the author, "They'll have to carry me in a pine box to get me out of the mountains."[56]

In the 1980s a teenaged boy (whose family was faced with imminent departure for the flatlands in search of employment) expressed this personal identification with the highlands even more excruciatingly. "We're part of these mountains," he said, "and when we go, there's something that will die here—we'll die. We'll be someone else as soon as we leave."[57] While recognizing the economic necessity of leaving, he described the existence of nonmountaineers as "graveyard lives."[58] Catherine Albanese accurately and poignantly described this as a relationship of "endearment" between the highlanders and their home.[59] During World War II and the postwar decades homesick mountaineers frequently left industrial employment in the Midwest and returned to the highlands.[60] The pain of forced relocation from the Mount Rogers and Grayson Highlands area has already been told in a preceding chapter. By now the rootedness of Appalachian people, wherein monetary value of land seems utterly irrelevant, should be legendary.[61] Some have managed to stay, and others have managed to return.

Robert Lee Combs, born in 1916 near Fairwood (in Grayson County, Virginia) was one of the few who managed to stay in the area his entire life. Interestingly and typically, he distinguished his home of over seventy-five years (Konnarock) from his birthplace (Comer's Creek), even though the two places were less than ten miles apart and definitely both in the shadow of the Mount Rogers–Whitetop plateau complex of mountains.[62] Combs traced his ancestry to the mixed-blood Cherokee Chief Ned Sizemore, apparently associated with what came to be called the Whitetop-Laurel subdivision of the Cherokees.[63] During the 1830s the Sizemore family, residents of the Whitetop area, had apparently offered safe haven to some of the Cherokee who fled into the Smokies rather than being forced to Oklahoma along the notorious Trail of Tears. These scattered peoples eventually formed the Eastern Band of the Cherokee Tribe, now based in western North Carolina. In any case, the oral tradition in Combs's family included remembrance of these historic Indians making stone projectile points in the Konnarock valley. Combs felt he and other area descendants of the Sizemores might be continuing an older attachment to the place. "Something that draws you back," he said. "It's a mystery."[64] Of the Konnarock area, Alice Hayes told interviewers in the mid-1990s, "There are a lot of people that never want to leave Konnarock. Everybody that left there for jobs are getting to retirement age, and they are all coming back."[65]

The Linear Church

The Appalachian Trail, or AT, has become one of the more famous hiking routes in the United States and beyond. Many hikers begin their journey in Georgia, when the southern spring is just arriving. They slowly make their way, over 2,000 miles north, relying on the weather to remain comparatively tolerable as they progress, until they arrive in Maine. Through-hiker Earl Shaffer named his 1948 book about the experience *Walking with Spring.*[66] The route traverses fourteen states and relies heavily upon public lands. The Virginia portion of the trail is the longest and altogether constitutes around a fourth of the entire mileage. Exact mileage figures depend upon when they are measured, for the AT has shifted its route numerous times in major and minor ways over the decades.

The JNF portion constitutes a unique microcosm of the overall AT experience, especially appreciated by through-hikers. Just before encountering the Mount Rogers area, northbound hikers usually stop in Damascus, Virginia, which seems to have a near-unanimous reputation as the most accommodating place for hikers along the entire trail.[67] Since 1987 Damascus has hosted the Trail Days celebration in May, when most northbound hikers are in that area. Further up the trail Pearisburg competes with Damascus as the most accommodating place for hikers.[68] Most through-hikers are going north, and their arrival in Damascus often represents a successful first leg of the journey. After Damascus almost all hikers note their impending climb of Virginia's highest peak, Mount Rogers. The feral ponies of the Grayson Highlands State Park frequently inspire photographs and descriptions, and around this part of the journey through-hikers mark their first 500 miles. This highland plateau of Virginia, encompassing Mt. Rogers, Whitetop, the Grayson highlands, and other peaks, tend to evoke thoughts of Montana, Wyoming, or other parts of the Rocky Mountains. For hikers who have walked up through the Smokies it is quite a remarkable surprise. Pearl Dawson, a 2002 hiker, had been to Montana and agreed with the comparison. She rated her day in the highlands as the best so far. "We had the bluest skies today [with] sun. I have never seen such a vast change in terrain since we started. One minute you're hiking thru grassland [with] steer, and the next you're in the woods again only to come out in the open 15 minutes later. I loved the rock outcroppings. The park is full of wild ponies."[69]

Often it is late May at this point, and thus late spring. As through-hiker Mark Hughes observed in May 2006, "The woods here in Southern Virginia are quite green, all the trees and ground cover are leafed out, and "the long green tunnel" is starting to emerge, which is the moniker for hiking in [Virginia]."[70] Atkins represents the halfway point between Damascus and Pearisburg. After Pearisburg, many hikers photograph and note Sinking Creek Valley's

Vista afforded from McAfee's Knob, in the JNF's old New Castle District, widely claimed to be the most photographed scene along the Appalachian Trail. Photograph by Teresa Chen Shaw.

Keffer Oak, a huge ancient tree that was a sapling in the early 1700s, around the time the first European Americans were making their way into southwestern Virginia. Various hikers describe it as the largest or second largest tree along the entire AT. Next it is on to Dragon's Tooth and the Catawba Valley. Finally, nearing the northern edges of the JNF, one of the AT's unique meccas reportedly inspires more photographs than any other part of the trail: McAfee's Knob.[71] According to 2006 hiker Ken Berry, "Virginia showcased some of her stellar beauty today. The magnificent views and the warm sun made McAfee Knob and Tinker Cliffs a highlight of Virginia thus far. While sitting out on tongues of rock I could see the valley below and the surrounding mountains and it was relaxingly beautiful. Cows were specks of black down in the valley."[72]

The Appalachian Trail experience is, of course, not limited to southwestern Virginia. For the observant hiker, however, each section of the trail remains singular, noted for unique natural features as well as particular phases in the pilgrimage. Resolve is tested at different places for different sojourners. Through-hiker Nate Olive repeatedly reflected an interest and knowledge in geology and botany. As he approached Bastian in early May 2001, he wrote,

The diversity here in Virginia simply amazes me. Grand vistas are separated by stretches of rolling trail pastures. Hot, still air trades places with cool winds around every other corner. Towns vary from simple collections of red barns and white houses to developed centers of business and commerce. The mountain ranges vary drastically from base layers of rock formed sedimentarily to foundations of ancient volcanic origin. Majestic stands of spruce are separated by mature poplars, and pine stands fight to gather momentum. Every step is as varied and different as every day, and any given moment holds a new surprise.[73]

To call the AT experience a "pilgrimage" is not intended to diminish the magnitude of the word in the least. On the contrary, it carries the same depth for many hikers that it held for medieval Christians on the way to a European shrine or for modern Muslims going to Mecca. In fact, for a large portion of the postsecular populace, this hike presents one of the few pilgrimages that might appeal to their spiritual instincts—anti-institutional instincts not so very unlike those of the traditional mountaineers who insisted that religion must be experienced, not taught. Furthermore, pilgrimages "may even be gaining in symbolic significance in some societies," wrote Simon Coleman and John Elsner in 1995, for "given that they constitute a potential means through which to react against an apparently dehumanised secular world."[74] Coleman and Elsner further point out that nineteenth-century Baptist minister Thomas Cook promoted "morally uplifting" tourism, sometimes (but not necessarily) to the biblical Holy Land itself.[75] We might see Cook as representing something of a transition between old and new notions of what constitutes "sacred" regarding travel.

Coleman and Elsner describe how even modern vacations and museum visits have come to resemble pilgrimages, replete with numinous overtones involving visitors intent on actually touching the moon rock in the Air and Space Museum or the Rosetta Stone in the British Museum.[76] Seen in this light, certainly hiking large distances of the AT easily qualify as pilgrimage, at least for the receptive sojourner.[77] As Victor and Edith Turner wrote, "If mysticism is an interior pilgrimage, pilgrimage is exteriorized mysticism."[78] Maurice J. Forrester Jr. wrote of the AT hikers that "this itinerant community has created over the years a mystical fellowship that becomes at times otherworldly in its pervasive sense of apartness."[79] A 2003 hiker identifying herself only as "Ann from Tennessee," regarded the AT experience as a "spiritual, physical, and mental cleansing."[80] Thomas Crutcher, a 1981 through-hiker and member of the Spiritual Life Institute, would agree. He explained the spiritual aspect most eloquently:

The AT is a pilgrimage and religious experience in the best and broadest sense. Almost all of the hikers would agree with this, but they may not use explicitly religious terminology. For me, it was life-changing; it brought me to a life with a great deal of silence, solitude, and meditation. The wilderness taught me a lot about myself and what was really important; both the outer and inner landscapes come into play. If you agree with the maxim that grace perfects nature, then whatever we can do to humanize ourselves sets the stage for grace to work. We need beauty, silence, and wilderness to become our truest, best selves. The challenge is to bring that vision back to the everyday world and try to transform it. We all need an irrepressible zest for life, and experiences like the AT can foster that.[81]

Spiritual experiences and awakenings along the trail are innumerable, as are religious associations.[82] Through-hikers often refer to Mount Katahdin (the pilgrimage's terminus in Maine) as the "Holy Mountain."[83] E. Jeffery Coons identified a tree near the Georgia–North Carolina line as an avatar after camping near it and having a mystical experience not rationally explainable.[84] Hikers attest to supernatural experiences and willingly suspend "profane" concepts of logical explanation.[85] Melissa Norman, a 2002 through-hiker, was well into Virginia on May 22 headed toward Burke's Garden when

> I entered this enormous beautiful meadow. It was absolutely marvelous with long green grass, a pond and a view 5 small [mountain] ranges long. All the aches and pains in my feet and knees left me immediately and I had this crazy urge to scream. I haven't really had the urge to yell but this surge of emotion just ran completely over me and I screamed hello to my mother and waved at everyone who was looking down at me. At that moment I felt closer to heaven than ever. I cried and cried for no other reason than because I felt so free and bound by nothing and then laughed a good laugh at myself for crying.[86]

The parallels between traditional postpagan pilgrimages and the AT through-hike are numerous: the initiate leaves structured society behind, enters a new pilgrimage society (though remaining very much an individual pilgrim), and aims toward the difficult pilgrimage's goal. The pilgrim "plays and prays" with said pilgrimage society and allows for (white) magic.[87] The pilgrims/hikers are "ascetic and stoic."[88] As scholars Coleman and Elsner explained it, "a Pilgrimage is not just a journey; it also involves the confrontation of travelers with rituals, holy objects, and sacred architecture."[89] The natural architecture for AT hikers is really the forested mountains, though shelters and hostels along the route offer a more formal manifestation of the concept. Rituals include the maintenance of gear and the basics of preparing and consuming food as well as making camp, and of course the now-requisite adoption of a trail name in addition to or in substitution for one's regular, societal name.[90] Holy objects can range from reluctantly discarded boots too worn out for further use, Georgia soil carried for deposit in Maine, and any number of idiosyncratic objects carried along the hike.[91] Some hikers have had their ashes scattered along the trail.[92]

When considered from a certain perspective and despite modern amenities, AT hikers are doing something on the land not seen since prehistoric times: long-distance walking. The introduction of the Spanish horse radically revolutionized Indian life, and naturally the most ancient mode of transportation has declined ever since. In this way AT hikers are indeed reaching back to something fundamentally primeval. If something unique is triggered in the human mind through weeks and weeks of walking, the hikers obviously recapture this.

Anthony J. Barrett walked through southwestern Virginia during the spring of 1999. He enjoyed a stop in Damascus for the annual trail days, climbed Mount Rogers at dawn, and enjoyed seeing the deer and ponies in the Grayson Highlands Park area. The comparatively pristine state of Catawba Valley (the view from McAfee's Knob), despite the Highway 311 link to nearby Salem and Roanoke, remained in his memory years later.[93] Barrett described himself as an adventurer rather than a spiritual seeker. "However, by the end of the journey," he wrote, "I was slowing down to savor the contemplative and more spiritual aspects of the trail experience."[94]

If Catherine Albanese is correct and the distinction between the Calvinistic strains of Appalachian religion and mainstream American Protestantism have indeed diminished in recent decades,[95] then perhaps the numinous associations with the traditional religion have as well. If so, the more spiritually oriented Appalachian Trail through-hikers may have come to represent the people who most associate the mountains with religious experience. Still, this in no way diminishes other, more traditional intimacies with the land.

Chapter 12

Old Commons Meets the New

For almost two centuries before JNF management, the highland woodlands in southwestern Virginia had functioned as a de facto commons. The early development of land speculators and absentee ownership of large tracts certainly fostered such informal use of the forested slopes, for the numerous landless tenants could not have survived otherwise. They used these highlands for livestock foraging (particularly swine), firewood gathering, ginseng digging, game hunting, and foraging for nuts, berries, and medicinal plants.[1] But outside of the southern Appalachians, concepts of the commons was prominent in the South and even in early New England.[2] During the colonial and early republic periods, fencing was prohibitively expensive in labor and materials all over the eastern seaboard. It remained so in the South even after the Civil War. Farmers found it was more practical to fence animals *out* of comparatively small vegetable gardens rather than inside expansive pastures. Thus the open range persisted and acquired deep customary roots.[3] As one western North Carolina woman put it during the 1970s, "The way I see [it], the good Lord put this land here for all of us to use and ain't nobody got the right to fence it off and try to tell people they can't pull galax" or other commons usage.[4] In some ways, the development of the commons itself runs counter to the Lockean advocacy that Americans once regularly took to heart.

As a seventeenth-century philosopher, John Locke championed democracy and property rights in his native England, and his conceptions of property influenced Americans more than any other Enlightenment or pre-Enlightenment philosopher.[5] He wrote the Carolina Constitution, strongly inspired both English and American Whigs, and profoundly influenced the Founding Fathers. Locke's original phrase, "Life, Liberty, and Property,"[6] is included in our constitutional rights and earlier manifested itself as our self-perceived rights of "Life, Liberty, and the Pursuit of Happiness" that Jefferson wrote in the preamble of the Declaration of Independence.[7] For Locke, people and their property ownership *preceded* the creation of government. In fact, it was the property-owning people who decided to *create* government in order to recognize and enforce their "God-given" property rights.[8] No government had a right to violate these rights. These ideas held great appeal along the nineteenth-century frontier and persisted in many parts of the twentieth-century American West, South, the Ozarks of Missouri and Arkansas, as well as the southern highlands. As illustrated earlier, the government soon asserted its authority otherwise, perhaps most strikingly in its Fifth Amendment eminent domain rights. Between hydropower impoundments and recreational land acquisitions, the southern highlands became among the more widespread regions to feel this power.

Lockean ideas of unregulated land use have clearly worn thin in many cases, even if they live on (sometimes fiercely) in libertarian sentiment. Libertarian concepts of unfettered property rights meshes fairly naturally with the deep tradition of independence that so many scholars and writers have associated with the southern highlands. But now southern mountaineers have seen the environmental and economic ravages that can accompany unregulated industrial exploitation of natural resources. Highlanders previously averse to zoning regulations might consider their possible benefits when contemplating the continuing trend in second home development. In this scenario the JNF and other public lands become something of a "mediator for the commons."

By the late nineteenth century, the main problem with the de facto commons of southwestern Virginia became one of an enlarged and growing population. A commons may function quite well with a low population base; too much human pressure and competition for resources tends to create conflict, environmental degradation, and poverty. In the midst of this deteriorating situation came the JNF as a new, albeit radically different, sort of commons. The population pressure has only increased since that time, and thus the JNF stands at the nexus of a modern and more complex commons conflict.

Postwar discussion about the commons began with Garrett Hardin's seminal 1968 article, "Tragedy of the Commons."[9] Early-twentieth-century southwestern Virginia might have served as an example of Hardin's initial thesis, which held that individual self-interest would inevitably override the common good. Hardin later modified his thesis and said his title should have been "Tragedy of the *Unmanaged* Commons."[10] But, in a way, this revision merely emphasized his original claim that government regulation was one of only two solutions to the tragedy. The other solution was private property and elimination of the commons.

Many scholars and writers responded to Hardin's idea. Like Frederick Jackson Turner's frontier thesis, the response was overwhelmingly one of detraction, and yet the detraction really only modified Hardin's thesis without ever debunking it completely. The detractors offered historic examples of functioning commons that contradicted Hardin's thesis, described examples where government regulation had actually caused its own harm, and generally disputed Hardin's cynical absolutism regarding the nature of human self-interest within a larger social dynamic.[11] Obviously government regulation is the JNF situation, and this situation is as flawed and problematic as the super structure encompassing any modern commons. Unstoppable "invasive" species and globalization of air quality, for example, became unignorable facts many decades ago.

One basic dynamic pertaining to the modern JNF commons is an enduring local orientation toward using forest resources, which conflicts most saliently with nonlocal interests more oriented toward the wilderness preservation or even outdoor recreation side of the spectrum. The former tends to see the latter as elitists who do not rely upon the land for a supplement to living. Wilderness advocates or even tourists primarily concerned with aesthetics worry that unchecked local use may alter the environment in ways offensive to the preservation ideal. Naturally, the JNF negotiates such conflicts, but it also deals with its own, sometimes cumbersome burden of a national policy not always suited to local conditions, particularly regarding timber harvesting.[12] Clearly, the highland woodlands of southwestern Virginia and other parts of the southern Appalachians can no longer function as the de facto commons of yore. For better or worse, some sort of governmental mediation is likely unavoidable to prevent self-interested exploitation of the public forest. The nature of the mediation will remain the eternal nexus of dispute and debate.[13]

No doubt the preceding chapter's emotional expressions of attachment to place do not characterize all local people. As in other locales, some are bound to think of southwestern Virginia's forested mountains as "home" in a more mundane sense. Furthermore, the post–World War II era has witnessed a general erosion of former ties to the land, which involved more cohesive community and social networks as much as intimacy with the land per se.[14] Without question, however, many local residents would rather use the forest for its resources—firewood, game, herbs, and the occasional logging contract—rather than commune with nature vis-à-vis the luxurious tourism of the privileged outsider.[15] Even more stark is the contrast between native-born desire for pragmatic forest use compared to idealistic wilderness preservation ideas advocated by outsiders living in urban or suburban environments.[16] This local preference is rooted deeply in preindustrial traditions and is what scholar Kathryn Newfont has called a "working commons" or "commons environmentalism," which centers heavily around hunting-and-gathering practices.[17] This cultural tradition has an ongoing, uneasy relationship with the worldview of newer arrivals. The aftermath of the 1970s demographic reversal persists as people continue to escape more intensely populated urban and suburban areas to live in the rural private lands surrounding the JNF and other southern Appalachian national forests. The national forest offers an aesthetic attraction to those wealthy enough to build homes in former farmland abutting the woodlands.[18] Likely such transplants would prefer the national forest be a national park instead. Retired hobby farmer Dick Austin perhaps epitomizes this attitude.

Compared to Appalachian poverty, Austin stands out as a wealthy transplant who acquired his Chestnut Ridge farm in Scott County, near the Clinch Ranger District. In 2009 he decried JNF management as "abominable" but could not offer any evidence to support this claim.[19] Were the evidence available, the JNF would have ended up in federal court, as the Monongahela did in the 1970s. With the organization of Forest Service Employees for Environmental Ethics and numerous environmental watchdog groups, there are many more people monitoring Forest Service timber cuts now than in the 1970s.[20] Austin's claim remains unsubstantiated. Timber cutting critics also tend to ignore things like the 27,000-plus acres the JNF added to wilderness in 2009 (see appendix C), including the 3,200-acre Stone Mountain Wilderness set aside in Lee County.

State officials with the Virginia Department of Forestry have their own task in getting loggers to comply with environmental protection laws, particularly regarding protection of the water quality of streams associated with their logging operations.[21] The overwhelming majority of southwestern Virginia's forests are in private hands, making the state the only agency of environmental law enforcement. This is true for all the forest lands in the east, where national forests constitute a small subset of a given state's woodlands. In 1975 Virginia pioneered one of the most progressive state law systems for protecting the environment during private logging. There were provisions for stream protection, erosion mitigation, and general watershed protection. Laws on the books are one thing, enforcement entirely another. Numerous violations followed passage of the 1975 regulations, and the Virginia Forest Watch continues to report violations annually.[22] Richmond attorney Robert Jackson Allen advocates that Virginia imitate the Forest Service by incorporating multiple use and sustained yield into its existing law code.[23] This is a wonderful ideal, but the expense of enforcement remains problematic.

This is not necessarily to condemn local loggers as antienvironmentalists, but their cultural attitude and their actual practices cannot be understood outside of the broad context of the region's pervasive poverty—something frequently (and ironically) overlooked by transplanted "limousine liberals" who have never known deprivation and who come into the region with advanced degrees. Practicing good environmental stewardship often costs more in the short term, and the long view remains a luxurious indulgence. The final insult arrives when the transplants blithely pretend that they speak for the mountain people. Reverend Gable, in his effort to help the Mount Rogers people, was very sensitive to such arrogance and avoided it. Others have not been so sensitive, nor so humble. Perhaps the only solution to any cultural dispositions toward the environmental lies in education. Here the

Abingdon-based Appalachian Sustainable Development pursues the noble endeavor of environmentally protective forestry and agricultural practices.[24] The local base of operation is important here, for there is much to be said for a decentralized approach to these issues.[25]

This would be a good time to review, briefly, the history of forestry in southwestern Virginia. First, there was the pioneering phase during which farmers cleared the valley floors of trees. Then, preindustrial salt and iron manufacturing intensified this forest clearing beyond farmland. Later, industrial logging stripped entire mountainsides of trees, and the JNF came into existence as a direct response. No doubt the JNF's forest management has been controversial, and federal officials have made mistakes. Watchdog groups like the Clinch Coalition have fulfilled a legally sanctioned function by monitoring JNF timbering.[26]

Unfortunately, a Catch-22 involves frequent public actions against Forest Service timber contracts, which in turn, escalates federal bureaucratic expenses. But aside from a few nascent sustainable forestry efforts,[27] the Virginia Department of Forestry oversees the most widespread effort to enforce environmental protection laws. So, in this sense, JNF critics like Austin, a self-styled environmental theologian, merely impose their Edenic ideology on the most convenient target, the JNF. On the other hand, in a strange lack of awareness of local poverty, writer Chris Bolgiano (an East Coast transplant to the Shenandoah Valley area) criticizes the native culture for its supposed overemphasis on utilitarianism regarding the land and its resources.[28] Both Austin and Bolgiano bring mainstream environmental ideas that, sadly, too often remain the indulgence of those independent of the land for a living. The fact remains, jobs are scarce in southwestern Virginia, and logging jobs appeal to some more than coal mining jobs.

Ironically, loggers often end up spending far more time in the forests than those who purport to protect them. More egregiously, wealthy transplanted cultural imperialists pretend to speak for native Appalachian culture without comprehending the first thing about economic poverty. It is difficult to say which is worse, their Marie Antoinette style of arrogant disconnectedness or their myopic self-righteous dogmatism. Regarding the JNF, the myopic perspective usually manifests itself in moral outrage over timber and road-building policy while ignoring environmental elephants in the room, such as nonpoint source pollution. The former is an exceedingly convenient target; the latter is as amorphous as it is ubiquitous. Everyone contributes.

Increasingly, scientists are proving nationwide that nonpoint source pollution derived from agricultural, suburban, and urban runoff is the major culprit regarding water quality degradation—much less so is erosion from abusive timber harvesting.[29] Responsible timber harvesting becomes insignificant when considering the larger picture of water quality degradation.

Organic runoff from animal manure chokes oxygen out of streams.[30] Every car that leaks oil contaminates streams and rivers, for chemistry's first law (Law of Conservation of Mass) tells us that this oil remains toxic indefinitely. All chemical fertilizers and pesticides, and now pharmaceutical traces in human urine (not to mention household chemicals), are all showing up in places like Chesapeake Bay.[31] But this is an extremely broad societal problem, and critics often cannot single out simple culprits. In this context loggers can become the easy target.[32]

In 2007 the total national forest acreage in Virginia was less than 18 percent of the state's total woodlands,[33] which would make the JNF's holdings around 5 percent. While federal managers obviously need to comply with all environmental laws regarding this acreage, the greater surrounding lands clearly present the greater concern regarding water quality. Obviously, best-practice logging that does minimal damage to watersheds remains crucial, and this presents serious challenges in steep terrain such as that found throughout the JNF. But there is also no denying the critical need of wood products industries in the depressed economy of Appalachia in general. The semi-viable alternative to wood products industries, oft touted, has been tourism.

The Mount Rogers NRA may never have become the concentrated recreational development that some of its designers anticipated, but the area has certainly witnessed its share of tourism. The usual tension arises among competing and conflicting interests: locals in need of economic stimulus, intruders unaware of their violations of privacy, the limitations upon traditional forest practices and frustrations inherent in dealing with a federal bureaucracy, and crowds of disrespectful tourists ruining the natural mood of the landscape by treating it like an amusement park attraction.[34] Hostility between the native-born and those seeking a second, seasonal home in the mountains is many generations old.[35] Everyone wants a piece of the paradise. In the thoughtful book *Ecotourism in Appalachia,* Al Fritsch and Kristin Johannsen cautiously advocate a model of green traveling in the highlands that has gained significant ground elsewhere in the world.[36] Certainly this approach holds great potential over the more garish forms of tourism promoted in earlier eras, but Fritsch and Johannsen are realists, not boosters. Tourism alone will not save the Appalachian economy. They write, "we believe that well-planned, community-based tourism can play an important role as one component in a more diversified Appalachian economy."[37] Many residents of Craig County apparently agree with this assessment.[38]

Selective logging on JNF land and environmentally friendly forestry on private land, in most scenarios, is going to be a part of that economy. But, generally, people in southwestern Virginia and other parts of the South appear to be shifting away from valuing their forests for commodity goods, like timber, to noncommodity benefits like hiking, sightseeing, and picnicking.

There seems to be a trend toward a biocentric view of the woodlands, with humans but a single entity within a larger ecological whole.[39] Unbeknownst to the critics of the JNF and other national forests, this shift is mirrored even more strongly within the Forest Service agency itself, at least in the personal opinions of employees.[40] Ironically, in a position diametrically opposed to its founding vision, a substantial number of Forest Service employees have even taken on a national ecological value system, which they see as compromised by local economic interests.[41] Obviously a great deal has changed since the Holston Working Circle specifically catered to local, small-scale sawmills and loggers in the aftermath of industrial timber stripping. Economics obviously concern local people the most, for good reason. Visitors and wealthy transplants have the luxury of focusing on other things, like the love of nature. Along with the particular local residents who possess a deep feeling for the mountains and trees, there is another group that stands out as religiously attached and attracted to these highland forests: Appalachian Trail hikers, particularly through-hikers going from terminus to terminus in Maine and Georgia. Any group must feature some diversity, but the physical demands and hardships of this outdoor experience tends to self-screen the general population, leaving behind a core of people who share some similar ethos. This is especially true for long-distance hikers who tend to see the experience as a pilgrimage.

Population pressures alone mean the modern commons cannot exist without conflict, and sometimes the cultural perceptions contrast profoundly. Maine resident and AT hiker Anthony Barrett (a retired petroleum employee who lived in Cairo, Egypt, for six years) observed this contrast in May 1999. He had stopped by Damascus's Trail Days but noted that the traditional trout rodeo took place on the same day. Apparently, before Trail Days exploded into a much-attended event, town leaders scheduled the two events together to attract worthwhile attendance. "There wasn't a more incongruous sight," Barrett wrote. "Streets filled with hippy-looking bearded hikers in shorts and stern rednecks in camo [i.e., camouflage] fishing gear."[42] Later that day the fishers formed their entourage of pickup trucks that followed the trout truck out of town toward some stream,[43] and later the hikers dispersed along the trail on foot. Incongruous indeed.

Still, the persistence of the commons illustrates some interesting aspects of perspective on land. As Albert Fritsch wrote in the aforementioned *Appalachia: A Meditation*, "Land can be divided and staked off and bordered with fences. This has its good points and bad points."[44] Fritsch elaborated by describing the benefits of sharing forest resources (including timber) and the necessity of controlling livestock. He concluded by advocating a commons-type of approach: "Appalachia is our to-be-shared land."[45] Regarding the highlands of northern Georgia, Lacy Hunter Nix no doubt spoke for many native-born southern Appalachian people when she wrote, "these mountains are a gift, a heritage, to be preserved and passed on to the next generation. We may disagree on the method of preservation. We may disagree on the best use of the land. We may even disagree on whether these outsiders are helping or hurting our precious home, but we all agree that these mountains and this place are a gift that we must treasure and protect."[46]

Memories of open range and the era preceding federal land management in the area naturally generate some conflict and resentment,[47] but so do second or seasonal home developers and investors. Ironically, as seen in the Konnarock area, some local residents have come to see the Forest Service as actually preserving the commons, albeit in a regulatory fashion that is bound to chafe old Lockean ideas of unfettered land use. This sentiment is even more pronounced in Craig County, which features more national forest acreage (54 percent) than any other Virginia county. The adjoining privately owned land is overwhelmingly rural, and multigenerational residents as well as more recent arrivals want to keep it that way. They see the JNF as an insurance policy against development.[48]

Linda Jilk, of Blacksburg, hiked the local national forest trails with her boyfriend, later got married in the forest, and subsequently continued to hike with their dog and children.[49] Jilk's sort of appreciation for the JNF is perhaps more typical around the old Blacksburg Ranger District, where forest-generated employment is negligible and Virginia Tech and Radford University college students, as well as numerous urban and suburban transplants, value the recreational aspects of the woodlands. Further southwest, Reverend Bill Gable periodically returns to the Konnarock area (where he lived in the 1970s) from his Abingdon home, down in the valley. He wrote in 2009, "Now that I don't live there all the time when I go back even just for a few hours I get a personal sense of well-being that's hard to describe. I don't know of any other place that does that to me."[50] Naturally there are mixed feelings about the JNF. Much misperception regarding property taxes persists, despite the aging PILT legislation that addressed this issue.[51]

Another element of the local populace has little respect for the JNF or any concept of public value in the land. The archaeological site looters on the Clinch District clearly illustrate this attitude, as do timber thieves in general. Beyond obvious opportunism, moonshine manufacturers of old and marijuana farmers of new are probably more or less indifferent to the commons. Aside from their obvious illegality, poachers' detriment to the public interest probably must be gauged against the scarcity or overpopulation of a given species, which opens an entire array of questions regarding ecology and the human role that has disrupted previous predator and prey dynamics. Clearly there are carryovers from Lockean property philosophy in most parts of the United States, and certainly among the people who live in proximity to the JNF.[52]

America is approaching the situation Europe and England have faced for centuries: the frontier is long gone, neighbors are here to stay, no paradise is undiscovered, and all future trends promise a larger population and an increase in competing and potentially conflicting property demands. In some ways, the JNF has evolved into an intriguing entity amidst this shifting American worldview regarding land. If nostalgia for an imagined paradise lost is overcome, and if some grassroots democratic participation is embraced, then perhaps the JNF commons possesses the potential to function as the most flexible neighbor in the emerging highland woodlands of southwestern Virginia. The religious reverence for this public land remains strong. Writer Liza Field, who spent summer retreats on the Glenwood District near Arcadia, wrote,

> This hemlock-steeped, castle-rocked place, plunged through with North Creek and Jennings Creek, their stone baths and crawdads and mountain walls, provided my family a low-cost vacation spot, a sanctuary from city life. . . . This enchanted forest saturated our hearts and returned home with us, providing the imagined scenery of scripture, Christmas carols, frontier history, Kipling, Mark Twain—the whole ancient human book of forests and wildlife, darkness and stars and journeys. . . . It was here that we acquired reverent behavior toward land and water. Something alive, a kind of majestic presence about the place, made us realize that it mattered how we lived.[53]

Field lived in a city, like so many who have enjoyed and continue to enjoy the JNF. She and her family appreciated their pilgrimages to the Glenwood District as a sanctuary from the "civilized" world. She and her family felt a sense of deep if informal stewardship for this place. She wrote, "In a world where such 'outdoor cathedrals' will only grow more rare, and whose populations are suffering costly surges in depression, violence and stress-disorders, whose children have rarely seen a star, heard a whippoorwill, or waded in a creek, perhaps this deeper valuing of a live forest will prove the more accurate accounting method."[54] For every Liza Field there must be many more who agree with her sentiment, even if they never publish their feelings. Local residents remain notoriously taciturn about such matters. A rare testimonial came from Smoky Mountains native Beuna Winchester, a Park Service guide. She said, "My love for the mountains is just as dear as love for life."[55]

Epilogue

In the Chinese philosophy of Daoism, there is an appreciation of things for their contrast regarding other things. This is most famously depicted in the familiar yin-and-yang symbol. Thereby daylight is appreciated through night, summer through winter, the female principle through the male, and so on. It is also how valleys and lowlands render mountains for what they are, and this is especially true in the valley and ridge province of the southern Appalachians, which includes much of southwestern Virginia. Valleys render magnificent views of mountains, and ridgetops afford views of valleys and other mountains. This topography becomes all the more remarkable when considering how public and private ownership tend to follow elevation patterns. In North Carolina lawmakers saw a necessity to pass legislation limiting ridgetop development.[1] This seems less likely in southwestern Virginia, where the public already controls many ridgetops.

The highest reaches of the southwestern Virginia mountains have evoked awe, religious regard, aesthetic appreciation, and respect from a great many people throughout the millennia. Briefly they were overly exploited, albeit comparatively mildly, by a swelling agrarian population struggling to survive. Even more briefly they were pillaged by industrialists whose main or sole concern was quick monetary gain.[2] But it is their very elevation that now renders them a public commons, and largely they became so in response to industrial abuse.

The first Forest Service chief, Gifford Pinchot, embraced the idea that national forests should serve "the greatest good, for the greatest number, for the longest run." Pinchot's supporters, President Theodore Roosevelt and Secretary of Agriculture James Wilson, agreed with this perspective. The phrase was lifted from the founder of Utilitarian philosophy, Jeremy Bentham. Whether or not Pinchot consciously followed Bentham's philosophy, he and other early national forest supporters certainly embraced a pragmatic sort of forest conservation centered squarely on commodity products. A great deal has changed in the agency's first century.[3]

If Utilitarianism characterized the earlier agency, ecological forestry is eclipsing it now. There are even elements of pantheism or natural religion, especially among wilderness advocates and supporters of noncommodity forest values in general. Pragmatism lives on most prominently among certain segments of the multigenerational local population, who still rely on the woodlands for game, firewood, pulpwood employment, tourism dollars, and the enduring tradition of gathering ginseng, morel mushrooms, sassafras, and other foraging items. All local customs aside, the national picture has long shifted toward environmentalism. Consider the following excerpt from the 1969 National Environmental Policy Act, which pertains to the JNF and every federal land management agency. According to Congress, the purpose of the act was "to declare a national policy which will encourage productive and enjoyable harmony between man and his environment; to promote efforts which will prevent or eliminate damage to the environment and biosphere and stimulate the health and welfare of man; to enrich the understanding of the ecological systems and natural resources important to the Nation."[4]

This forty-year-old proclamation seems astonishingly idealistic, and yet in retrospect it clearly provided one of the turning points in American environmental history. For the Forest Service, it became one of the legal stipulations that began turning the agency away from almost exclusive timber commodity emphasis to a greater orientation toward ecological forestry. For the JNF, as of 2009 the trend has included an accumulation of over 96,000 acres reserved in wilderness, or about 13.7 percent of the original, premerger JNF acreage. This is hardly an insubstantial amount.

☙

From a distance perhaps much of the forested mountainous land of southwestern Virginia seems pristine, untouched, or even ignored by humans. But obviously, over the centuries, these lands have experienced a large amount of direct and indirect human influence. Biologically and botanically, this forest is significantly altered from the woodland that preceded it even one century earlier. It only faintly resembles the forest of 400 years ago. The American chestnut tree, eastern wolves, eastern elk, and passenger pigeons are among the many life forms that have diminished or disappeared completely. On the other hand, oak trees and white tail deer are much more numerous than in any past time. Because of the changing flora, and for other reasons, topsoil depths and their chemical compositions have changed. Without question, this land and its flora and fauna, will continue to change, both through its own evolution and in relation to human action or nonaction. What is unprecedented is the knowledge, awareness, and even self-consciousness in regard to human influence on this land.

The prehistory of southwestern Virginia remains very incompletely told, especially considering that prehistory encompasses the huge majority of the overall span of human occupation. The historic period of southwestern Virginia involves less than 400 years. The prehistoric era covers more than 9,000 years. Human interaction with the environment during these millennia was quite varied in itself, ranging from highly nomadic hunting to swidden horticulture. Aboriginal interaction remained distinct from even the earliest interaction of peoples transplanted from the eastern hemisphere. Native Americans remained keenly attuned to the mountain woodland environment and less technologically removed; their mode of survival absolutely depended upon such an orientation.

Even before Eastern Hemisphere peoples or their descendants populated southwestern Virginia, they induced indirect effects on the land through the white tail deer pelt trade. Subsequent direct and indirect effects included the introduction of exotic flora and fauna, combined with unprecedented alteration of the landscape through forest clearing and regularized agriculture. In the eighteenth century alone human population grew to levels never before existent in southwestern Virginia, and this factor obviously increased the magnitude of an already more concerted influence on the environment.

As the threatening wildness of southwestern Virginia diminished, a contingent of Romantics became the first group of European American people to dwell on the highland woodland's scenic aspects. Their luxurious period of tourism, the antebellum nineteenth century, became a singular interim following the wilderness conquering era and preceding post–Civil War and early-twentieth-century heavy industrialization. During these decades southwestern Virginia supported varying levels of prosperity or poverty until population growth, economic changes, and an evolving national conservation mood coalesced to begin bringing large portions of land under an unprecedented collective consideration. The JNF has existed ever since.

In general, the JNF inherited a timber-poor and damaged environment. These were not the "lands that nobody wanted,"[5] but economic reality had caused many industrialists to move on to greater opportunities, while local residents faced a similar situation on individual bases. That local people loved the land became apparent when a core population stayed rooted in the hills even amidst the most dynamic decades of out-migration. It is only human nature to love and desire to live in a beautiful land, all the more so when that land is home for all living generations, and where several generations of ancestors are buried.

The issues, controversies, and basic history of the JNF is largely indicative of the greater history of federal forest lands in southern Appalachia as well as the eastern United States. How humans have treated this land, and how they have come to regard it, reflects many of the salient themes, problems, and challenges of modern forest economy, recreational usage, and ecological concerns. The JNF has been something of an ongoing conservation experiment within an evolving democratic context. In the emerging environmental political scene of the 1960s and 1970s, a shift in the American temperament affected all federal and state land management agencies. The history of the Mount Rogers NRA captured many aspects of this scenario. New directions for participatory public land management began and helped solidify an environmental movement at the core of American culture.

In all, the JNF has undergone some enormous changes during only the last three decades, perhaps propelling itself out of the contentious 1960s and 1970s toward an entirely new era of ecological forestry. Considerations for wildlife and their habitats have grown from basic game species to nongame species, to special care for threatened, rare, and endangered species. Add rare or endangered flora to these fauna, and the latest turn toward ecosystem management naturally arises. Biologists have risen to new prominence within the greater agency, and specialized botanists have joined the diversity of staff on individual national forests. The spectrum is further expanded to include hydrologists and air quality specialists, which reflect original watershed concerns and postwar air pollution problems. And where absolute fire suppression once seemed like an admirable goal, forest managers now use fire as an ecological agent. The importance of the JNF's wilderness areas in the scientific world will probably become most comprehensible through studies of "island biogeography," which would apply to these areas of older growth surrounded by lands featuring younger forests.[6] Certainly by the early 1990s original JNF employees who began their work during the Depression era could scarcely recognize the remnants of the old agency.

Multiple use constantly arises as a focal point of debate among those interested in national forest management. Public ignorance of the Forest Service mission and its limitations, as defined by federal laws, remains very common. The basic concepts surrounding multiple use remain about as challenging as they were fifty years ago, when the Multiple Use Sustained Yield Act ushered in a long trend toward greater congressional decision making of Forest Service policy.[7] The greatness of multiple use lies in its recognition of and attention to diverse interests. Its shortcomings lie in the difficulties of determining what kind of emphasis to place on which resource in which circumstances.

Certainly the JNF's employees themselves do not always agree on how national forest operations should commence. As recently as the 1990s, tensions continued among various divisions of JNF staff. Archaeologists ("dead Indian hunters") and timber staff ("timber beasts") tend to see the forest and its resources in such different ways that a certain absence of camaraderie is bound to result. Generational and philosophical differences continue. Older, often retired wildlife biologists tended also to be deer hunters, while a younger generation might include vegetarians.

Jim Loesel, secretary of the watchdog group Citizens Task Force on National Forest Management, observed in 1997 that the Forest Service had reached a point where it was "looking for allies." Old alliances with the private timber industry had eroded for some time, and increasingly after World War II the Forest Service had begun to operate as a mediator among contending interest groups. Loesel recognized the mid-1990s as an "exceptionally complex" situation for JNF management, one involving a certain amount of harmonious public input, but also signs of increasingly polarized public groups, perhaps especially timber harvesting and wilderness advocates.[8]

A century of national forest stewardship may have proven that the JNF best functions as a protector of ecology and culture, with only a highly diminished role for natural resource extraction. The Pioneer Forest essay in a following appendix may offer some insight into future eastern national forest timber management. This approach could be harmonized with the agency's historical focus on logging while allowing the JNF and other eastern national forests to function as hybrids, with a mission somewhere between those of the U.S. Forest Service proper and the National Park Service. The most central question might be administrative costs. If the Forest Service could provide an example of reducing costs and pioneering innovative management, this agency would indeed once again become a model for the rest of the federal bureaucracy.

The JNF continues to undertake a highly ambitious, complex, and multifaceted endeavor in the administration and management of this modern-day commons. The very diversity of this endeavor and its successes deserve no small appreciation. The JNF and many other national forests continue to face unprecedented demands upon their resources; it can be no easy task to maintain a healthy interplay between extractive, aesthetic, cultural, economic, and ecological concerns. Buried amidst the sometimes overwhelming bureaucracy of the modern era, the JNF tradition of community interaction struggles to survive. Gone are the moonshiners who used to build their stills on national forest land, as are the days of national forest tenants. Rangers no longer issue 202-C timber permits, but local people can still get permits to gather firewood, and (as recently as the 1980s) someone building their own log cabin could still buy individual trees. The public constituency remains ever evolving, but perhaps the best aspects of JNF community relations tradition will endure.

Given the serious and increasing threat to the Appalachian bioregion, the public lands stand out even more as islands of comparative mitigation. It is ironic that environmental critics can see nothing good in the management of the Clinch Ranger District, when only a few miles to the north some of the most devastating environmental change is occurring in the way of mountaintop

JNF critics tend to ignore how the Forest Service provides a sanctuary against such drastic environmental alteration as mountaintop removal. This satellite image (circa 2009) focuses on land abutting the north side of the Clinch Ranger District. Enhanced from Google satellite images.

removal.[9] For all the flaws that might be found in JNF management, the national forest seems like a haven in comparison. So far, the sacred land is fairly safe.

For all the inescapable practical considerations, religious regard for the land should never be underestimated. It is, after all, central to the modern environmental movement, even if Green Calvinists do not see themselves as such. Religious regard, by its very nature, does not usually appeal to logic or rational thought, and yet its political manifestations make it a potent force that the Forest Service has come to encounter on a regular basis for the past half century or so. Forested mountains of any sort might be enjoyed from afar, but the JNF controls the timber management destiny on its lands. Even more important, those who wish to commune more closely with the woodlands absolutely depend upon the public access that the JNF and AT provide. Critics of national forest management might do well to imagine southwestern Virginia without the commons.

The highland woodlands in southwestern Virginia now receive more intense human regard than at any time in recorded history. The intensity of perception is easily matched by the variety and multiplicity of views, and the JNF remains an ongoing experiment in democratic land management. One thing for sure, never before has any entity, governmental or otherwise, legally designated nearly 100,000 acres in southwestern Virginia as off-limits for development or extractive management. This seems like more than a token nod to the values of a natural world not widely beheld since prehistoric times. A movement back to selective-age German silviculture would offer a further possible step, radical as it may seem, toward ecological forestry. In any case, the Jefferson National Forest is now well established as caretaker to forested mountains that many have always sensed was sacred ground. Many more sense this now.

Appendix A

Pioneer Forest: A Case Study in Modern Selective Forest Management

Gifford Pinchot would be delighted. The German school of select-cut forestry and uneven age timber management that informed Pinchot and a couple of generations of American foresters is alive and well. It has continued and become even further refined on approximately 154,000 acres of forestlands in the Missouri Ozarks.[1]

The visionary behind the Pioneer Forest is Leo A. Drey, born in St. Louis in 1917. In 1951 Drey bought around 30,000 acres on the advice that this was the minimal acreage he needed to sustain an economically viable private forest. In only a few years, however, Drey expanded his holdings to become the largest private landowner in the state.

The eastern Missouri Ozarks were famous for their white oak trees, and a number of enterprises harvested this wood for barrel staves, particularly for storing and shipping bourbon whiskey. Pioneer Cooperage Company, and later National Distillers Company, concentrated their efforts on tens of thousands of acres mainly in Shannon, Reynolds, and Carter counties. In 1954 Drey purchased some 90,000 acres from National Distillers. As he has said many times, he has been in "over his head" ever since.

The Pioneer's staff has remained around half a dozen people. In 1956 a general manager, forester, and three rangers ran the operation. In the 1990s the staff consisted of a chief forester, biologist, forest manager, and three ranger-technicians.

National Distillers, Drey explained, had very superior timber management practices. According to Drey, "The two people in charge, Ed Woods and Charlie Kirk, were really recognized as being preeminent in the field of foresters. They'd been at it a long time. They're the ones who established the model that we're still following, this individual tree selection." Woods and Kirk, of course, came out of the earlier school of forestry then focused on selective cutting. Other professional foresters predicted that a maple forest would come to dominate the area, since hypothetically the absence of direct sunlight afforded by a clear-cut would encourage shade-tolerant maples. "Of course, we've been at that for almost fifty years," Drey observed wryly, "and demonstrated that you *can* regenerate an oak forest. And we *have* generated an oak forest through individual tree selection."

So just as silviculturists nationally were turning away from selective cutting to clear-cutting, the self-described "bullheaded" Drey began his operation adamantly opposed to clear-cutting. It is a direction from which he never departed, and now, more than half a century later, the Pioneer Forest's continued profitability and environmental stewardship has vindicated his original vision. The Pioneer Forest now contains two research natural areas (as defined by the Society of American Foresters); one comprised of white oak, another of old growth eastern red cedar. Finally, the forest has a number of "forest reserves" that Drey and staff decided needed protection, even if they did not meet the standards for a research natural area.

Forest reserves "mainly revolve around unique plant communities that we wanted to protect at some level, but that are not something that's unique enough to put into a natural area," said forest manager Clint Trammel, who began working for the Pioneer Forest in 1970. Trammel stressed that these areas might qualify for research natural areas in the future. Such forest reserves might be situated around damp fens that support a distinct ecology. "Our logging activities are designed so that we don't impact those," Trammel explained. "We'll stay far enough back away from them that we don't change the water regime on those sites." Another forest reserve includes an old stand of white oak and hickory. The older hickory, in particular, is unusual for the area, so the Pioneer Forest staff decided to leave it alone to see what would naturally evolve.

Trammel gets a little peeved at the overwhelming predominance of even-age timber management in the United States. "We have enough research on even-age management that we could probably fill ten warehouses. But we probably could not cover the top of a conference table with the research that's been done on uneven-age management in oak-hickory." Trammel said his research, including a master's thesis for University of Missouri, has revealed remarkably similar economic returns from even- and uneven-age management. "So you can't justify doing even-age management because of the economic returns. You have to have some other reason to justify it," said Trammel, who decries the emphasis on fiber production over returns to the landowner.

Trammel offered a more detailed explanation. "When you say, for example, that a stand of red oak or black oak is mature at seventy years, and you go in and cut it all down and start over again, what you're doing is forgetting the fact that probably 15–20 percent of that stand could grow for another twenty or thirty years, and in that period of time add four or five or six inches in diameter to the size of the tree. That extra twenty or thirty years is what really begins to put on the high value wood that is the most desirable. It's free of knot. It's the highest percentage of high-grade lumber that comes out of a tree. And if you cut those trees down at an early stage in their life cycle, what you're basically doing is short-circuiting that value return that you get from the longer growth periods."

Before World War II, professional forestry in America focused on selective-cut silviculture. During the housing construction boom of the late 1940s and the 1950s, clear-cutting or even-age management gained favor in Douglas fir forests of the Pacific Northwest. But this approach to timber management soon spread like gospel throughout the United States, in deciduous and coniferous forests alike. Since then, selective cutting has drawn the much-repeated criticism of "high grading," or the culling of only the best trees, thus leaving an inferior grade of forest and seed stock. Pioneer Forest staff have grown somewhat inured to this criticism but offer much evidence to the contrary.

Pioneer Forest timber sales are designed to remove trees that have maximized their growth and are succumbing to stronger, taller trees who are gaining dominance. Trees still accumulating high value are left. Such an approach makes no sense to suppliers of wood fiber for chip mills, of course, whereas the Pioneer Forest tries to emphasize high quality timber production for a lumber market.

One of their timber management goals lies in bringing back white oak and shortleaf pines by logging out red, black, and scarlet oaks that took over after early-twentieth-century logging targeted the former species, leaving the latter to dominate the culled environment. They do not seek a complete conversion through total eradication of red oak, but rather seek to change the relative composition until it more closely approximates the preindustrial logging environment. "And we're doing it basically through a natural selection system by preserving good quality white oak and shortleaf pine stands and thinning extensively in some of our red oak stands," Trammel explained. "We're gradually making that conversion."

In addition to maximizing forest volume through selective cutting, the Pioneer actually low grades some stands for occasional firewood sales. Gatherers of firewood for personal use get a free permit from the Pioneer Forest office in Salem, Missouri. Those selling firewood to others have to gain a timber contract. In the latter case, each crooked or diseased tree is hand marked, as in any Pioneer sale. "Most of our firewood comes out of tops and residue behind a logging sale," Trammel said. "But once in a while we'll go through and mark a stand of timber. If for some reason we think it's too thick we'll have some stands thinned that way." Trammel admits that a dwindling number of people are willing to engage in the manual labor of gathering firewood.

Trammel laughed at the suggestion that they might not use skidders in their operations. "The skidder is not the problem," he explained. "The problem is the guy driving the skidder." Fortunately, over the years certain loggers (particularly family operations) have come to appreciate the sort of operation expected of them on a Pioneer timber sale. Chief Forester Terry Cunningham and staff consistently award these loggers the cutting contracts, and in recent years there has been very little gratuitous damage caused by skidders.

Like appraising real property, practicing good medicine, or any number of other endeavors, marking timber in a selective cut is a combination of art, science, and logical and instinctive reasoning. "You've still got to look at every tree on the tract to decide which ones you want to paint and which ones you don't," Trammel explained. Their approach is to thin the poorest trees first, then the smaller of the better trees. Their goal is to end up with about fifty square feet of basal area containing sawlog-quality timber and to avoid leaving any given stand until it requires 100 percent restocking. Timber inventorying is a regular duty.

Don Stevens, a National Park Service historian, once dismissed Drey's motives as mere "conservation and profit." But this is clearly mistaken. Since the 1950s Drey could have liquidated his timber resource through clearcutting and reaped much greater, albeit short-term, cash profit. Instead, his priority has always been responsible land stewardship and environmental profit. Pragmatic forestry, in fact, has always been but one component in Drey's overall environmental vision. His designation of research natural areas and forest reserves demonstrate this fact. There are many other examples.

During the late 1980s Drey proposed reintroducing the once indigenous black bear, mountain lion, and elk onto his lands. Missouri officials, however, would not give their approval. In the interest of science, the Pioneer Forest has also hosted many dozens of research projects, particularly those of forestry and wildlife graduate students. But it was the 1990 Natural Streams Act ballot measure that perhaps showed Drey at his most idealistic.

The law would have enacted land use restrictions along fifty-four Missouri stream embankments and prohibited ATV use, restricted damming, and imposed other constraints designed to prevent erosion and protect water ecology. Drey's Pioneer Forest, of course, has long practiced this approach in erosion protection by avoidance of cutting standing timber in proximity to stream and river embankments. Such land use stipulations are standard procedure in places like Oregon, Wisconsin, and Vermont. From Drey's perspective, the act would have protected landowners' interests without threatening their rights. The opposition—including the Farm Bureau, Missouri Department of Conservation, ATV groups, and others—saw the act as a threat to personal rights and property rights. In any case, the ballot measure went down to resounding defeat, but not after both sides spent many hundreds of thousands of dollars in their respective campaigns.

While riparian protection and selective cutting evoke their own commonsensical logic, Trammel remains astounded by the philosophical resistance he has encountered from professional foresters. "I think the problem is, they feel like it's some kind of indictment against what they themselves are doing, rather than seeing it as an alternative approach to forest management." Trammel wishes that the general run of forest managers would instead see selective cutting as a viable alternative that could at least be practiced in conjunction with clear-cutting. "Rather than seeing it as another tool that they could use, they fear that they're going to be criticized for doing what they're doing if they admit that what we're doing is working. It's a psychological thing."

Is the Pioneer Forest unique in the United States? Almost. Since 1854 the Menominee Indians of Wisconsin have maintained 220,000 acres of forest on their 235,000-acre reservation forty-five miles west of Green Bay.[2] The Menominee, the Pioneer, and others practicing sustainable forestry have organized themselves in the Forest Trust and the Forest Stewards Guild.

In 1984 Henry Carey established the Forest Trust, which "views sustainable forestry as a fragile calculus between the natural systems of the Earth's ecology and those of the human economy." In 1997 the Forest Trust sponsored the creation of the Forest Stewards Guild, which now contains private forest owners from many parts of the nation. The Forest Stewards Guild's mission is "to promote ecologically responsible resource management that sustains the entire forest across the landscape." The Pioneer Forest, of course, is a member of the Forest Stewards Guild, and Trammel has met and shared ideas with other members from various parts of the United States.

"Anyone with any size piece of property can take what we're doing on Pioneer Forest and use the same approach managing their land," Trammel insisted. Such forest owners may not generate a substantial profit for themselves, but they will be able to pay for the selective cut, pay their property taxes, and, most important of all, keep a standing forest.

In 1998 forest manager Clint Trammel took me on a jeep tour of the Pioneer Forest. We drove past several timber sales, and at twenty-five miles per hour it was difficult to distinguish any difference in forest appearance. Without being told that we were passing a timber sale, I would not have known it. Trammel grew excited when we approached the white oak research natural area (formally named the Current River Natural Area). "I love to drive down this road," he said. "You walk down on that hillside there and you can see what this forest *can* grow. I think a lot of what we're looking at today is not anything close to what the Ozarks can grow. You can see some really outstanding specimens that are beginning to show up in the forest."

"What I really like about working here on Pioneer, in addition to just the philosophy that's used, is that I have personally marked and seen harvested the same tracts, twice. And before I retire, I'll see some of those tracts cut a third time. Now, how many people can say that? And when I walk away from that stand of timber for the third time, [aesthetically] it's going to look just like it did the first time I walked into the stand of timber."

During the 1990s sustainable forestry blossomed into an international phenomenon as foresters began expanding their scientific endeavors into multifaceted ecocultural approaches to woodlands management. These new approaches attempt to achieve environmental, social, and economic health instead of the earlier, narrower silvicultural focus on wood products output. For many of these endeavors, such as Trillium Corporation's operation in southern Chile, the long-term results remain to be seen. On the other hand, a positive legacy of many decades of sustainable forestry is already evident on the Pioneer Forest, Collins Pine operations in Oregon, and other smaller woodlands.

Drey established a foundation that will ensure that the Pioneer Forest and its forestry practices continue into the indefinite future. As *Audubon* magazine concluded in 1988, Drey "happens to have more money than he needs and the wisdom and generosity to spend it on land preservation. Missouri is lucky. Every state should have a Leo Drey." Conservation and profit? More like a streak of pragmatic forestry within a greater, idealistic environmental vision.

☙

The Pioneer Forest is a relevant case study for timber management in all mid-Atlantic southeastern national forests, including the Jefferson National Forest.[3] Pioneer Forest has relevance for these public forests for at least two reasons: all the aforementioned territories contain highland topography and feature somewhat similar mixed evergreen and deciduous forests. But the differences are more striking. Pioneer Forest is a private entity that makes a profit even while eschewing clear-cutting silviculture. The eastern national forests are obviously more complicated (and unfortunately more bureaucratic) entities financed with public monies. Nevertheless, eastern national forest timber managers would do well to learn something about the old-fashioned selective-cut silviculture still practiced on the Pioneer Forest. After all, if administrative costs could be kept reasonable, a return to selective cutting could solve several problems for the eastern national forests: greater protection of watersheds, the end of below-cost timber sales, the end of protests from environmentalists, ecological foresters, recreationists, tourists, and adjoining landowners who object to clear-cutting—and last but not least, a continuation of the traditional utilitarianism of the U.S. Forest Service that makes the Department of Agriculture its appropriate base. Selective cutting of low-grade timber could contribute to the glue-lamination wood products industry and solve the traditional criticism of high-grading. But some high-grade trees could also contribute to fine woodworking products, such as furniture and timber frame beams.

If the U.S. Forest Service once abandoned selective cutting for clearcutting, is it so unreasonable to conceive of a return to its original silviculture (albeit, now even more sophisticated)? Eastern national forests could never come near replicating the Pioneer Forest's economic model—but it might come close in the timber and engineering departments with a staff proportional to the harvesting volume and methodology. Rangers would have to relearn their predecessors' skills at individual tree-marking, supplement such practice with the latest ecosystem management scientific thinking, and remain subject to the scrutiny of groups like Forest Service Employees for Environmental Efforts, not to mention numerous local environmental watchdog groups such as the Clinch Coalition (Wise, Virginia) and Citizens Task Force on National Forest Management (Roanoke, Virginia).

Selective cutting on the JNF could serve as a stellar example for neighboring private timber owners, which would also be consistent with the founding idea of national forests as models for private timber managers to emulate. The Forest Service serving as a model of timbering conservation for private woodland owners goes back to its 1905 founding legislation up to contemporary times and has been reiterated with various interim legislative measures.[4] Selective cutting would also continue to fulfill the utilitarian aspect of deeply ingrained multiple-use philosophy of the Forest Service but, to an important extent, also secure the general interest of eastern national forests as a commons. The most radical (and perhaps ultimately unrealistic) aspect of this idea would be the proposed reduction in federal bureaucracy. Perennial political rhetoric to the contrary notwithstanding, governmental bureaucracies at all levels tend to grow rather than diminish. If the U.S. Forest Service retooled its eastern lands with a selective woodland management mission and reduced staff accordingly, they would become a model federal bureau—as the overall agency actually was during the 1950s.

If the decades-long resistance to allowing an ecological role for wildfire in timber management was any indication, probable agency resistance (not to mention political resistance from pulp producers) to this idea obviously presents formidable obstacles. Still, there is reason to believe that this revolutionary change would find a good deal of public support, and thus possibly all eastern national forests could continue functioning in a partially public-subsidized fashion not unlike how national parks are financed. Over 13.6 percent of the JNF is now protected wilderness; as recently as the 1970s this would have struck many as a radically improbable vision for the future. Is the idea of returning to selective timber management really any more radical than that?

Appendix B

Payment in Lieu of Taxes

Forest taxation is an old and vexing problem. Long before the advent of American national forests, various governments throughout the world struggled to create an equitable taxation system for land bearing a long-term yield product. During the early 1930s, Fred Fairchild and other forest economists extensively studied forest taxation in the United States and various European countries in an effort to provide some answers to this problem. Among their discoveries, they found that both Sweden and parts of Switzerland provided special tax allocations for forest bearing lands, allowed for a smaller tax base percentage from timber lands, or appraised rural forest property at lower rates compared to rural farmland. They also found that European forests in general tended to have been under such management for the long term and that timber-generated tax revenues were steadier and more predictable. They discovered that "the tax systems of all the [European] countries studied in one way or another recognized and make some allowance for the peculiar nature of forest property." Fairchild made similar recommendations for forest property taxation in the United States and suggested an adjusted property tax for forest lands, a reduced tax for deferred-yield forests, and a differential timber taxation for second-growth forests, which would provide "an adjustment of the property tax to the normal degree of income deferment."[1]

The issue that arose with American national forests lay in providing fair compensation to state or county governments for land the Forest Service reserved or acquired, thus taking it out of the regular property tax rolls. Originally, the federal government gave states bearing national forest land 10 percent of all timber sales, mineral sales, and other resource receipts from national forests. States, in turn, distributed these monies to their respective counties based on national forest acreage contained in each. The Forest Service raised that figure to 25 percent in 1908, and these payments became known as "payments in-lieu of taxes."[2] The policy, however, was never without its critics. As early as 1920, John Ise felt compelled to defend public timber ownership and federal payment in lieu of taxes:

> Much of the reasoning on the subject [of taxation] takes only a short-time view of the matter. If the government timber were turned over to private exploitation, there can be no doubt that most of it would be exploited

more rapidly than it is now, and that some of the communities involved would enjoy an era of . . . "prosperity"; but if such a policy resulted in the speedy and wasteful destruction of timber, such a "prosperity" would be short-lived. The government, in its forest reserve policy, is aiming at long-time results.[3]

Many timber companies of the late nineteenth and early twentieth centuries bought land for the trees, not the land itself, and grew eager to dispose of the acreage as soon as they finished harvesting trees. They certainly did not wish to pay land taxes on cutover land, and with increasing taxes during the 1920s, and especially during the Great Depression, county governments found themselves in serious financial straits owing to high rates of property tax delinquency.[4] In fact, during the Great Depression, national forest purchases provided counties with a certain amount of desperately needed income.[5] Of this phenomenon, Fairchild wrote in 1935,

> There are those who believe that the poorest land would better be in public ownership and who therefore incline to regard with complacency the reversion of such land through tax delinquency.
>
> Tax delinquency which advances to the point where titles are surrendered on a large scale is prima facie evidence of serious economic or political maladjustments, and there can be no adequate solution of the problem apart from important—perhaps drastic—changes in the organization and financing of local government.[6]

But as the national economy recovered after World War II, some of the 25 Percent Fund's flaws grew apparent. For fiscal year 1958, the 25 Percent Fund included $31,165 for counties in the JNF territory, which then amounted to around 620,000 acres. Thus, counties received little more than five cents per acre. As the county historically possessing the largest percentage of JNF land (about 99,000 acres in 1958), Craig County had the greatest financial stake in the 25 Percent Fund. Nevertheless, in 1958 the local *New Castle Record* observed that "all indications are that income . . . will continue to increase" as the forest improved through protection, fire control, and multiple-use management.[7] No one anticipated the clear-cutting controversy of the 1960s and 1970s that would eventually limit Appalachian national forest timber harvests and, concurrently, receipts derived from such sales.

Even before the clear-cutting controversy erupted, receipts from Appalachian national forests could not keep pace with increasing costs for roads and schools, and various county governments throughout the eastern United States began to complain. Also, the decline of timber sales' receipts compensation accompanied a transformation in the timber industry, in which mechanization precluded, to an important extent, employment of significant numbers of local people in wood-related production. Larger sales to pulpwood manufacturers replaced many of the earlier, smaller sales to local individuals and sawmills.

At best, the system proved unequal. The heavily forested Pacific Northwest received tremendous amounts of income from their national forest receipts, but the relatively timber-poor Appalachian region suffered.[8] Finally, public outcry over clear-cutting on national forests ensured that no substantial growth in timber harvesting (and thus timber sales' receipts) would take place. For all these basic reasons, the Public Land Law Review Commission produced a study in 1970 entitled *One Third of the Nation's Land,* in which they agreed that some county and state governments were not receiving as much income through 25 Percent Fund payments as they would through private ownership land taxation.[9] As they reported, "If national interest dictates that lands should be retained in Federal ownership, it is the obligation of the United States to make certain that the burden of that policy is spread among all the people of the United States and is not borne only by those states and governments in whose areas the lands are located."[10]

Congress addressed this problem in 1976 by enacting Public Law 94–565, or the Payment in Lieu of Taxes Act (PILT). The PILT legislation allocated a minimum amount of revenue for counties (about seventy-five cents per acre) based upon the actual amount of federally owned land, or "entitlement acres," in that county.[11] Not only did counties receive base amounts for their "entitlement acres," they also began receiving compensation for road building and reforestation costs in the national forest, which the Forest Service had previously deducted from timber receipts before calculating the 25 percent in lieu payment.[12]

An astounding unawareness of PILT payments has persisted ever since Congress passed this legislation. National forest timber cutting advocates omit PILT fund data in order to lobby for more national forest timber sales, which, without PILT, would mean more revenue to localities. *With* PILT payments, however, increased timber revenues become irrelevant. Enormous clear-cuts would be required in the Appalachian region in order for the 25 Percent Fund to match PILT's seventy-five cents per acre, and clearly the public would not tolerate such intensive national forest timber cutting. Congress does not fully fund the PILT allocation every year (at sixty cents per JNF acre, 1987 rendered the lowest compensation up to the mid-1990s), but between 1979 and 1991, combined JNF 25 Percent Fund monies and PILT payments have averaged seventy-nine cents per JNF acre. Such payments compare favorably or even surpass corporate property taxes.

In 1980 corporations in Bland County paid seventy-six cents per acre, while combined JNF 25 Percent Fund and PILT payments totaled eighty-one cents per acre (individuals paid an average of ninety-five cents per acre). Even more dramatic, 1980 Wise County corporations paid an average of only forty-eight cents per acre, individuals paid fifty-nine cents, while combined JNF 25 Percent Fund and PILT monies totaled ninety-four cents per acre.[13] Obviously, Congress rectified Appalachian national forest property tax inequities with the PILT legislation.

Appendix C

Miscellaneous Statistics and Recent Budget Data

Table 1. Acreage Statistics for Entire JNF

1936	140,000
1940	480,000
1945	580,000
1950	610,000
1955	620,000
1960	540,000*
1964	546,636
1965	550,000
1970	610,000
1975	680,000
1980	690,000
1985	700,000
1992	705,192

* Boundary change—acreage transferred to George Washington National Forest.

Note: Elevation ranges from 600 feet at the James River to 5,729 feet at Mount Rogers, the highest point in Virginia.

Table 2. County Acreage

COUNTY	1957	1964	1980	1991
Craig	98,800	112,462	114,932	114,664
Bland	19,000	21,247	70,455	72,084
Giles	48,200	50,872	61,507	62,974
Smyth	59,100	61,611	71,631	72,745
Wythe	45,300	48,508	56,511	57,020
Botetourt	52,700	56, 272	64,582	65,421
Grayson	11,600	13,446	32,105	32,846
Wise	26,900	29,020	29,600	35,796
Scott	30,300	30,912	34,174	34,122
Pulaski	16,900	18,763	19,291	19,288
Montgomery	14,800	17,805	19,211	19,070
Monroe, WV	—	0	18,241	—
Washington	16,200	17,328	21,364	22,081
Rockbridge	—	19,158	21,182	21,192
Lee	9,000	9,895	11,873	11,303
Dickenson	8,600	9,003	9,003	8,235
Bedford	—	18,792	18,074	18,762
Tazewell	5,000	5,256	7,097	9,427
Carroll	4,300	4,325	5,564	5,722
Roanoke	1,600	1,845	2,900	3,079
Letcher, KY	0	0	845	—
Pike, KY	—	116	116	—

Table 3. Timber Sale Statistics, 1946—1954

Year	Extract (MMBF*)	Saw timber (MMBF)	Other products (MMBF)	Total (MMBF)
1946	9.8	3.9	5.4	19.1
1947	5.4	8.1	2.8	16.3
1948	4.9	4.9	3.2	13.0
1949	5.1	5.5	2.0	12.6
1950	1.8	4.0	1.6	7.4
1951	3.3	4.6	1.7	9.6
1952	1.7	1.6	2.4	5.7
1953	0.6	4.3	3.0	7.9
1954	200 tons	5.7	3.2	8.9 (excluding extract)

* MMBF = Millions of Board Feet

Note: After 1954 JNF Timber Staff ceased to record "extract" as a category, which reflected the last of the salvaging operations in regard to the American Chestnut tree (killed by the Chestnut Blight from the 1910s to the 1930s), the primary species used in extract manufacture.

Table 4. Timber Sale Statistics, 1955–1992

Year	Saw timber (MMBF*)	Other products (MMBF)	Total (MMBF)
1955	—	—	8.4
1956	—	—	11.0
1957	8.3	3.5	11.8
1958	7.9	3.8	11.7
1959	6.8	6.2	13.0
1960	4.3	9.5	13.8
1961	6.1	9.1	15.2
1962	6.2	8.3	14.5
1963	9.7	11.3	21.0
1964	8.9	17.6	26.5
1965	8.3	16.7	25.0
1966	9.7	17.2	26.9
1967	7.6	26.5	34.1
1968	6.9	23.5	30.4
1969	14.5	24.3	38.8
1970	6.0	16.3	22.3
1971	4.5	7.4	11.9
1972	6.8	8.8	15.6
1973	7.4	14.5	21.9
1974	5.5	13.3	18.8
1975	9.0	10.1	19.1
1976	0.9	4.3	5.2
1977	3.6	9.3	12.9
1978	5.3	10.7	16.0
1979	5.5	7.4	12.9
1980	4.0	13.4	17.4
1981	7.1	15.5	22.6
1982	7.2	15.9	23.1
1983	9.6	16.6	26.2
1984	8.5	17.4	25.9
1985	10.9	20.4	31.3
1986	7.6	14.6	22.2
1987	6.0	14.1	20.1
1988	8.5	17.7	26.2
1989	10.2	21.5	31.7
1990	8.2	15.5	23.7
1991	7.8	15.6	23.4
1992	—	—	17.1

*MMFB = Millions of Board Feet

Note: In fiscal year 1992, JNF pulpwood accounted for 50 percent of wood sales; 33 percent went for sawtimber, and the remaining 17 percent went for firewood. During that year, 2,100 acres were clear-cut, 800 were thinned, and 300 experienced group selection.

Source: Hank Sloan and Karen Goode, "Timber Sale Program Information Reporting System Report (TSPIRS), Fiscal Year 1992."

Table 5. Cut Method for GW/JNF's Combined and Volume Harvested

Fiscal year	Clear-cut (acres)	Thinning	Group selection	Shelter-wood	Salvage	Total acreage	Total volume (MMFB*)
1988	4,703	889	37	83	—	5,712	69.2
1989	3,976	289	173	123	—	4,561	62.9
1990	4,496	454	223	149	—	5,322	62.5
1991	4,465	971	553	288	—	6,277	69.4
1992	2,606	738	1,627	754	—	6,105	57.3
1993	2,013	649	1,250	1,241	—	5,668	60.6
1994	1,451	725	681	834	—	4,723	57.3
1995	723	844	194	1,623	1,038	4,422	55.7
1996	405	372	207	1,253	945	3,182	37.6

Table 6. Suitable Timber Base

	George Washington National Forest	Jefferson National Forest
Total national forest acreage	1,061,000	690,254
1997 suitable land base for timber harvesting	350,000	339, 600

Note: 1990s statistics supplied by Jim Sitton, Timber Staff Officer for George Washington—Jefferson National Forest, July 30, 1997.

Table 7. Pre-Merger Budget Figures

Fiscal year (FY 1993 missing)	1990	1991	1992	1994	1995
General purpose	332,500	635,000	845,200	70,729	86,400
Recreation facility construction	169,000	461,000	230,000	662,000	75,000
Recreation road construction	213,000	622,900	503,900	378,390	565,600
Timber road construction	369,600	324,500	380,000	136,365	190,700
Trail construction	29,000	362,000	180,000	244,000	225,000
Cooperative work / other	121,495	124,444	207,864	273,559	228,835
Cooperative Work / Knutson-Vandenburg funds	326,655	361,751	387,529	567,473	485,822
Job Corps capital	75,452	80,469	84,612	125,440	—
Center operations / VST (vehicle service table?)	2 mil. + 763,787	2 mil. + 962,660	3 mil. + 247,733	3 mil. + 630,214	—
Job Corps program direction	73,800	69,156	60,000	74,666	—
		Construction / rehabilitation 115,885	Construction / rehabilitation 41,033	Construction / rehabilitation 126,347	
Forest fire protection	593,098	634,436	584,097	432,559	391,200
					emergency suppression and rehabilitation 400,000

Fiscal year (FY 1993 missing)	1990	1991	1992	1994	1995
Federal highway advertising	5,000	4,500	—	11,100	9,000
Hazardous waste program	13,000	11,000	0	9,500	5,500
LWCF land acquisition	199,213	256,089	311,000	304,235	284,754
Cooperative law enforcement	75,000	101,000	75,000	70,000	127,994
Facilities maintenance	67,267	101,762	185,000	248,552	253,406
General administration	1 mil. + 342,964	1 mil. + 471,927	1 mil. + 566,267	1 mil. + 606,896	1 mil. + 702,513
Lands	78,746	106,089	127,124	120,509	Real estate management 161,370
Land line location	287,287	251,781	330,079	452,000	170,316
Common variety	14,204	10,994	—	—	—
Leasable	101,784	110,498	—	—	—
Geology	14,003	8,296	131,652	111,177	165,000
Locatable	1,022	1,360	—	—	—
Road maintenance	509,300	655,338	584,527	417,594	561,400
Range	80,719	155,386	29,277	32,835	25,306
			Vegetation management 77,032	Vegetation management 84,881	Vegetation management 6,825
Reforestation	260,800	195,100	9,783	70,700	—
Recreation	1 mil. + 390,000	2 mil. + 808,000	1 mil. + 737,700	1 mil. + 377,299	1 mil. + 123,320

Fiscal year (FY 1993 missing)	1990	1991	1992	1994	1995
Recreation (cont.)			cultural resource management 84,000	cultural resource management 245,000	Heritage resources 140,000
Watershed	313,418	418,078	Soil, water, air operation 317,77	Soil, water, air operation 400,042	Soil, water, air operation 78,554
			Soil and water resource improvements 21,882	Soil and water resource improvements 21,120	Watershed improvements 31,939
			Soil inventories 115,050	Soil inventories 80,840	
Timber	1 mil. + 079,648	1 mil. + 186,304	Sales preparation 575,000	Sales preparation 288,627	Sales management 1 mil. + 040,000
			Harvest advertising 204,000	Harvest advertising 100,536	

Fiscal year (FY 1993 missing)

	1990	1991	1992	1994	1995
Timber (cont.)					Forestland vegetation management 145,000
			Silvicultural examination 128,000	Silvicultural examination 166,145	
			timber resource inventory planning 126,000	timber resource inventory planning 165,384	
			TSI 61,000	TSI 156,500	
			nursery and genetic improvement 24,000	nursery and genetic improvement 27,000	
Trails	275,000	251,000	326,000	192,000	—
Wildlife and fish	373,863	452,253	Wildlife habitat 182,769	Wildlife habitat 286,725	Wildlife habitat 174,145
			Inland fish 135,661	Inland fish 152,896	Inland fish 84,820

Fiscal year (FY 1993 missing)	1990	1991	1992	1994	1995
Timber purchaser road construction	219,100	152,300	255,100	137,000	191,538
Office and Management Forest Service Quarters	15,000	—	12,000	30,000	20,000
Reforestation Trust Fund	225,402	233,100	261,217	—	123,000
SCSEP	255,537	263,244	279,744	306,828	328,106
Timber salvage sales	229,000	0	0	233,000	98,000
Ecosystem planning, inventory, monitoring	0	0	0	0	1 mil. + 20,000
Noxious weed control	0	0	2,000	6,000	0
Threatened and endangered sensitive species	0	0	156,507	226,610	249,843
Wilderness management	0	0	239,000	161,595	282,000
Resource conservation and development	0	0	0	5,000	5,000
Economic recovery / economic action program	0	0	0	Economic recovery 5,000	Economic action program 141,938
Total	12 mil. + 494,164	15 mil. + 959,600	15 mil. + 464,611	15 mil. + 329,588	11 mil. + 401,144
Misc.		0	National Forest Service drug control 35,000	Student pay 296,020	Gifts 2,000
			Undefined code 6,500		

Table 8. George Washington and JNF Budget Data

Fiscal year	1996	1997
General purpose road construction	15,739	0
Recreational road construction	464,995	332,859
Timber road construction	288,640	288,902
Trail construction	267,000	314,000
Cooperative work / other	132,377	60,000
Cooperative work / Knutson-Vandenburg funds	986,512	1,241,500
Fire presuppression and fuels	1,467,048	1,374,016
Federal highway administration	12,800	13,720
Hazardous waste program	8,500	8,500
LWCF land acquisition	303,615	125,190
Law enforcement	233,078	226,234
Facilities maintenance	266,019	289,176
General administration	2,624,400	2,184,400
Minerals and geology	180,000	145,000
Lands (real estate management)	380,260	356,470
Land line location	321,648	333,133
Road maintenance	1,080,124	1,088,062
Range	Grazing management 113,335	Grazing management 88,914
	Range vegetation management 40,082	Range vegetation management 189,393
Recreation	2,850,000	2,749,000
	Heritage resources 339,000	Heritage resources 269,000
		Recreation facility construction 400,000
		Recreation fee collections 60,000

Fiscal year	1996	1997
Watershed	Watershed improvements 161,047	Watershed improvements 187,802
	Soil, water, air operations 110,092	Soil, water, air operations 110,350
Timber sales management	1,684,000	1,826,000
Wildlife and fish	Wildlife habitat management 303,173	Wildlife habitat management 353,235
	Inland fish habitat management 222,022	Inland fish habitat management 186,597
	Anadromous fish habitat management 11,500	Anadromous fish habitat management 42,060
Timber purchaser road construction	300,000	264,522
Office and Management Forest Service Quarters	2,500	12,000
Reforestation Trust Fund	232,000	289,000
Senior Community Service Employment Program (DOL)	821,664	858,186
Timber salvage sales	400,000	500,000
Ecosystem planning, inventory, monitoring	1,629,708	1,639,801
T&E species habitat management	260,000	238,796
Wilderness management	357,000	362,000
Forestland vegetation management	245,000	241,000
Research Conservation and Development	4,500	gifts 2,038
Economic action program	6,500	5,000
Total	19,125,878	19,255,851

Note: Budget data supplied to the author by Jim Loesel of the Citizens Task Force.

Table 9. JNF Wilderness Areas, Established 1975–1988

Wilderness area	Acreage	Ranger district
Beartown	6,375	Wythe (now 5,609 acres)
Kimberling Creek	5,580	Wythe
Lewis Fork	5,730	Mount Rogers NRA (now 5,926 acres)
Little Dry Run	3,400	Mount Rogers NRA (now 2,858 acres)
Little Wilson Creek	3,855	Mount Rogers NRA (now 5,458 acres)
Mountain Lake	10,753	Blacksburg (now 16,511 acres)
Peters Mountain	3,326	Blacksburg (now 4,531 acres)
Thunder Ridge	2,450	Glenwood (now 2,344 acres)
James River Face	9,000	Glenwood (now 8,886 acres)
Barbers Creek	5,700	New Castle (now 5,382 acres)
Shawvers Run	3,665	New Castle (now 5,686 acres)

Table 10. Additional JNF Wilderness Areas, Established 2009

Wilderness area	Acreage	Ranger district
Brush Mountain East	3,743	Eastern Divide (old New Castle)
Brush Mountain	4,794	Eastern Divide (old Blacksburg)
Garden Mountain	3,291	Eastern Divide (old Wythe)
Hunting Camp Creek	8,470	Eastern Divide (old Wythe)
Raccoon Branch	4,223	Mount Rogers NRA
Stone Mountain	3,270	Clinch

Note: Total wilderness acreage 96,562—or, about 13.7 percent of the maximum premerger JNF acreage. Original acreages derived from JNF Wilderness files. Updated acreage and new wilderness areas derived from Omnibus Public Land Management Act of 2009. For descriptions of these and all other national wilderness areas, also see http://www.wilderness.net.

Notes

Preface

1. See Will Sarvis, "A Difficult Legacy: Creation of the Ozark National Scenic Riverways," *Public Historian* 24 (Winter 2002): 31–52; and Will Sarvis, "Old Eminent Domain and New Scenic Easements: Land Acquisition for the Ozark National Scenic Riverways," *Western Legal History* 13 (Winter/Spring 2000): 1–37.

2. A magnificent exception may be found in Richard West Sellars, *Preserving Nature in the National Parks* (New Haven: Yale Univ. Press, 1997).

Introduction

1. To no surprise, observers have reached no consensus as to what geography even constitutes southern Appalachia. There is much agreement about differentiation of subregions, such as the Cumberland Plateau or the Ridge and Valley area that comprises much of the JNF. At the very least a major distinction lies between any coal- or non-coal-bearing region of Appalachia, but beyond that scholars and government bureaucrats hold wide and divergent views of what lands constitute Appalachia or even southern Appalachia. For a sampling of maps and literature regarding what constitutes geographical Appalachia, see Anthony P. Cavender, *Folk Medicine in Southern Appalachia* (Chapel Hill: Univ. of North Carolina Press, 2003), 6–7; Wilma A. Dunaway, *The First American Frontier: Transition to Capitalism in Southern Appalachia, 1700–1860* (Chapel Hill: Univ. of North Carolina Press, 1996), 2, 26, 110, 144, 310; Paul Salstrom, "Newer Appalachia as One of America's Last Frontiers," in *Appalachia in the Making: The Mountain South in the Nineteenth Century,* ed. Mary Beth Pudup, Dwight B. Billings, and Altina L. Waller (Chapel Hill: Univ. of North Carolina Press, 1995), 76–102; John B. Rehder, *Appalachian Folkways* (Baltimore: Johns Hopkins Univ. Press, 2004), 3–15; Ann DeWitt Watts, "Does the Appalachian Regional Commission Really Represent a Region?" in *Appalachia: Social Context Past and Present,* ed. Bruce Ergood and Bruce E. Kuhre, 2nd ed. (Dubuque, IA: Kendall Hunt, 1983), 225–33; John Alexander Williams, *Appalachia: A History* (Chapel Hill: Univ. of North Carolina Press, 2002), 9–14.

2. A compilation of the JNF effort at Saltville appeared in USDA Forest Service–Southern Region and Archeological Society of Virginia, "Upland

Archeology in the East," Symposium no. 5, Special Publication No. 38, pt. 5 (Richmond: Archeological Society of Virginia, 1996). In the past the Smithsonian Institution has sponsored some of the paleontology work of Jerry McDonald in the Saltville area (see *North American Bison: Their Classification and Evolution* [Berkeley: Univ. of California Press, 1981]). Thomas Mays argues that a mass grave of black Civil War soldiers lies buried in the Saltville vicinity. See Thomas Davidson Mays, "The Price of Freedom: The Battle of Saltville and the Massacre of the Fifth United States Colored Cavalry" (master's thesis, Virginia Polytechnic Institute and State Univ., 1992).

3. Preservations Technologies employed the author on a few occasions on sites in the Roanoke Valley and near Altavista. An example of a cost-sharing study, this one with Western Carolina University, is Anne Frazer Rogers, ed., *The Jaybird Branch Project: Report of Investigations* (Cullowhee: Western Carolina Univ., 1982), on file with JNF Cultural Resources Division.

4. See Henry Glass, "The Appalachian Log Cabin," *Mountain Life and Work* 39 (Winter 1963): 5–14; Henry Glass, "The Types of the Southern Mountain Cabin," in *The Study of American Folklore,* 3rd ed., ed. Jan H. Brunwand (New York: Norton & Co., 1986), 529–62.

5. On the topic of highland farming in this region, generally see Phil Gersmehl, "Factors Leading to Mountaintop Grazing in the Southern Appalachians," *Southeastern Geographer* 10 (Apr. 1970): 67–72; Sara M. Gregg, "Uncovering the Subsistence Economy in the Twentieth-Century South: Blue Ridge Mountain Farms," *Agricultural History* 78 (Autumn 2004): 417–37; John F. Hart, "Land Rotation in Appalachia," *Geographical Review* 67 (Apr. 1977): 148–66; John S. Otto, "Forest Fallowing among the Appalachian Mountain Folk: An Ethnohistorical Study," *Anthropologica* 30, no. 1 (1988): 3–22; and John S. Otto, "Forest Fallowing in the Southern Appalachian Mountains: A Problem in Comparative Agricultural History," *Proceedings of the American Philosophical Society* 133 (Mar. 1989): 51–63.

6. See, for example, William G. Robbins, *American Forestry: A History of National, State, and Private Cooperation* (Lincoln: Univ. of Nebraska Press, 1985) chap. 2; William G. Robbins, *Lumberjacks and Legislators: Political Economy of the U.S. Lumber Industry, 1890–1941* (College Station: Texas A&M Univ. Press, 1982), 246–47; Albert C. Worrell, *Principles of Forest Policy* (New York: McGraw-Hill, 1970), 28.

7. For one formulation, analysis, and very interesting assessment to this complexity, see Clayton R. Koppes, "Efficiency, Equity, Esthetics: Themes in American Conservation," in *The Ends of the Earth: Perspectives on Modern Environmental History,* ed. Donald Worster (New York: Cambridge Univ. Press, 1988), 230–51.

8. The "contact period" for eastern Virginia begins in 1607 with the establishment of the Jamestown colony. For other areas (such as the interior of Virginia) contact remained contingent upon the specific area and specific peoples involved, plus varying degrees and consequences of direct or indirect exposures.

9. Michael Frome, *The Forest Service,* 2nd ed. (Boulder, CO: Westview Press, 1983), 237; Robin R. Gottfried, "Observations on Recreation-Led Growth in Appalachia," *American Economist* 21 (Spring 1977): 44; George L. Hicks, *Appalachian Valley* (New York: Holt, Rinehart & Winston, 1976), 52–53; Shelley Smith Mastran and Nan Lowerre, *Mountaineers and Rangers* (Washington, DC: GPO, 1983), 164–66. New Castle Ranger Bob Boardwine remembered an incident on his district that reflected this scenario. A local individual complained at one point over too much Forest Service road construction, which that person felt needlessly damaged the environment. Not long afterward, however, when outsiders moved in and prevented this individual from crossing their land to reach national forest hunting grounds, that same person approached Boardwine asking for more Forest Service road construction (Bob Boardwine, pers. comm. to author, Jan. 28, 1992).

10. William E. Shands and Robert G. Healy, *The Lands Nobody Wanted* (Washington, DC: Conservation Foundation, 1977), 80.

11. For a description of this ongoing land acquisition process, see the *New River Current* (of the *Roanoke Times and World News*), July 13, 1990. Former JNF forester Art Hadacek recalled that during the 1960s–1980s many Forest Service personnel from western national forests expressed amazement at the JNF's complexity. Not only did the JNF handle complicated land acquisitions, juggle numerous extractive resource permits (involving timber, coal, iron, and natural gas), but also met the intensive recreational needs—hunting, fishing, hiking, and camping—of both local people and throngs of eastern urbanites. See Art Hadacek interview, Dec. 16, 1991.

12. For an earlier episode, see the *Galax Gazette,* Jan. 8, 1979, and the *Scott County Herald Virginian,* Apr. 14, 1982. For a later episode, see *Roanoke Times and World News,* Mar. 17, 1991, June 25, 1991.

13. Harold K. Steen, *The U.S. Forest Service: A History* (Seattle: Univ. of Washington Press, 1976), 238–44.

14. Timber acreage statistics supplied to the author by JNF timber staff officer Jim Sitton, July 30, 1997. See Appendix C for more detailed information.

1. Prehistoric Southwestern Virginia

1. James B. Griffin, "Eastern North American Archaeology," *Science* 156 (Apr. 14, 1967): 176; J. Sanderson Stevens, "A Story of Plants, Fire, and People: The Paleoecology and Subsistence of the Late Archaic and Early Woodland in Virginia," in *Late Archaic and Early Woodland Research in Virginia: A Synthesis,* ed. Theodore R. Reinhart and Mary Ellen N. Hodges (Richmond: Archeological Society of Virginia, 1991), 188–89.

2. See Edward V. McMichael, "Environment and Culture in West Virginia," *Proceedings of West Virginia Academy of Science* 33 (1961): 146–50; Burton L. Purrington, "Ancient Mountaineers: An Overview of the Prehistoric Archaeology of North Carolina's Western Mountain Region," in *The Prehistory of North Carolina,* ed. Marck A. Mathis and Jeffrey J. Crow (Raleigh: North Carolina Division of Archives and History, 1983), 132–35.

3. Not to be confused with modern horticultural activities. Here I adopt the usage preferred by many archaeologists to distinguish comparatively less-intense aboriginal planting practices (sometimes generalized as "slash and burn") with later Euro and Afro-American practices, which included actual plowing and annual cultivation of the same tracts. For a usage example, see Carl O. Sauer, *Sixteenth Century North America: The Land and the People as Seen by the Europeans* (Berkeley: Univ. of California Press, 1971), 286–87.

4. Keith T. Egloff, "The Late Woodland Period in Southwestern Virginia," in *Middle and Late Woodland Research in Virginia: A Synthesis,* ed. Theodore R. Reinhart and Mary Ellen N. Hodges (Richmond: Archeological Society of Virginia, 1992), 215; Helen C. Rountree, "Powhatans and Other Woodland Indians as Travelers," in *Powhatan Foreign Relations, 1500–1772,* ed. Helen C. Rountree (Charlottesville: Univ. Press of Virginia, 1993), 30, 33; Helen Horbeck Tanner, "The Land and Water Communication Systems of the Southeastern Indians," in *Powhatan's Mantle: Indians in the Colonial Southeast,* ed. Peter H. Wood, Gregory A. Waselkov, and M. Thomas Hatley (Lincoln: Univ. of Nebraska Press, 1989), 10, 16.

5. See Maureen S. Meyers, "The Mississippian Frontier in Southwestern Virginia," *Southeastern Archaeology* 21 (Winter 2002): 178–92.

6. Ronald J. Mason, "The Paleo-Indian Tradition in Eastern North America," *Current Anthropology* 3 (June 1962): 239, 253. Also see J. Mark Witthofski and Theodore R. Reinhart, eds., *Paleoindian Research in Virginia: A Synthesis,* 2nd ed. (Courtland, VA: Archeological Society of Virginia, 1994).

7. Don W. Dragoo, "Some Aspects of Eastern North American Prehistory: A Review 1975," *American Antiquity* 41 (Jan. 1976): 6, map; R. C. Dunnell, "Prehistory of Fishtrap, Kentucky," *Yale University Publications in Anthropology,* no. 75 (New Haven: Yale Univ., 1972), 73; R. Barry Lewis, ed., *Kentucky Archaeology* (Lexington: Univ. Press of Kentucky, 1996), 22, 35; Purrington, "Ancient Mountaineers," 107–8.

8. By 1988 archaeologists had documented an interesting distribution of fluted Paleo-Indian points in and around Rich Valley in what is now Washington, Smyth, and Tazewell counties. See E. Randolph Turner, "PaleoIndian Settlement Patterns and Population Distribution in Virginia," in *PaleoIndian Research in Virginia: A Synthesis,* Special Publication No. 19 of the Archeological Society of Virginia, ed. J. Mark Wittkofski and Theodore R. Reinhart (Richmond: Archaeological Society of Virginia, 1989), 80.

9. Robert J. Wenke, *Patterns in Prehistory: Humankind's First Three Million Years,* 3rd ed. (New York: Oxford Univ. Press, 1990), 218–19. Also see P. S. Martin and H. E. Wright Jr., eds., *Pleistocene Extinctions: The Search for a Cause* (New Haven: Yale Univ. Press, 1967).

10. The presence or absence of bison in southwestern Virginia has been a matter of some dispute and involves several disciplines. A sophisticated paleontological analysis of bison in southwestern Virginia may be found in Jerry N. McDonald, *North American Bison: Their Classification and Evolution* (Berkeley:

Univ. of California Press, 1981), 104, 251–56. A summary and sometimes dismissal of historic sources claiming eye witness accounts of bison or buffalo in southwestern Virginia may be found in Frank G. Roe, *The North American Buffalo: A Critical Study of the Species in Its Wild State*, 2nd ed. (Toronto: Univ. of Toronto Press, 1970), 228, 245–47. Also see Samuel Cole Williams, *Adair's History of the American Indians* (London, 1755; New York: Argonaut Press, 1966), 27; Joel A. Allen, *The American Bisons: Living and Extinct* (Cambridge: Cambridge Univ. Press, 1876; New York: Arno Press, 1974), 85–87, 92, 225; Rountree, "The Powhatans and Other Woodland Indians as Travelers," 45. For examples of primary reports of buffalo in southwestern Virginia, see Robert G. Albion, ed., *Philip Vickers Fithian: Journal, 1775–1776: Written on the Virginia-Pennsylvania Frontier and in the Army around New York* (Princeton: Princeton Univ. Press, 1934), 147; Adelaide L. Fries, ed., *Records of the Moravians in North Carolina* (Raleigh: State Department of Archives and History, 1968), 1:50–51; Louis B. Wright, ed., *The Prose Works of William Byrd of Westover: Narratives of a Colonial Virginian* (Cambridge: Harvard Univ. Press, 1966), 402–4.

11. Dragoo, "Some Aspects of Eastern North American Prehistory," 11–12; Wenke, *Patterns in Prehistory*, 219.

12. Wenke, *Patterns in Prehistory*, 561.

13. Michael B. Barber, "Human Prehistory beyond the Blue Ridge: A Brief Introduction," Jefferson National Forest Cultural Resources Division, Roanoke, p. 25; Dunnell, "The Prehistory of Fishtrap," 73; Keith Egloff and Deborah Woodward, *First People: The Early Indians of Virginia* (Richmond: Virginia Department of Historic Resources, 1992), 12, 22.

14. Jay F. Custer and Dennis C. Curry, "Prehistoric Settlement-Subsistence Systems in Grayson County, Virginia," *Quarterly Bulletin of the Archeological Society of Virginia* 41 (Sept. 1986): 126. In his study of the Gilbert Site in Tazewell County, Emory Jones wrote, "the high percentage of broken Archaic Period projectile points demonstrates that the site was a convenient, perhaps sought out, stopping place for early hunter-gatherers to rest and to refurbish or replace any damaged or worn-out hunting equipment" (Emory Eugene Jones Jr., "The Gilbert Site, Tazewell County, Virginia," *Quarterly Bulletin of the Archeological Society of Virginia*, 44 (Dec. 1989): 222. Also see Howard A. MacCord Sr., "The Dalton Site, Pulaski County, Virginia: A Report on Phase III (Date Recovery) Excavations," *Quarterly Bulletin of the Archeological Society of Virginia* 39 (Dec. 1984): 216; Howard A. MacCord, "The Flannery Site, Scott County, Virginia," *Quarterly Bulletin of the Archeological Society of Virginia* 34 (Sept. 1979): 29–30.

15. Barber, "Human Prehistory beyond the Blue Ridge," 15; Michael B. Barber, pers. comm. to author, Feb. 24, 1993.

16. Lewis, *Kentucky Archaeology*, 46; Wenke, *Patterns in Prehistory*, 561.

17. Custer and Curry, "Prehistoric Settlement-Subsistence Systems in Grayson County, Virginia," 127.

18. JNF archaeologist Michael B. Barber stresses that the Daughtery's Cave site is the only extensive occupational Archaic site in southwestern Virginia that archaeologists have scientifically examined, and that examination of other sites could significantly modify the current picture. Michael Barber, pers. comm. to author, Feb. 24, 1993. Also see Barber, "Human Prehistory beyond the Blue Ridge," 17; Egloff and Woodward, *First People*, 13.

19. Barber, "Human Prehistory beyond the Blue Ridge," 20, 22; Joseph R. Caldwell, "Eastern North America," in *Prehistoric Agriculture*, ed. Stuart Struever (Garden City, NY: American Museum of Natural History, 1971), 367.

20. For general discussions of Woodland cultural development, see Dragoo, "Some Aspects of Eastern North American Prehistory," 18–19; Griffin, "Eastern North American Archaeology," 175; and especially Charles Hudson, *The Southeastern Indians* (Knoxville: Univ. of Tennessee Press, 1976), 55–66, 77–80, 95, 327.

21. Barber, "Human Prehistory beyond the Blue Ridge," 18; Egloff, "The Late Woodland Period in Southwestern Virginia," 187; Clarence R. Geier, "Development and Diversification: Cultural Directions during the Late Woodland/ Mississippian Period in Eastern North America," in *Middle and Late Woodland Research in Virginia: A Synthesis*, ed. Theodore R. Reinhart and Mary Ellen N. Hodges (Richmond: Archeological Society of Virginia, 1992), 279, 291. For similar developments in neighboring western North Carolina, see Purrington, "Ancient Mountaineers," 136; for eastern Kentucky, see Dunnell, "Prehistory of Fishtrap," 74–75.

22. Joseph L. Benthall, "The Litten Site: A Late Woodland Village Complex, Washington County, Virginia," *Quarterly Bulletin of the Archeological Society of Virginia*, 26 (Sept. 1971): 34; Jones, "The Gilbert Site," 222–23; Howard A. MacCord Sr., "The Brown Johnson Site—Bland County, Virginia," *Quarterly Bulletin of the Archeological Society of Virginia*, 25 (June 1971): 268; Howard A. MacCord Sr., "The Sullins Site, Washington County, Virginia," *Quarterly Bulletin of the Archeological Society of Virginia* 36 (Dec. 1981): 120.

23. William T. Buchanan Jr., "The Hall Site, Montgomery County, Virginia," *Quarterly Bulletin of the Archeological Society of Virginia* 35 (Dec. 1980): 97–99. Buchanan surmises that the Hall Site probably reflects a single family unit. The Litten Site in Washington County divulged various types of pottery and projectile point styles resembling traits from other groups, such as the Saponi and Occaneechee of Central Virginia, and peoples of the eastern Tennessee Mississippian (Dallas) culture. See Benthall, "The Litten Site," 33. The Crab Orchard site in Tazewell County also reflected trade with southern and distant northern peoples with copper and marine shell bead artifacts. See Keith Egloff and Celia Reed, "Crab Orchard Site: A Late Woodland Palisaded Village," *Quarterly Bulletin of the Archeological Society of Virginia*, 34 (Mar. 1980): 147.

24. Dunnell, "Prehistory of Fishtrap," 74–75. Wenke recognizes this kind of ceramic development as part of a cultural phenomenon occurring at various

times with various peoples throughout the world. See Wenke, *Patterns in Prehistory*, 563.

25. Barber, "Human Prehistory beyond the Blue Ridge," 18; Egloff and Woodward, *First Peoples*, 23; Lewis, *Kentucky Archaeology*, 81; Douglas C. McLearen, "Late Archaic and Early Woodland Material Culture in Virginia," in *Later Archaic and Early Woodland Research in Virginia: A Synthesis*, ed. Theodore R. Reinhart and Mary Ellen Hodges (Richmond: Archeological Society of Virginia, 1991), 114, 125.

26. Caldwell, "Eastern North America," 368; Keith T. Egloff, *Ceramic Study of Woodland Occupation along the Clinch and Powell Rivers in Southwestern Virginia*, Research Report Series no. 3 (Richmond, VA: Department of Conservation and Historic Resources, Division of Historic Landmarks, 1987).

27. Sebert L. Sisson, "Pot Rock Cliff Shelter, Carroll County, Virginia," *Quarterly Bulletin of the Archeological Society of Virginia*, 34 (Sept. 1979): 56. Bott observed that shell-tempered ceramics predominated at the Hansonville Site in Russell County and noted "the importance of the temporal and regional relationships between shell and limestone tempered ceramics." See Keith Edward Bott, *44RU7: Archaeological Test Excavations at a Late Woodland Village in the Lower Uplands of Southwestern Virginia* (Richmond: Virginia Division of Historic Landmarks, 1981), 12. For analysis of ceramics and their possible indications for a New River site, see William T. Buchanan, *The Trigg Site, City of Radford, Virginia* (Richmond: Archeological Society of Virginia, 1984).

28. Egloff, *Ceramic Study*, 6–8, 48, 49.

29. MacCord, "Brown Johnson Site," 264.

30. MacCord, "The Flannery Site," 27.

31. Dragoo, "Some Aspects of Eastern North American Prehistory," 18; Douglas C. McLearen, "Virginia's Middle Woodland Period: A Regional Perspective," in *Middle and Late Woodland Research in Virginia: A Synthesis*, ed. Theodore R. Reinhart and Mary Ellen N. Hodges (Richmond: Archeological Society of Virginia, 1992), 56; Purrington, "Ancient Mountaineers," 139; Wenke, *Patterns in Prehistory*, 565–67, 569.

32. Dennis Blanton, "Middle Woodland Settlement Systems in Virginia," in *Middle and Late Woodland Research in Virginia: A Synthesis*, ed. Theodore R. Reinhart and Mary Ellen N. Hodges (Richmond: Archeological Society of Virginia, 1992), 68, 69.

33. Blanton, "Middle Woodland Settlement Systems in Virginia," 75, 77, 81, 82; McLearen, "Virginia's Middle Woodland Period," 53–55.

34. Barber, "Human Prehistory beyond the Blue Ridge," 19, 20; Lewis, *Kentucky Archaeology*, 117. Also see Caldwell, "Eastern North America," 368.

35. Egloff and Woodward, *First People*, 29–30. Also see Egloff and Reed, "Crab Orchard," 146–47. The Crab Orchard village complex epitomizes prehistoric horticulture in southwestern Virginia. The Indians there were possibly influenced by the nearby Saltville Valley, an important Indian hunting ground

and a place where Indians possibly made salt to exchange with other groups (Michael B. Barber, pers. comm. to author, Feb. 24, 1993).

36. Egloff and Reed, "Crab Orchard Site," 146–47.

37. Egloff and Woodward, *First People,* 25, 27. Similar Mississippian evidence appears in adjacent areas of Kentucky during this period. See Lewis, *Kentucky Archaeology,* 86.

38. Caldwell, "Eastern North America," 361; Egloff, "Late Woodland Period in Southwestern Virginia," 213.

39. Dragoo, "Some Aspects of Eastern North American Prehistory," 20–21; Geier, "Development and Diversification," 279, 281.

40. Barber, "Human Prehistory beyond the Blue Ridge," 22–23; Joseph L. Benthall, *Archeological Investigation of the Shannon Site* (Richmond: Virginia State Library, 1969), 145–48; Roy S. Dickens Jr., *Cherokee Prehistory: The Pisgah Phase in the Appalachian Summit Region* (Knoxville: Univ. of Tennessee Press, 1976), 14, 172–88, 191–92, 201, 206, 210–14; Dunnell, "Prehistory of Fishtrap," 76; Egloff, *Ceramic Study,* 3; Lewis, *Kentucky Archaeology,* 150, 177; MacCord, "Flannery Site," 30; Jacquelyn G. Piper, "An Interpretation of Mount Rogers National Recreation Area" (master's thesis, Univ. of South Florida, 1977), 147; Purrington, "Ancient Mountaineers," 144–45; Ralph S. Solecki, "An Archeological Survey of Two River Basins in West Virginia," *West Virginia History* 10 (July 1949): 319–432.

41. Egloff and Woodward, *First People,* 32. For the actual report see Lucien Carr, "Report on the Exploration of a Mound in Lee County, Virginia," in the *Tenth Annual Report of the Peabody Museum* (Cambridge: Salem Press, 1877), 75–94.

42. Carr, "Report on the Exploration of a Mound," 79–83; Egloff and Woodward, *First People,* 32.

43. The rock shelter occupations themselves reflect "short-term exploitative camps" used during hunting and gathering activities. Many of these rock shelters are located on the Jefferson National Forest's Clinch Ranger District and easily comprise, to date, the Forest's most significant archaeological sites. In 1981 Anne Frazer Rogers and her field crew from Western Carolina University studied eight of these rock shelters in Wise County. See Rogers, "The Jaybird Branch Project: Report of Investigations," 5, 6, 39. Neighboring Kentucky rock shelters may have been occupied in similar sporadic fashion or year round, the latter possibly coinciding with abandonment of area bottomlands and rise of hillside horticulture. See Lewis, *Kentucky Archaeology,* 86, 110.

44. Barber, "Human Prehistory beyond the Blue Ridge," 22–23; Michael B. Barber, pers. comm., Feb. 24, 1993; Jeffrey L. Hantman, "Between Powhatan and Quirank: Reconstructing Monacan Culture and History in the Context of Jamestown," *American Anthropologist* 92 (Sept. 1990): 684; Helen C. Rountree, "Summary and Implications," in Rountree, *Powhatan Foreign Relations,* 216–17.

45. Jon Muller, "The Southern Cult," in *The Southeastern Ceremonial Complex: Artifacts and Analysis,* ed. Patricia Galloway (Lincoln: Univ. of Nebraska Press, 1989), 11–26.

46. Former JNF archeologist Mary Louise Arend, pers. comm. to author, June 10, 1992. And, as Douglas Mclearen noted, where quartzite dominated Savannah River points in general, Indians living in southwestern Virginia also used locally available rhyolite and limestone or chert in addition to quartzite. See Mclearen, "Virginia's Middle Woodland Period," 95, 97, 98.

47. Blanton, "Middle Woodland Settlement Systems in Virginia," 75, 77; Helen C. Rountree, *The Powhatan Indians of Virginia* (Norman: Univ. of Oklahoma Press, 1989), 32, 71, 120.

48. Barber, "Human Prehistory beyond the Blue Ridge," 26–27; Michael B. Barber interview, June 3, 1992; Michael Barber, "Continued Archaeological Reconnaissance of the Coeburn Exchange, Wise County, Virginia" (Roanoke, VA: Jefferson National Forest, 1985), 7, 8, 50, 57; Bott, *44RU7,* 37. Also see Geier, "Development and Diversification," 290–91.

49. See Shepard Krech III, *The Ecological Indian: Myth and History* (New York: Norton, 1999).

50. One of the more prominent examples lies in J. Donald Hughes, *American Indian Ecology* (El Paso: Texas Western Press, 1983). Also see Robert F. Berkhofer Jr.'s false dichotomy in "Cultural Pluralism versus Ethnocentrism in the New Indian History," in *The American Indian and the Problem of History,* ed. Calvin Martin (New York: Oxford Univ. Press, 1987), 35–45; Peter Heinegg, "Lessons from the Indians: Ecological Piety," *North American Review* 163 (Spring 1978): 66–69; Calvin Martin, "Fire and Forest Structure in the Aboriginal Eastern Forest," *Indian Historian* 6 (Fall 1973): 38–42, 54; and Chris Vecsey and Robert W. Venables, *American Indian Environments: Ecological Issues in Native American History* (Syracuse: Syracuse Univ. Press, 1980), though the latter focuses more on the postcontact era.

51. One of the more discerning earlier assessments of the topic, as well as a very useful historiographic analysis, lies in J. Baird Callicott, "American Indian Land Wisdom? Sorting out the Issues," *Journal of Forest History* 33 (Jan. 1989): 35–42. Other interesting observations dealing with varied environmental attitudes may be found in Cornelius J. Jaenen, "Thoughts on Early Canadian Contact," and Frederick Turner, "On the Revision of Monuments," both in Martin, *The American Indian and the Problem of History,* 55–56, 116.

52. Indications of this abound; following are a few printed primary sources pertaining to the southern highlands: Williams, *Adair's History of the American Indians,* 27; "The Indians of Virginia . . . . 1689," *William and Mary Quarterly,* 3rd ser., 16 (Apr. 1959): 230–43; Louis B. Wright, ed., *The History and Present State of Virginia by Robert Beverley* (Chapel Hill: Univ. of North Carolina Press, 1947), 202, 210; Stanley Pargellis, "An Acct of the Indians of Virginia," *William and Mary Quarterly,* 3rd ser., 16 (Apr. 1959): 228–29.

53. Robert Heizer addresses this general topic with much wisdom and within a worldwide context. See Robert F. Heizer, "Primitive Man as an Ecological Factor," *Kroeber Anthropological Society Papers* 13 (Fall 1955): 1–31. Also see Michael P. Hoffman, "Prehistoric Ecological Crises," in *Historical Ecology: Essays on Environment and Social Change,* ed. Lester J. Bilsky (Port Washington,

NY: Kennikat Press, 1980), 33–42; Cornelius J. Jaenen, "Thoughts on Early Canadian Contact," in Martin, *The American Indian and the Problem of History,* 55–66.

54. Krech, *Ecological Indian,* 213.

55. The most extensive study on this topic lies in Erhard Rostlund, *Freshwater Fish and Fishing in Native North America* (Berkeley: Univ. of California Press, 1952). For other accounts focusing specifically on Southeastern Indians, see James Mooney, *Myths of the Cherokee and Sacred Formulas of the Cherokee* (Nashville: Charles and Randy Elder Booksellers, 1982), 422; Frank G. Speck, "The Ethnic Position of the Southeastern Algonkian," *American Anthropologist,* n.s. 26 (1924): 191; and Frank G. Speck, *Ethnology of the Yuchi Indians,* Univ. of Pennsylvania Museum of Anthropological Publications No. 1 (Philadelphia: Univ. Museum, 1909–1911), 23–24. Gene Wilhelm speculates that the use of plant piscicides originated in Europe and spread to the Western Hemisphere, but this question (as with so many involving prehistory) remains only hypothetical. See Gene Wilhelm Jr., "The Mullein: Plant Piscicide of the Mountain Folk Culture," *Geographical Review* 64 (Apr. 1974): 246–50. In any case, mountaineers certainly used poisons to stun fish, similar to earlier Indian practices. In addition to Wilhelm, see Rehder, *Appalachian Folkways,* 163; and Roy E. Thomas, comp., *Southern Appalachia, 1885–1915: Oral Histories from the Residents of the State Corner Area of North Carolina, Tennessee and Virginia* (Jefferson, NC: McFarland & Co., 1991), 74.

56. *Adair's History of the American Indians,* 248; Pargellis, "An Acct of the Indians of Virginia," 243; Stephen J. Pyne, *Fire in America: A Cultural History of Wildland and Rural Fire* (Princeton: Princeton Univ. Press, 1982), 74; Sauer, *Sixteenth Century North America,* 285; Stevens, "A Story of Plants, Fire, and People," 209; William L. Thomas Jr., *Man's Role in Changing the Face of the Earth* (Chicago: Univ. of Chicago Press, 1956), 115–33.

57. Clarence W. Alvord and Lee Bidgood, *The First Explorations of the Trans-Allegheny Region by the Virginians, 1650–1674* (Cleveland: Arthur H. Clark Co., 1912), 73; C. G. Holland, "The Ramifications of the Fire Hunt," *Quarterly Bulletin of the Archeological Society of Virginia* 33 (June 1979): 134–40; John Lawson, *New Voyage to Carolina* (Chapel Hill: Univ. of North Carolina Press, 1967), 215–16; Pargellis, "An Acct of the Indians of Virginia," 243; Speck, "The Ethnic Position of the Southeastern Algonkian," 191; Williams, *Adair's History of the American Indians,* 248.

58. For a fascinating essay on Indian concepts of time, space, and metaphysics, see Benjamin Lee Whorf, "An American Indian Model of the Universe," in *Teachings from the American Earth: Indian Religion and Philosophy,* ed. Dennis Tedlock and Barbara Tedlock (New York: Liveright, 1975), 121–29. Another excellent essay offering indications of an Indian worldview is N. Scott Momaday, "Native American Attitudes to the Environment," in *Seeing with a Native Eye,* ed. Walter H. Capps (New York: Harper & Row, 1976), 79–85. Also see Sam D. Gill, *Beyond "The Primitive": The Religions of Nonliterate Peoples* (Englewood Cliffs, NJ: Prentice-Hall, 1982), 6, 16, 18–19, 81.

59. As late as 1929 Robert Mason claimed a persistence of tradition among the Cherokee. See Robert L. Mason, "The Myths of the Cherokees," *American Forests and Forest Life* 35 (1929): 259–62, 300.

60. For a contemporary early eighteenth-century observation of this phenomenon, see Lawson, *A New Voyage to Carolina*, 214. For a comprehensive secondary study, see Gregory A. Waselkov, "Indian Maps of the Colonial Southeast," in Wood et al., *Powhatan's Mantle*, 292–343.

61. Mooney, *Myths of the Cherokee*, 445; Sokyo Ono, *Shinto: The Kami Way* (Rutland, VT: Charles Tuttle Co., 1962), 7–8; Floyd H. Ross, *Shinto: The Way of Japan* (Boston: Beacon Press, 1965), 49; Noel W. Schutz Jr., "The Study of Shawnee Myth in an Ethnographic and Ethnohistorical Perspective" (Ph.D. diss., Indiana Univ., 1975), 101.

62. Williams, *Adair's History of the American Indians*, 137–43; Mooney, *Myths of the Cherokee*, 295–97. An interesting exception to generally restrained hunting practices, which included avoidance of killing young animals, John Lawson encountered Indians in the North Carolina highlands who enjoyed eating fawns as a distinct delicacy. See Lawson, *New Voyage to Carolina*, 58, 182.

63. Calvin Martin, who admirably criticizes much "Indian-white" biased history of the past, rightfully distinguishes between Euro-American and Indian worldviews and accurately emphasizes the mystic component of the native perspective. Martin, however, ends up substituting a "biological perspective" for earlier Eurocentric approaches to aboriginal thinking and thus perpetuates the problem of translation between two culturally distinct worldviews. (See Martin, *American Indian and the Problem of History*, 6–34, esp. 8, 9, 15, 24, 27, 28, 29, 30). In the final analysis all intellectual disciplines possess irrevocable limitations, and probably none of them will ever really approximate the Native American prehistorical perspective.

64. J. Baird Callicott, "Traditional American Indian and Western European Attitudes toward Nature: An Overview," *Environmental Ethics* 4 (Winter 1982): 305; Heizer, "Primitive Man as an Ecological Factor," 4–7; Charles Hudson, "Cherokee Concept of Natural Balance," *Indian Historian* 3, no. 4 (1970): 54; William C. McCleod, "Conservation among Primitive Hunting Peoples," *Scientific Monthly* 43 (Dec. 1936): 562–66; Pargellis, "An Acct of the Indians of Virginia," 240; Ruth E. Suddeth, "The Myths of the Cherokees," *Georgia Review* 10 (Spring 1956): 85. The Shawnee apparently extended such anthropomorphism even further in relation to their "Female Deity." See C. F. Voegelin, "The Shawnee Female Deity," *Yale University Publications in Anthropology*, no. 10 (New Haven: Yale Univ., 1970).

65. Lawson, *A New Voyage to Carolina*, 58.

66. Mooney, *Myths of the Cherokee*, 263–65, 294. The Shawnee apparently held a similar regard for snakes. See Schutz, "Study of Shawnee Myth," 196–97, 201.

67. James Mooney, "The Cherokee River Cult," *Journal of American Folklore* 13 (Jan.–Mar. 1900): 1–10; Williams, *Adair's History of the American Indians*, 239. Also see James Mooney, *Sacred Formulas of the Cherokee* (Washington, DC: Bureau of American Ethnology, 1891), 341, 370, 394.

68. H.W.F. Saggs, *Civilization before Greece and Rome* (New Haven: Yale Univ. Press, 1989), 269–71, 279, 283.

69. Mooney, *Myths of the Cherokee,* 330–33, 337–41, 475n78, 477n81; Mooney, *Sacred Formulas,* 311, 312, 341, 374.

70. Ibid.

71. Mooney, *Myths of the Cherokee,* 479; Mooney, *Sacred Formulas,* 361–62, 366.

72. Neal Salisbury, "American Indians and American History," in Martin, *The American Indian and the Problem of History,* 50.

73. Some insightful observations of this may be found in Salisbury, "American Indians and American History," 46–54, and Clarence J. Glacken, *Traces on the Rhodian Shore: Nature and Culture in Western Thought from Ancient Times to the End of the Eighteenth Century* (Berkeley: Univ. of California Press, 1967), 494. For an interesting example of coexisting African tribal and "state-oriented" societies, see Paul Bohannon, *Africa and Africans* (Garden City, NY: Natural History Press, 1964), 188–205. An introduction to the traditional Chinese cyclical view of human events can be found in Colin A. Ronan, *The Shorter Science and Civilisation in China: An Abridgement of Joseph Needham's Original Text,* vol. 1 (Cambridge: Cambridge Univ. Press, 1978), or see the original multivolume project begun by Joseph Needham, *Science and Civilisation in China* (Cambridge: Cambridge Univ. Press, 1954–).

74. Krech, *Ecological Indian.* Krech draws his overarching conclusions from specific topics such as native impact on buffaloes, deer, and beaver, the Pleistocene extinctions, and aboriginal influence on environmental change in the Arizona desert.

2. Agricultural Settlers

1. Through the admittedly limited medium of colonial and postcolonial writings, the emergence of identifiable Indian groups or "tribes" begins to take shape. Cautious archaeologists stop short of assigning historic-era tribal affiliations to prehistoric cultural patterns. Strictly speaking, as the archaeological record meets the historic record, only speculation may link previously unnamed Indian peoples with what Euro-Americans came to identify as Indian "tribes." See Bennie C. Keel, *Cherokee Archaeology: A Study of the Appalachian Summit* (Knoxville: Univ. of Tennessee Press, 1976), 213.

2. Gary C. Goodwin, *Cherokees in Transition: A Study of Changing Culture and Environment Prior to 1775* (Chicago: Univ. of Chicago Press, 1977), 107; William G. McLoughlin, *Cherokee Renascence in the New Republic* (Princeton: Princeton Univ. Press, 1986), 7, 9, 27.

3. Goodwin, *Cherokees in Transition,* 142.

4. Williams, *Adair's History of the American Indians,* 2n1.

5. James Mooney, *The Siouan Tribes of the East,* Bureau of American Ethnology, Bulletin 22 (Washington, DC: GPO, 1894), 83; Schutz, "The Study of Shawnee Myth," 305, 463–64; John R. Swanton, *The Indians of the Southeastern United States,* Bureau of American Ethnology, Bulletin 137 (Washington, DC: GPO,

1946), 184; John Witthoft and William A. Hunter, "The Seventeenth Century Origins of the Shawnee," *Ethnohistory* 2 (Winter 1955): 53; Peter H. Wood, "The Changing Population of the Colonial South: An Overview by Race and Region, 1685–1790," in Wood et al., *Powhatan's Mantle*, 85–86.

6. The Tutelo were likely the group who aided Thomas Batts and Robert Fallam during their 1671 journey of exploration from eastern Virginia into the mountains. But by 1700 or so the Tutelo seem to have moved south to join the Tuscarora Indians of the North Carolina piedmont. During the next century the Tutelo gradually moved north, until the League of the Iroquois eventually adopted them in 1753. See Edward P. Alexander, *The Journal of John Fontaine: An Irish Huguenot Son in Spain and Virginia 1710–1719* (Charlottesville: Univ. Press of Virginia, 1972), 12–13, 90–109; Benthall, *Archeological Investigation of the Shannon Site*, 146–48; David I. Bushnell Jr., "The Five Monacan Towns in Virginia, 1607," *Smithsonian Miscellaneous Collection* 82, no. 12 (Washington, DC: GPO, 1930), 5; Horatio Hale, "The Tutelo Tribe and Language," *Proceedings of the American Philosophical Society* 21 (Mar. 2, 1883): 2–4, 46–47; Gertrude P. Kurath, "The Tutelo Fourth Night Spirit Release Singing," *Midwest Folklore* 4, no. 2 (1954), 99; Lawson, *A New Voyage to Carolina*, 53; Mooney, *Siouan Tribes of the East*, 38; Swanton, *The Indians of the Southeastern United States*, 200–201, 800. Most researchers have concluded or accepted the Tutelo as a Siouan language people, stemming back to the reports of the early western Virginia explorers Batts, Fallam, and Lederer. This is the conclusion reached by Virginia State Archaeologist Keith Egloff, who also supported this hypothesis with ceramic evidence. See Keith T. Egloff, *Ceramic Study of Woodland Occupation along the Clinch and Powell Rivers in Southwest Virginia*, Research Report Series No. 3 (Richmond, VA: Department of Conservation and Historic Resources, Division of Historic Landmarks, 1987), 3; also see Barber, "Human Prehistory beyond the Blue Ridge," 22–23. An argument of them as an Algonquian people is presented in Carl F. Miller, "Revaluation of the Eastern Siouan Problem with Particular Emphasis on the Virginia Branches— The Occaneechi, the Saponi, and the Tutelo," *Bureau of American Ethnology Bulletin* 164, Anthropological Papers No. 52 (Washington, DC: GPO, 1957).

7. Verner W. Crane, *The Southern Frontier 1670–1732* (1929; Ann Arbor: Univ. of Michigan Press, 1956), 61–62; J. Joseph Bauxar, "Yuchi Ethnoarchaeology, Part I: Some Yuchi Identifications Reconsidered," *Ethnohistory* 4, no. 3 (Summer 1957): 296; Thomas M. N. Lewis and Madeline Kneberg, *Hiwassee Island: An Archaeological Account of Four Tennessee Indian Peoples* (Knoxville: Univ. of Tennessee Press, 1946), 12, 15–16; Marvin T. Smith, "Aboriginal Population Movements in the Early Historic Period of the Interior Southeast," in Wood et al., *Powhatan's Mantle*, 29; Swanton, *The Indians of the Southeastern United States*, 14, 212–15.

8. Ray Allen Billington, *Westward Expansion: A History of the American Frontier*, 4th ed. (New York: Macmillan, 1974), 59; Crane, *The Southern Frontier 1670–1732*, 40, 61; J. Leitch Wright Jr., *The Only Land They Knew: The Tragic Story of the American Indians in the Old South* (New York: Free Press, 1981),

96–97. Alvord and Bidgood, *First Explorations,* remains a standard interpretation regarding the Batts and Fallam journey. See Alvord and Bidgood, *The First Explorations of the Trans-Allegheny Region by the Virginians, 1650–1674.* Also see William M. Darlington, *Christopher Gist's Journals: With Historical, Geographical and Ethnological Notes and Biographies of His Contemporaries* (Pittsburg: J. R. Weldin & Co., 1893), 18–20. For an alternate interpretation of this journey, and for a thorough examination of early Virginia westward travels in general, see Alan V. Briceland, *Westward from Virginia: The Exploration of the Virginia-Carolina Frontier 1650–1710* (Charlottesville: Univ. Press of Virginia, 1987), 8–14.

9. Goodwin, *Cherokees in Transition,* 83, 87, 96.

10. For a good source indicating not only the deerskin trade dynamics, but also Euro-American and Indian relations, see Kathryn E. Holland Braund, *Deerskins and Duffels: The Creek Indians Trade with Anglo-America, 1685–1815* (Lincoln: Univ. of Nebraska Press, 1993). The late seventeenth century witnessed competition between the British and French for monopolizing the Cherokee trade. In this particular contest, the British eventually gained the advantage, though the Cherokee invariably suffered in the long run as internal divisions and split loyalties only further eroded any effectual bargaining they might have exercised as a unified group. The great Yamassee War of 1715–16 is one example of struggle directly associated with the new economy and its weaknesses. The much-noted rivalry between the Cherokee and Shawnee was probably related to this trade before evolving a step further into cross-alliances among the Euro-Americans and their motherland loyalties during the French and Indian War. This latter conflict coincided with the first substantial wave of settlers to establish themselves in southwestern Virginia. See David H. Corkran, *The Creek Frontier 1540–1783* (Norman: Univ. of Oklahoma Press, 1967), 50–51; Crane, *The Southern Frontier 1670–1732,* 162; Fries, *Records of the Moravians in North Carolina,* 1:285, 290, 300, 307, 373, 389; Goodwin, *Cherokees in Transition,* 83, 86–87, 95, 98, 142, 144, 148; Richard L. Haan, "The 'Trade Do's Not Flourish as Formerly': The Ecological Origins of the Yamassee War of 1715," *Ethnohistory* 28 (Fall 1981): 341–58; Keel, *Cherokee Archaeology,* 216; Wood, "The Changing Population of the Colonial South," 85. For studies of the northern North American Indian–Euro-American fur trade, see Calvin Martin, *Keepers of the Game: Indian-Animal Relations and the Fur Trade* (Berkeley: Univ. of California Press, 1978) and Arthur J. Ray, *Indians in the Fur Trade: Their Role as Trappers, Hunters, and Middlemen in the Lands Southwest of the Hudson Bay* (Toronto: Univ. of Toronto Press, 1974).

11. Goodwin, *Cherokees in Transition,* 98, 105, 110, 124, 142; McLoughlin, *Cherokee Renascence,* 16–18; Williams, *Adair's History of the American Indian,* 244.

12. Goodwin, *Cherokees in Transition,* 107, 141; McLoughlin, *Cherokee Renascence,* 19–20, 26.

13. Barber, "Human Prehistory beyond the Blue Ridge," 22–23; Lewis et al., *Hiwassee Island,* 16.

14. Alvord and Bidgood, *First Explorations,* 56; Alan V. Briceland, *Westward from Virginia: The Exploration of the Virginia-Carolina Frontier 1650–1710* (Charlottesville: Univ. Press of Virginia, 1987), 2–3, 171–95.

15. Thomas Perkins Abernethy, *Three Virginia Frontiers* (Baton Rouge: Louisiana State Univ. Press, 1940), 41–42, 57; Kenneth P. Bailey, *The Ohio Company of Virginia and the Westward Movement 1748–1792: A Chapter in the History of the Colonial Frontier* (Glendale, CA: Arthur H. Clark Co., 1939), 17; David E. Johnston, *A History of Middle New River Settlements and Contiguous Territory* (1906; Radford: Commonwealth Press, 1969), 12; F. B. Kegley, *Kegley's Virginia Frontier* (Roanoke: Southwest Virginia Historical Society, 1938), 56–57; Robinson, *The Southern Colonial Frontier, 1607–1763,* 158–60.

16. Naturally, the Cherokee did not recognize the treaty between the colonists and Iroquois signed at Fort Stanwix and continued to attack settlers. Similar to their French and Indian War experience, southwestern Virginia experienced mostly British-instigated Indian harassment during the Revolutionary War. See Archer Butler Hulbert, *Historic Highways of America* (Cleveland: Arthur H. Clark Co., 1903), 6:22; Oren F. Morton, *A History of Monroe County, West Virginia* (1916; Baltimore: Regional Publishing Co., 1980), 43; Robert Douthat Stoner, *A Seed-Bed of the Republic: A Study of the Pioneers in the Upper (Southern) Valley of Virginia* (Roanoke: Roanoke Historical Society, 1962), 103. Raids remained in the Kentucky mountains after diminishing to the western flatlands. See R. Barry Lewis, *Kentucky Archaeology,* (Lexington: Univ. Press of Kentucky, 1996), 189.

17. For an interesting study that includes focus on the Great Valley, see Kenneth W. Noe, *Southwest Virginia's Railroad: Modernization and the Sectional Crisis* (Urbana: Univ. of Illinois Press, 1994).

18. William C. Pendleton, *History of Tazewell County and Southwestern Virginia, 1748–1920* (Richmond, VA: W. C. Hill Printing Co., 1920), 238.

19. Ronald D. Eller, *Miners, Millhands, and Mountaineers: Industrialization of the Appalachian South, 1880–1930* (Knoxville: Univ. of Tennessee Press, 1982), 21; Wilson Goodridge, *Smyth County History and Traditions* (Kingsport, TN: Kingsport Press, 1932), 170–71; Morton, *History of Monroe County, West Virginia,* 257–58.

20. For colonial legal precedents regarding favored and unfavored animals, see David S. Hardin, "Laws of Nature: Wildlife Management Legislation in Colonial Virginia," in *The American Environment: Interpretations of Past Geographies,* ed. Lary M. Dilsaver and Craig E. Colten (Lanham, MD: Rowman & Littlefield, 1992), 137–62.

21. Albert E. Cowdrey, *This Land, This South: An Environmental History* (Lexington: Univ. Press of Kentucky, 1983), 114–15; Joseph Ewan and Nesta Ewan, eds., *John Banister and His Natural History of Virginia 1678–1692* (Urbana: Univ. of Illinois Press, 1970), 43; Goodwin, *Cherokees in Transition,* 132. For primary accounts related to wolf bounties in southwestern Virginia, see William Preston, receipt book 1759–1760, folder 345, Preston Family Papers,

Virginia Tech Special Collections, Blacksburg, VA; William Preston, Mar. 28, 1762, folder 390, Preston Family Papers; *Abingdon Virginian,* Dec. 25, 1868.

22. Alfred W. Crosby, *The Columbian Exchange: Biological and Cultural Consequences of 1492,* new ed. (Westport, CT: Praeger, 2003), 66.

23. Edgar Anderson, *Plants, Man and Life* (Berkeley: Univ. of California Press, 1969), 8, 156; Lyman Carrier and Katharine S. Bort, "The History of Kentucky Bluegrass and White Clover in the United States," *Journal of the American Society of Agronomy* 8 (July–Aug. 1916): 256–66; Eva Crane, *A Book of Honey* (New York: Charles Scribner's Sons, 1980), 103–27; Alfred W. Crosby, *Ecological Imperialism: The Biological Expansion of Europe, 900–1900,* new ed. (New York: Cambridge Univ. Press, 2004), 155–58, 188–90; Everett Oertel, "Bicentennial Bees: Early Records of Honey Bees in the Eastern United States," *American Bee Journal* 116 (five parts, Feb.–June, 1976); Carl O. Sauer, "The Settlement of the Humid East," *Climate and Man, 1941 Yearbook of Agriculture* (Washington, DC: GPO, 1941), 158–66.

24. Charles Rufus Boyd, *Resources of South-West Virginia* (New York: John Wiley & Sons, 1881), 210; Kathleen Bruce, *Virginia Iron Manufacture in the Slave Era* (New York: Century Co., 1931), 24, 137, 276–77; Charles B. Dew, *Iron-Maker to the Confederacy* (New Haven: Yale Univ. Press, 1966), 2–3; Robert A. Rutland, "Men and Iron in the Making of Virginia—Part I (1619–1860)," *Iron Worker* 40 (Summer 1976): 17; Richard H. Shallenberg, "Evolution, Adaptation, and Survival: The Very Slow Death of the American Charcoal Iron Industry," *Annals of Science* 32 (1975): 352; James M. Swank, *History of the Manufacture of Iron in All Ages* (Philadelphia: American Iron & Steel Assoc., 1892), 267–68.

25. Bruce, *Virginia Iron Manufacture,* 119, 119n54, 137, 212, 213, 276; Dew, *Iron-Maker to the Confederacy,* 78, 100; Shallenberg, "Evolution, Adaptation, and Survival," 353; Jeffrey C. Turner, "Cloverdale Furnace: A Century of Iron Manufacture in Botetourt County, Virginia" (master's thesis, Virginia Polytechnic Institute and State Univ., 1984), 65, 66; Thomas L. Watson, *Mineral Resources of Virginia* (Lynchburg, VA: J. P. Bell Co., 1907), 403.

26. Boyd, *Resources of South-West Virginia,* 78, 291; William Jolliffe, *The Buchanan and Clifton Forge Railroad: A Necessity to Virginia: Developing the Coal and Iron of the James and Kanawha River Valleys* (Lynchburg: Bell, Browne & Co., 1875), 18.

27. As early as 1795, a Grayson County, Virginia, ironmaker named John Blair employed at least nine slaves at his ironworks. In December 1864, upper James River basin Tredegar furnaces employed 609 male slaves, 200 of which worked at the Catawba and Glenwood furnaces. See Bruce, *Virginia Iron Manufacture,* 454–57; Dew, *Iron-Maker to the Confederacy,* 78; Ronald L. Lewis, *Coal, Iron, and Slaves: Industrial Slavery in Maryland and Virginia, 1715–1865* (Westport, CT: Greenwood Press, 1979), 26–27, 33; Rutland, "Men and Iron in the Making of Virginia—Part I (1619–1860)," 6, 13; Watson, *Mineral Resources of Virginia,* 452. By the late nineteenth century, amidst the coal

and coke fired iron and steel age, some people began to criticize charcoal iron manufacture for its voracious consumption of trees. See Robert A. Rutland, "Men and Iron in the Making of Virginia—Part II," *Iron Worker* 40 (Autumn 1976): 7.

28. C. B. Hayden, "On the Rock Salt and Salines of the Holston," *American Journal of Science* 44 (1843): 178; Charles Lanman, *Letters from the Alleghany Mountains* (New York, 1848), 158; Thomas L. Preston, *Historical Sketches and Reminiscences of an Octogenarian* (Richmond, VA: B. F. Johnson, 1900), 57–59; Rev. H. Ruffner, "Notes of a Tour from Virginia to Tennessee, in the Months of July and Aug., 1838," *Southern Literary Messenger* 5 (Jan. 1839): 47; Francis Smith to S. D. Ingham, in S. D. Ingham, "Salt Works—United States," House Doc. 55, 21st Cong., 1st sess., 1830; James Sanders deposition, Apr. 6, 1850, *Saltworks Lawsuit*, Emory and Henry College archives, Emory, VA; James McNew deposition, Apr. 8, 1850, *Saltworks Lawsuit*; Jonas S. Kelly deposition, Nov. 1871, *Saltworks Lawsuit*; J. P. Lesley, "Report to the President and Directors of the Virginia and Tennessee Railroad Company, 1854," Norfolk & Western Collection, Virginia Tech Special Collection; "Statement Relating to Plan of Organizing Preston Salt Works under Charter and Cost of Same," Nov. 1, 1853, Campbell-Preston-Floyd Papers, Virginia Historical Society, Richmond; Thomas L. Preston to Anna Maria Preston, Nov. 30, 1846, Dec. 2, 1846, Preston Davis Papers, Univ. of Virginia Alderman Library Archives, Charlottesville; Wyndham Robertson, "Some Notes on the Holstein (Va.) Salt and Gypsum," *Virginias* 3 (Feb. 1882): 21; "Estimate of Business and Returns for the Year," Nov. 1, 1853, Campbell-Preston-Floyd Papers, Virginia Historical Society; Alexander M'Call to C. O. Sanford, 1848, published letter, Norfolk and Western Collection, Virginia Tech Special Collections, Blacksburg; J. P. Hale, *History of the Great Kanawha Valley* (Madison, WI: Brant, Fuller & Co., 1891), 1:225; U.S. Census, Smyth County, VA, 1850. Also see the description of denuded woodlands in the *Lynchburg Daily Virginian,* June 22, 1855.

29. *Abingdon Virginian,* July 31, 1874; Smyth County census, 1870, 1880 (the 1880 figures for Saltville were recorded in the 1890 U.S. Census); Ella Lonn, *Salt as a Factor in the Confederacy* (New York: Walter Neale, 1933), 35; William M. E. Rachal, "Salt the South Could Not Savor," *Virginia Cavalcade* 3 (Autumn 1953): 4–7; C. S. Sargent, "Fuel Consumption in the Virginias and U.S.," *Virginias* 4 (Jan. 1883): 3; Goodridge Wilson, *Smyth County History and Traditions* (Kingsport, TN: Kingsport Press, 1932), 195–96.

3. Turnpikes and Romance in the Mountains

1. Although Virginia is curiously omitted, in general see Kevin E. O'Donnell and Helen Hollingsworth, *Seekers of Scenery: Travel Writing from Southern Appalachia, 1840–1900* (Knoxville: Univ. of Tennessee Press, 2004).

2. Edward Halsey Foster, *The Civilized Wilderness: Backgrounds to American Romantic Literature, 1817–1860* (New York: Free Press, 1975), 8, 21, 22, 46.

3. Ibid., 13–15.

4. The masterful overview of this story in a broader context of Western civilization may be found in Marjorie Hope Nicolson, *Mountain Gloom and Mountain Glory: The Development of the Aesthetics of the Infinite* (New York: Cornell Univ. Press, 1959; Seattle: Univ. of Washington Press, 1997).

5. Thomas Cole, "Essay on American Scenery," *American Monthly Magazine* 7 (Jan. 1836), 1.

6. Ibid., 2; also see 5.

7. Ibid., 3–4; also see 12.

8. Ibid., 5–11.

9. Ibid., 12. The landscape art aspect of this phenomenon is also documented in Charles Lanman, "Our Landscape Painters," *Southern Literary Messenger* 16 (May 1850): 272–80; Karol Ann Peard Lawson, "An Inexhaustible Abundance: The National Landscape Depicted in American Magazines, 1780–1820," *Journal of the Early Republic* 12 (Fall 1992): 303–30; Kenneth Myers, *The Catskills: Painters, Writers, and Tourists in the Mountains 1820–1895* (Yonkers: Hudson River Museum of Westchester, 1987); and Barbara Novak, *Nature and Culture: American Landscape Painting 1825–1875* (New York: Oxford Univ. Press, 1980).

10. The great work on historic and evolving American mentalities toward the environment remains Roderick Nash, *Wilderness and the American Mind,* 4th ed. (New Haven: Yale Univ. Press, 2001). Also see Alfred Kazin, *A Writer's America: Landscape in Literature* (New York: Alfred A. Knopf, 1988), chap. 2; David Morse, *American Romanticism* (Totowa, NJ: Barnes & Noble Books, 1987), 1:3; and Donald Worster, *Nature's Economy: A History of Ecological Ideas* (New York: Cambridge Univ. Press, 1977), 81–89. For general references see Foster, *The Civilized Wilderness,* and John F. Sears, *Sacred Places: American Tourist Attractions in the Nineteenth Century* (New York: Oxford Univ. Press, 1989).

11. Sources on early American tourism and tourist attractions include Alf Evers, *The Catskills: From Wilderness to Woodstock* (Garden City: Doubleday & Co., 1972), 387–88; Kenneth A. Johnson, "Origins of Tourism in the Catskill Mountains," *Journal of Cultural Geography* 11, no. 1 (1990): 5–16; Philip G. Terrie, "The New York Natural History Survey in the Adirondack Wilderness, 1836–1840," *Journal of the Early Republic* 3 (Summer 1983): 186–206; Philip G. Terrie, "Romantic Travelers in the Adirondack Wilderness," *American Studies* 24, no. 2 (1983): 59–75; Roland Van Zandt, *The Catskill Mountain House* (New Brunswick, NJ: Rutgers Univ. Press, 1966).

12. Not all of these turnpikes lay along the valley floors. In what became the Glenwood and New Castle ranger districts of the JNF, the remnants of turnpikes that traversed mountain ridges still remain. Because of topography, these turnpikes tended to be the most expensive to maintain and therefore less profitable. Despite more than a century of abandonment, however, their paths remain and are easily traveled by foot or horseback, and sometimes even by off-road vehicle. Among other things, such persistence would seem

to illustrate the environmental impact of road construction not followed by human efforts to restore top soil and flora.

13. Turnpikes often improved upon older roadbeds by eliminating curves, avoiding hazardous stream crossings, or sidestepping courses that had become severely eroded through use and weather. A rough horse and wagon route between Fincastle and Sweet Springs preceded the Sweet Springs and Price's Mountain Turnpike by at least thirty-seven years. In 1824 and 1826 Allegheny County, Virginia, surveyors referred to this route as the "road leading from the Sweet Springs to Fincastle" or the "Waggon Road." In 1837 James Shanks completed the Sweet Springs and Price's Mountain Turnpike along this general course, though evidence of the old road remained. See Allegheny County Surveyor's Book No. 1, Allegheny County Courthouse, Covington, Virginia: Joseph Pinnell (also Pennell), Apr. 3, 1824, 46; John Shawver, Apr. 3, 1826, 88; and Margaret Shawver, Apr. 26, 1838, 308.

14. In July 1847 the Farish Ficklin & Co. stage line offered a 100-mile run from Lynchburg to the White Sulphur Springs, with an alternate route through Fincastle to the Sweet Springs. See *Lynchburg Virginian,* July 12, 1847; Wellington Williams, *Appletons' Southern and Western Travellers' Guide* (New York: D. Appleton & Co., 1850).

15. Oliver Wendell Holmes, "Stagecoach and Mail: From Colonial Days to 1820" (Ph.D. diss., Columbia Univ., 1956), iii–iv, 271–72; Edward Graham Roberts, "The Roads of Virginia, 1607–1840" (Ph.D. diss., Univ. of Virginia, 1951), 155.

16. On the other hand, by 1845 stage travel had drastically reduced the ostler's profession at White Sulphur Springs, where "large stables built for the accommodation of visiter's [sic] horses . . . once a source of great profit, have since the improvement of the roads and the general adoption of stage coaches, become almost empty." The resident ostler commented in 1845 that where they had once handled as many as four hundred horses, he now maintained only a dozen (*Lynchburg Virginian,* July 28, 1845). The ostler's lament clearly reflected how turnpike and stage travel had usurped the former era characterized by rough road and horseback travel.

17. E. S. Abdy, *Journal of a Residence and Tour in the United States of North America* (London, 1855), 2:310; Paulding, *Letters from the South,* 1:198–99; "Viator," "The White Sulphur Springs," *New England Magazine* 3 (Sept. 1832): 225. John Inscoe describes a similar situation in the mountains of western North Carolina during the same period. See John C. Inscoe, *Mountain Masters, Slavery, and the Sectional Crisis in Western North Carolina* (Knoxville: Univ. of Tennessee Press, 1989), 30. Robert Mitchell generally describes this post-frontier developmental phenomenon, and his designations of greater agricultural specialization and growth in service industries especially apply to this case. See Robert D. Mitchell, *Commercialism and Frontier: Perspectives on the Early Shenandoah Valley* (Charlottesville: Univ. Press of Virginia, 1977), 161–62.

18. Based on figures compiled in Ernest Cline Snyder, "Fincastle, Virginia, 1772–1942: A Study in Small Town Life" (master's thesis, Virginia Polytechnic

Institute, 1942), 25, tables 5–6. John Inscoe describes the gateway phenomenon on a larger scale for Asheville, North Carolina, of the same period. See Inscoe, *Mountain Masters,* 31–32, 37.

19. "Visit to the Virginia Springs, during the Summer of 1834, No. 1," *Southern Literary Messenger* 1 (May 1835): 477.

20. Beginning July 19, 1833, the "proprietors of the line from Lynchburg to Fincastle, Sweet Springs and White Sulphur or to Lewisburg" announced three additional trips, totaling six per week, charging $5.00 as far as Fincastle, another $4.50 to Sweet Springs, and 9 cents per mile for any intermediate travel (*Lynchburg Virginian,* Aug. 15, 1833). In 1839 the Galbraith stage company offered a line from Lynchburg to Natural Bridge and on to Dagger's Springs (*Lynchburg Virginian,* July 22, 1839). The S. D. Patteson & Co. stage line made a triweekly run to Fincastle during the summer of 1839 (*Lynchburg Virginian,* July 15, 1839), and two years later G. M. Bruce's stage line traveled to White Sulphur Springs via Natural Bridge (*Lynchburg Virginian,* Feb. 25, 1841, Aug. 23, 1841). Also see *Lexington Gazette,* July 17, 1835; June 10, 1836; Aug. 27, 1837; June 17, 1841; Aug. 20, 1846; Aug. 22, 1850; June 8, 1854; *Lynchburg Virginian,* Sept. 14, 1853.

21. Domestic manufacture of stagecoaches began in the northern United States during the late 1700s, primarily around New York. Where six New York carriage shops operated in 1788, twenty-eight functioned by 1810. By 1820 stage builders had improved upon the eighteenth century's cruder stage wagon to create the Jersey wagon and finally the post coach, or "oval Jersey stagecoach, the first distinctive style of stagecoach in America." During the late 1820s, manufacturers in various New England villages made carriages and shipped them to New York for distribution. Between the 1820s and 1850s various new models appeared, including adaptations of Old World models, such as barouches, and wholly American innovations, such as rockaways and new rochelles. See Don H. Berkebile, *Carriage Terminology: An Historical Dictionary* (Washington, DC: Smithsonian Institution Press, 1978), 25, 239–41; Holmes, "Stagecoach and Mail," 266–67; Ezra M. Stratton, *The World on Wheels* (New York: Benjamin Bloom, 1972), 415, 420, 429, 433, 445, 455.

22. Durrenberger, *Turnpikes,* 78, 115–16, 121; Hunter, "Turnpike Movement in Virginia," 143, 151, 267, 269, 270; George Rogers Taylor, *The Transportation Revolution, 1815–1860* (New York: Harper & Row, 1951), 27–28. As the Natural Bridge Turnpike Company reported in 1840, "we have received no tolls from the stage contractors, and have recently instituted suit against one of them (Galbraith) and shall pursue the same course with the other, (Porter,) should he not make speedy payment." See Natural Bridge Turnpike Company 1840 Report, 25th Annual Report, Board of Public Works, Turnpike Companies, Virginia State Library and Archives, Richmond (hereafter referred to as BPW).

23. Fincastle and Blue Ridge Turnpike Company to the Board of Public Works, Nov. 12, 1843, BPW. Again in 1845 the company wrote "no dividend was declared for the half year ending 12th of Sept.; oweing [*sic*] to the fact that

heavy Expenses have been incur-d in repairs made upon the mountain section of the Road." Similar severe weather forced repairs the following year, precluding dividend payments. If rains damaged a mountain turnpike badly enough, or if the road deteriorated through neglect, the turnpike company often opened it free to the public (see Fincastle and Blue Ridge Turnpike Company, FY1845 and FY1846 reports to the Board of Public Works, BPW).

24. For a full variety of accounts see the following: Abdy, *Journal of a Residence and Tour*, 304, 306; "Virginia Springs," *Southern Literary Messenger* 3 (May 1837): 282; Count Francesco Arese, *A Trip to the Prairies and in the Interior of North America* (New York: Harbor Press, 1934), 30–31; William Burke, *The Mineral Springs of Western Virginia with Remarks on Their Use*, 2nd ed. (New York: Wiley & Putnam, 1846), 15–16; Edwin Bedford Jeffress, Diary, Aug. 1– Sept. 12, 1852, Virginia Historical Society, Richmond; Harriet Martineau, *Society in America* [1834–1836] (New Brunswick: Transaction Books, 1981), 138–39; Peregrine Prolix (Philip Houlbrooke Nicklin), *Letters Descriptive of the Virginia Springs* (Philadelphia: H. S. Tanner, 1837; Austin: AAR/Tantalus, 1978), 72. For comparative examples of turnpike conditions, see Narry E. Cole, *Stagecoach and Tavern Tales of the Old Northwest* (Cleveland: Arthur Clark Co., 1930), and for the general picture see Holmes and Rohrbach, *Stagecoach East*, 47, 50, 54.

25. *Lynchburg Virginian,* May 29, 1854. By the late nineteenth century an entire new network, including Craig Healing, Allegheny, Blue Ridge, and Roanoke Red Sulphur, arose directly in conjunction with expanding railroad travel. See Charlotte Lou Atkins, "Rockbridge Alum Springs: A History of the Spa: 1790–1974" (master's thesis, Virginia Polytechnic Institute and State Univ., 1974), 30; Dorothy H. Bodell, *Montgomery White Sulphur Springs: A History of the Resort, Hospital, Cemeteries, Markers, and Monument* (Blacksburg, VA: Pocahontas Press, 1993), 1, 6–27, 54–60; Stan Cohen, *Historic Springs of the Virginias: A Pictorial History* (Charleston, WV: Pictorial Histories, 1981), 1, 7–8, 13, 33, 97, 177–78.

26. Durrenberger, *Turnpikes,* 154, 158; James River Project Committee of the Virginia Academy of Science, *The James River Basin: Past, Present and Future* (Richmond: Virginia Academy of Science, 1950), 685–86.

27. Alice Morse Earle, *Stage-coach and Tavern Days* (New York: Benjamin Bloom, 1900), 328, 340; Holmes and Rohrbach, *Stagecoach East,* 72, 150. John Inscoe recognized how tourism culturally "enhanced" mountain communities of western North Carolina, but also accurately described the limits of such interaction (see Inscoe, *Mountain Masters,* 37). Perhaps this limitation applied to a somewhat lesser degree in the case of western Virginia's springs and the upper Shenandoah Valley for two reasons: the less rugged and more accessible location on the "fringe" of Appalachia and the much more intense concentration of tourist traffic compared to that of western North Carolina.

28. As in everything regarding southern Appalachian history, much of this depends upon the area of the region. Here, the early tourist trade pertained mostly to the eastern edge of the Virginia mountains and did not pertain to

the coal field region, farther to the west (although Scott County's Natural Tunnel drew some visitors). In any case, many scholars have addressed the isolation topic. For a useful historiographical summation of a good deal of material, as well as a qualification of "isolation" in upper East Tennessee, see David C. Hsiung, "How Isolated Was Appalachia? Upper East Tennessee, 1780–1835," *Appalachian Journal* 16, no. 4 (1989): 336–49. For a detailed analysis of another part of western Virginia, see Ralph Mann, "Mountains, the Land, and Kin Networks: Burkes Garden, Virginia, in the 1840s and 1850s," *Journal of Southern History* 58, no. 3 (1992): 411–34. Probably the best single source documenting the postindustrial rise of negative Appalachian stereotypes, including isolationism, lies in Henry D. Shapiro, *Appalachia on Our Mind: The Southern Mountains and Mountaineers in the American Consciousness, 1870–1920* (Chapel Hill: Univ. of North Carolina Press, 1978). The disadvantageous, postindustrial economic position that Appalachia fell into is documented in both the old groundbreaking work of Ronald Eller and the new economic history of Paul Salstrom. See Eller, *Miners, Millhands, and Mountaineers,* and Paul Salstrom, *Appalachia's Path to Dependency: Rethinking a Region's Economic History, 1730–1940* (Lexington: Univ. Press of Kentucky, 1994). Finally, Wilma A. Dunaway's highly researched *The First American Frontier* challenges many former interpretations of the region but also ultimately argues against the isolation notion.

29. In addition to Shapiro, *Appalachia on Our Mind,* also see Katherine Ledford, "A Landscape and a People Set Apart: Narratives of Exploration and Travel in Early Appalachia," in *Confronting Appalachian Stereotypes: Back Talk from an American Region,* ed. Dwight B. Billings, Gurney Norman, and Katherine Ledford (Lexington: Univ. Press of Kentucky, 1999), 47–66; and Sandra L. Ballard, "Where Did Hillbillies Come From? Tracing Sources of Comic Hillbilly Fool in Literature," in Billings, *Confronting Appalachian Stereotypes,* 138–49.

30. Nash, *Wilderness and the American Mind,* 60, 67, 78, 79, 82. For an extensive treatment of these subjects, see chapter 3, "The Romantic Wilderness," and chapter 4, "The American Wilderness." Also see David Morse, *American Romanticism* (Totowa, NJ: Barnes & Noble Books, 1987), 1:3.

31. Dona Brown, *Inventing New England: Regional Tourism in the Nineteenth Century* (Washington, DC: Smithsonian Institution Press, 1995), chaps. 1–2; Esther Moir, *The Discovery of Britain: The English Tourists* (London: Routledge and Kegan Paul, 1964); Ian Ousby, *The Englishman's England: Taste, Travel and the Rise of Tourism* (Cambridge: Cambridge Univ. Press, 1990); Lynne Withey, *Grand Tours and Cook's Tours: A History of Leisure Travel, 1750 to 1915* (New York: William Morrow & Co., 1997), chaps. 1–4.

32. In a valuable volume on early American tourism, John Sears designates various categories of tourist attractions, including some cited above, as well as the monumentalism of Yosemite and the Wild West attraction of Yellowstone. See Sears, *Sacred Places.*

33. Dandridge Spotswood Diary, July 12–Aug. 18, 1848, Virginia Historical Society, Richmond.

34. William Edmonds Horner, *Observations on the Mineral Waters in the South Western Part of Virginia* (Philadelphia: J. Thompson, 1834), 27.

35. Anonymous, *The Peaks of Otter, A Monograph of the Religious Experience of a Young Man* (Philadelphia: Presbyterian Board of Publication, 1859), 25–26. For another romantic depiction of the Peaks of Otter, see Porte Crayon, "Virginia Illustrated," *Harper's New Monthly Magazine* 12 (Jan. 1856): 159.

36. "Virginia Springs," *Southern Literary Messenger* 3 (May 1837): 281. For other examples of varying romanticism, see Anonymous, *A Description of the Peaks of Otter* (Lynchburg: Virginia Job Office, 1853), 10–29, 71–74; Martineau, *Society in America,* Mark Pencil, *The White Sulphur Papers* (New York: Samuel Colman, 1839), 16–17; Prolix, *Letters Descriptive of the Virginia Springs,* 21, 29, 76–78; Dandridge Spotswood Diary, July 12–Aug. 18, 1848; Jane Tayloe Worthington, "Spring Memories among the Mountains," *Southern Literary Messenger* 12 (June 1846): 349–52; Jane Tayloe Worthington, "To the 'Far Blue Mountain,'" *Southern Literary Messenger* 13 (Jan. 1847): 41–42.

37. Prolix, *Letters Descriptive of the Virginia Springs,* 21, 29. For a mid-1830s romantic description of Hawk's Nest, in Fayette County (now West Virginia), see Harriet Martineau, *Society in America* (New Brunswick: Transaction Books, 1981), 138; Prolix also described the Natural Bridge in this style in 1836 (see pp. 76–78).

38. Pollard, *Virginia Tourist,* 106–7.

39. Ibid., 151.

40. John Esten Cooke, *The Last of the Foresters* (New York: Derby & Jackson, 1856), opening lines of chapter 38. This work was researched via the Gutenberg Project, which unfortunately does not include original pagination. See http://www.gutenberg.org/files/10560/10560–8.txt.

41. Ibid.

42. Archibald Henderson, *Conquest of the Old Southwest* (New York: Century, 1920), ix. Available through http://www.gutenberg.org/dirs/etext00/cnqsw10.txt.

43. Pollard, *Virginia Tourist,* 100–101.

44. Ibid., 101–2.

45. Anne Newport Royall, *Sketches of History, Life, and Manners, in the United States* (1826; New Haven: Johnson Reprint Corp., 1970), 41. Other writers even limited themselves when describing Hawk's Nest, the Peaks of Otter, and Natural Bridge. For examples, see "Ride to the Peaks of Otter," *Southern Literary Messenger* 7 (Dec. 1841): 850–53; "Scenery of Western Virginia," *Lexington Gazette,* Aug. 7, 1851; Horner, *Observations on the Mineral Waters in the South Western Part of Virginia,* 26–27, 28; Edwin Bedford Jeffress, Diary, Aug. 1–Sept. 12, 1852, Virginia Historical Society, Richmond.

46. Royall, *Sketches of History,* 19–20. Also see Foster, *Civilized Wilderness,* 26–27.

4. Industrial Logging Discovers Appalachia

1. "Virgin" has been commonly used to describe a forest never logged by humans. The term, however, is misleading, since natural agents (such as lightning-caused fires, volcanic explosions, hurricanes, tornadoes, ice storms, and landslides) can disrupt even old growth forest. Botanically speaking, even without such "disasters," the forest ecosystem is always evolving anyway. See Stephen H. Spurr and Burton V. Barnes, *Forest Ecology,* 3rd ed. (Malabar, FL: Krieger, 1992), 367–70.

2. H. B. Ayres and W. W. Ashe, "The Southern Appalachian Forests," U.S. Geological Survey, Professional Paper No. 37 (Washington, DC: GPO, 1905), 17; Charles Rufus Boyd, *Resources of South-West Virginia* (New York: John Wiley & Sons, 1881), 27, 42–43, 78, 123, 125, 224. "Southern Appalachia" in these paragraphs is defined as the "approximately 10,000 square miles between New River Gap in Virginia and Hiwassee River in western North Carolina and northern Georgia" that Ayres and Ashe used in the focus of their report. Of this region, they actually examined about 8,300 square miles in 1900 and 1901 (see Ayres and Ashe, "The Southern Appalachian Forests," 14).

3. Boyd also reported ravaged forests in the vicinity of Wythe County's iron furnaces and around Saltville. See Boyd, *Resources of South-West Virginia,* 110, 197.

4. Charles S. Sargent, *Report on the Forests of North America* (Washington, DC: GPO, 1884), 512.

5. Conventional locomotives were rod-driven and suited to flat land or mild slopes of generally no more than 3 percent.

6. Leslie Brooks, "Incline Logging," *Southern Lumberman* 114 (May 10, 1924): 49; Martin Hassinger, interview by Mary Kegley and Lionel Melancon, July 24, 1979, in Bristol, VA; Robert S. Lambert, "Logging the Little River, 1890–1940," *East Tennessee Historical Society's Publications* 33 (1961): 32–42; Ellis Lucia, *The Big Woods: Logging and Lumbering—from Bull Teams to Helicopters—in the Pacific Northwest* (New York: Doubleday, 1975), 21; Ivar Samset, *Winch and Cable Systems* (Boston: Martinus Nijhoff/Dr W. Junk, 1985), 27, 37–40. A well-documented and detailed overview of American logging history may be found in Michael Williams, *Americans and Their Forests: A Historical Geography* (New York: Cambridge Univ. Press, 1989).

7. 1910 Manufacturing Census / 13th Census of the United States 1910.2, Virginia and West Virginia, pp. 1262, 1312, 1316; Ayres and Ashe, "The Southern Appalachian Forests," 57–58, 78; Eller, *Miners, Millhands and Mountaineers,* 98, 104; Luther C. Hassinger, "The Lumber Industry of Southwest Virginia," *Historical Society of Washington County, Virginia Bulletin,* ser. 2, no. 4 (Spring 1967): 9; Lloyd D. Lewis, "An Abingdon Branch Accolade," *Trains* 44 (June 1984): 23; *Plow,* July 1976; W. M. Ritter Lumber Co., *The Romance of Appalachian Hardwood Lumber* (Columbus, OH: W. M. Ritter Lumber Co., 1940), 15.

8. 1910 Manufacturing Census/13th Census of the United States 1910.2, Virginia and West Virginia, 1262, 1312, 1316; 14th Census of the United States, State Compendium: Virginia, 127; State Compendium: West Virginia, 94–95;

Ronald B. Craig, *Virginia Forest Resources and Industries* (Washington, DC: USDA Misc. Pub. no. 681, Apr., 1949), 12; Eller, *Miners, Millhands and Mountaineers*, 104; National Forest Reservation Commission (NFRC), *1920 Annual Report* (Washington, DC: GPO, 1921), 19; Samuel P. Hays, *The American People and the National Forests: The First Century of the U.S. Forest Service* (Pittsburgh: Univ. of Pittsburgh Press, 2009), 3; John Henry Reeves, "The History and Development of Wildlife Conservation in Virginia: A Critical Review" (Ph.D. diss., Virginia Polytechnic Institute, 1960), 51, 56; Terry Seyden, "Jefferson National Forest—Rich in History," *Virginia Forests Magazine* 45 (Spring 1990): 4.

9. For an interesting promotional piece concerning this area, see Lewis M. Haupt, *Description of 100,000 Acres of Land in Giles, Craig, Monroe Counties, State of Virginia* (Philadelphia: Leisenring Steam Printing House, 1871). A copy may be found at the State Library in Richmond, VA.

10. Deed of Conveyance, Big Stony Railroad, box 2.116, Printed Materials 1896–1930, Norfolk and Western Collection, Virginia Tech Special Collections, Blacksburg (hereafter cited as N&W Collection). Also see the *10th Annual Report of the N&W Railway Co.* (fiscal year ending June 30, 1906), 13; *11th Annual Report*, 14; *14th Annual Report*, 15–16; *15th Annual Report*, 21. All N&W annual reports located in N&W Collection.

11. Chief Engineer (Churchill) to (Vice President) N. D. Maher, Aug. 25, 1908, Oct. 2, 1908, box 2.59, file 977, N&W Collection.

12. Charles Churchill to L. E. Johnson, Feb. 15, 1905, box 2.65, file 1686, N&W Collection.

13. *Monroe Watchman*, Oct. 2, 1975. According the Oren F. Morton, timber was the Potts Valley Branch's main freight from the outset. See Morton, *A History of Monroe County*, 220.

14. Chief Engineer to N. D. Maher, Aug. 25, 1908, Nov. 11, 1908, box 2.59, file 977, N&W Collection.

15. Herman Harry, pers. comm. with author, Aug. 20, 1992, in Potts Creek Valley, VA; box 2.59, file 977, N&W Collection.

16. See the advertisement in *Monroe Watchman*, Jan. 30, 1913.

17. After an initial legal complication involving Washington Lumber Company and a chancery court case, the Craig-Giles Iron Company bought this tract at auction on July 24, 1912. By October 1912 Craig-Giles had resold the timber to Tri-State and given them eighteen years to harvest the wood, with an obligation to pay for at least 10 million board feet annually. See the *New Castle Record*, July 27, 1912; Craig County Deed Book P, 583; *Monroe Watchman*, Jan. 23, 1913.

18. *Monroe Watchman*, Jan. 23, 1913; *New Castle Record*, July 6, 1912.

19. For descriptions of other incline operations in the southern mountains, see C. S. Badgett, "Equipment for Incline Logging," *American Lumberman* 144 (Oct. 8, 1921): 54; Lambert, "Logging the Little River, 1890–1940," 32–42; J. P. Murphy, "Advantages of Incline and Mechanical Logging," *Southern Lumberman* 119 (Apr. 25, 1925): 36.

20. H. G. Cowling, "Logging Inclines," *Timberman* (Aug. 1926): 37.

21. For technical descriptions of counterbalance inclines, see H. G. Cowling, "The Counterbalance Incline," *Timberman* (June 1925): 54; S. J. Dumbolton, "A Counter-Balanced Incline," *Timberman* (June 1924): 51, 67.

22. See Leslie Brooks, "Incline Logging," *Southern Lumberman* 114 (May 10, 1924): 49–50. Unfortunately, Brooks did not specify the exact location of this operation. Starkey describes a Pocahontas County incline in operation by 1907 in Edith Kimmell Starkey, "Over the Mountain: Timbering at Braucher," *Goldenseal* 13 (Summer 1987): 35.

23. The efficacy of erecting a band mill in what turned out to be not a lucrative operation seems doubtful in retrospect, though numerous references to the band mill exist. See, for instance, Craig County Chancery Order Book 4, 318–19; Lucy Huffman interview, June 2, 1992; or R. L. Humbert, *Industrial Survey: Craig County Virginia* (Blacksburg: Virginia Polytechnic Institute Engineering Extension Service, 1930), 32 (copy located in Carol Newman Library, Virginia Polytechnic Institute and State Univ.).

24. For a broad assessment of this aspect of incline logging, see Badgett, "Equipment for Incline Logging," 54.

25. Frank J. Sheridan, *Italian, Slavic, and Hungarian Unskilled Immigrant Laborers in the United States,* Bulletin of the Bureau of Labor, no. 72 (Washington, DC: GPO, 1907), 434.

26. In addition to those involved with the Tri-State operation, Italians also helped build the Clinchfield Railroad in Southwest Virginia. See Margaret Ripley Wolfe, "Aliens in Appalachia: The Construction of the Clinchfield Railroad and the Italian Experience," in *Appalachia: Family Traditions in Transition,* ed. Emmet M. Essin (Johnson City: Eastern Tennessee State Univ. Press, 1975), 83–88. Italians also labored on the forest railroad of the Hassinger Lumber Company lands in southern Washington County, Virginia, near North Carolina (Martin Hassinger, pers. comm. with author, June 17, 1992, in Bristol, VA). A description of the Hassinger Operation may be found in the *Plow,* July 1976, 13–15 (copy in the Kelly Library, Emory and Henry College, Emory, VA).

27. Around the turn of the century, the Union Pacific Railroad employed between 500–1,000 such laborers, while the Great Northern employed some 1,500. See Robert F. Foerster, *The Italian Emigration of Our Times* (Cambridge: Harvard Univ. Press, 1924), 357–60. Also see Dominic Ciolli, "The 'Wop' in the Track Gang," in *A Documentary History of the Italian Americans,* ed. Wayne Moquin (New York: Praeger, 1974), 141–45; and Cesidio Simboli, "When the Boss Went Too Far," in *A Documentary History of the Italian American,* ed. Wayne Moquin (New York: Praeger, 1974), 146–49.

28. Kenneth R. Bailey, "A Judicious Mixture: Negroes and Immigrants in the West Virginia Mines, 1880–1917," *West Virginia History* 34 (Jan. 1973): 141–61; Barry J. Ward, "Italian-American Folk Poetry in the Industrialized Appalachian Mountains," *West Virginia History* 43 (Spring 1982): 285. Italians

also worked at the logging operation surrounding the Brancher area of West Virginia, described in Starkey, "Over the Mountain," 35.

29. Herman Harry, pers. comm. to author, Aug. 20, 1992.

30. Huffman interview, June 2, 1992.

31. Unfortunately for the historian, the Tri-State operation whisked in and out of Craig County between censuses, thus leaving no record of their laborers. By the time of the 1920 population census, not a single immigrant, Italian or otherwise, remained in the Johns Creek area. See the 14th Census of the United States, 1920—Population, Craig County, Virginia, Alleghany Magisterial District, Johns Creek Valley.

32. Huffman interview, June 2, 1992; *New Castle Record,* Jan. 22, 1916; *Monroe Watchman,* Jan. 27, 1916. Lifetime Johns Creek Valley resident Lucy Huffman specifically recalled the locomotive as a "Shay," though it is possible that this name was generally used to describe gear-driven logging locomotives. Huffman also remembered this accident and recalled that everyone remained puzzled over why more than one engineer was riding the locomotive.

33. 14th Census of the United States, State Compendium: Virginia, 127; State Compendium: West Virginia, 94–95.

34. R. L. Humbert, C. C. Beggs, et al., *Industrial Survey, Craig County, Virginia* (Blacksburg: Engineering Extension Division, Virginia Polytechnic Institute, 1930), 32; *Monroe Watchman,* Jan. 23, 1913.

35. Craig County Chancery Order Book 4, 319.

36. *37th Annual Report N&W Railway Company,* fiscal year ending Dec. 31, 1932, N&W Collection.

37. 13th Census of the United States, Population, 1910, Craig County, Alleghany Magisterial District.

38. 14th Census of the United States, Population, 1920, Craig County, Alleghany Magisterial District, Johns Creek Valley.

39. In 1920 one man owned such a sawmill in Sinking Creek Valley, neighboring Johns Creek. By 1930 fifteen such mills were in use. See 14th Census of the United States, 1920, Craig County, Alleghany Magisterial District, Sinking Creek Valley; Humbert et al., *Industrial Survey: Craig County Virginia,* 32.

40. Ovid Butler, ed., *The Biltmore Story: Recollections of the Beginning of Forestry in the United States* (St. Paul: American Forest History Foundation, 1955), 16–18, 25–26; Samuel Trask Dana and Sally K. Fairfax, *Forest and Range Policy: Its Development in the United States,* 2nd ed. (New York: McGraw-Hill Book Co., 1980), 115; E. H. Frothingham, "Scientific Research and Southern Appalachian Forests," *Lumber World Review* 51 (Nov. 10, 1923): 50, 51; E. H. Frothingham, "Forest Research in the Southern Appalachians," *Southern Lumberman* 108 (Dec. 23, 1922): 126. Originally called the Appalachian Forest Experiment Station, the name of this early scientific outpost is now called the Southeast Forest Experiment Station.

41. Frothingham, "Forest Research," 125.

42 Ibid., 51.

43. Ibid., 48, 51, 52.

44. USDA, Forest Service, *National Forest Manual: Regulations and Instructions* (Washington, DC: GPO, 1928–33), 61-L.

45. Frome, *The Forest Service,* 220–21; Steen, *The U.S. Forest Service,* 141, 326. Schiff argues that the first half of the twentieth century witnessed little significant influence from scientists on daily national forest forestry practices. Thus, despite an initial scientific basis for professional forestry management, foresters soon "divorced" themselves from the scientific realm to practice "practical forestry." See Ashley L. Schiff, *Fire and Water: Scientific Heresy in the Forest Service* (Cambridge: Harvard Univ. Press, 1962), 2, 7, 8, 115.

46. Frothingham, "Forest Research," 122.

5. JNF and the Rise of National Conservation

1. Although concerted conservation in America did not arise until the late nineteenth century, even during the colonial period the English government tried to regulate the New World's timber resources for ship construction. Old growth, tall, straight, and strong trees, were in particular demand for outfitting sailing masts, for what Europe had long since depleted North America had in abundance. In 1803 the young U.S. government continued this tradition by reserving 250,000 acres of the Louisiana Purchase's timber lands specifically for the U.S. Navy. As for preserving land, George Catlin, the great painter of a fleeting era in American Indian and western United States history, proposed such an idea as early as 1832, decades ahead of widespread popular sentiment. American wilderness interest gained momentum during the late eighteenth century among urban philosophers involved in the Romantic Movement. Other early contributors included nineteenth-century Transcendentalists such as Henry David Thoreau and Ralph Waldo Emerson. John Muir, of course, contributed to wilderness preservation during the 1870s and in 1892 helped form the Sierra Club, which has advocated environmental causes ever since. See Dana and Fairfax, *Forest and Range Policy,* 4–6; Roderick Nash, ed., *The American Environment: Readings in the History of Conservation* (Reading, MA: Addison-Wesley, 1968), 5–9; Nash, *Wilderness and the American Mind,* 44, 51, 85, 86, 89, 95; Worrell, *Principles of Forest Policy,* 26, 30, 81.

2. In particular, 1871 saw a rash of fire storms across the northern Midwest, including the notorious Peshtigo and Hinckley fires in the Wisconsin-Minnesota area. The Peshtigo fire alone burned over a million acres and killed over 1,500 people. Such raging infernos continued in following decades, with especially destructive fires occurring periodically, such as the Cochrane Fire of 1911. See Fred McClement, *The Flaming Forests* (Toronto: McClelland and Stewart, 1969), 17–21, 62–68. Ellis Lucia describes much of the early prevailing attitude on the part of logging companies in his book *The Big Woods.* Although very much a product of the era's phobia against fire, see Stewart H. Holbrook, *Burning an Empire: The Story of American Forest Fires*

(New York: Macmillan, 1944) for detail surrounding the Peshtigo fire and other major fires.

3. Organic Administration Act of 1897, 55th Cong., 1st sess., 1897, chap. 2, sec. "Surveying the Public Lands."

4. Culhane, *Public Lands Politics: Interest Group Influence on the Forest Service and the Bureau of Land Management* (Baltimore: Johns Hopkins Univ. Press, 1981), 4; Samuel Trask Dana, *Forest and Range Policy: Its Development in the United States* (New York: McGraw-Hill, 1956), 34, 40, 42, 84; Frome, *The Forest Service,* 18–19, 78–79; Donald J. Pisani, "Forests and Conservation, 1865–1890," *Journal of American History* 72, no. 2 (1985): 340–59; Steen, *The U.S. Forest Service,* 17, 53; Worrell, *Principles of Forest Policy,* 27. Other indications of early conservation interests established themselves around several prominent forest-oriented organizations, such as the American Forestry Association (created in 1889) and the Society of American Foresters (established in 1900).

5. Worrell, *Principles of Forest Policy,* 27.

6. Generally, see Samuel P. Hays, *Conservation and the Gospel of Efficiency: The Progressive Conservation Movement, 1890–1920,* rev. ed. (1959; Cambridge: Harvard Univ. Press, 1969) and Elmo R. Richardson, *The Politics of Conservation: Crusades and Controversies 1897–1913* (Berkeley: Univ. of California Press, 1962). Also see Donald Worster, *The Wealth of Nature: Environmental History and the Ecological Imagination* (New York: Oxford Univ. Press, 1993), 145.

7. Dana, *Forest and Range Policy,* 112.

8. "Message from the President of the United States Transmitting a Report of the Secretary of Agriculture in Relation to the Forests, Rivers, and Mountains of the Southern Appalachian Region" (Washington, DC: GPO, 1902), 39–40.

9. Ovid Butler, "The National Forests of the East," *American Forests* 41 (Sept. 1935): 508; John Ise, *The United States Forest Policy* (New Haven: Yale Univ. Press, 1920), 207, 208; L. F. Kneipp, "Uncle Sam Buys Some Forests," *American Forests* 42 (Oct. 1936): 443.

10. H. B. Ayers and W. W. Ashe, "The Southern Appalachian Forests," U.S. Geological Survey, Professional Paper No. 37 (Washington, DC: GPO, 1905), 17. Also see Leonidas C. Glenn, "Denudation and Erosion in the Southern Appalachian Region and the Monongahela Basin," U.S. Geological Survey, Professional Paper No. 72. (Washington, DC: GPO, 1911), 13–14. Ideas linking watershed conservation and hydroelectric development continued after passage of the Weeks Act and culminated with the creation of the TVA in 1933. Early JNF territory pertinent to TVA concerns included the Holston, Powell, and Clinch rivers, which formed the main headwaters for the Tennessee River, partly located in what was then the White Top purchase unit, later the JNF's Holston and Clinch ranger districts. By 1969 the JNF protected the direct water supplies of thirty-two communities, including important tributaries to the James and New rivers. See Ashe, "The Creation of the Eastern National Forests," 524; Willis Hawley, "Buying National Forests," *American*

Forests 31 (May 1925): 294–95; Charles Edgar Randall, "Jefferson's Forest," *American Forests* 75 (Apr. 1969): 61.

11. Examples include Ayres and Ashe, "The Southern Appalachian Forests"; E. H. Frothingham, "Ecology and Silviculture in the Southern Appalachians: Old Cuttings as a Guide to Future Practice," *Journal of Forestry* 15 (Mar. 1917): 343–49; Glenn, "Denudation and Erosion in the Southern Appalachian Region"; William L. Hall, "Influence of the National Forests in the Southern Appalachians," *Journal of Forestry* 17 (1919): 402–7; William L. Hall, *The Waning Hardwood Supply and the Appalachian Forests,* USDA Forest Service Circular No. 116, Sept. 24, 1907; Ise, *The United States Forest Policy,* 211, 212; M. O. Leighton and A. H. Horton, *Relation of the Southern Appalachian Mountains to Inland Water Navigation,* USDA Forest Service Circular No. 143, 1908; Hu Maxwell, "The Use and Abuse of Forests by the Virginia Indians," *William and Mary Quarterly* 19 (Oct. 1910), 73–103; "Message from the President of the United States Transmitting a Report of the Secretary of Agriculture in Relation to the Forests, Rivers, and Mountains of the Southern Appalachian Region" (Washington, DC: GPO, 1902), 39–40.

12. Weeks Act of 1911, Public Law 61-435, 61st Cong., 3rd sess., 1911.

13. Glenn, "Denudation and Erosion in the Southern Appalachian Region and the Monongahela Basin," 121; David M. Emmons, "Theories of Increased Rainfall and the Timber Culture Act of 1873," *Forest History* 15 (Oct. 1971): 6–14; Charles R. Kutzleb, "Can Forests Bring Rain to the Plains?" *Forest History* 15 (Oct. 1971): 14–21; George Perkins Marsh, *Man and Nature* (1864; Cambridge: Harvard Univ. Press, 1965), 171–82.

14. Hays, *The American People and the National Forests,* 32. Forest Service hydrology began with an emphasis on water quantities as affected by different land uses and then evolved into an emphasis on quality of water as related to its cycles and its nutrients' cycles within various forest ecosystems. The modern era now recognizes hydrology as one ecological component or medium among several that includes soil biota, insects, and biotic flora and fauna. See Wayne T. Swank and D. A. Crossley Jr., eds., *Forest Hydrology and Ecology at Coweeta* (New York: Springer-Verlag, 1988), vi, 17–31; Bill Morrison interview, Jan. 16, 1992. For an example of hydrological study on the JNF, see Mary Alice Schaeffer, "Effects of Land Use on Oxygen Uptake by Microorganisms on Fine Benthic Organic Matter in Two Appalachian Mountain Streams" (master's thesis, Virginia Polytechnic Institute and State Univ., 1993).

15. For the earlier debate, see Steen, *The U.S. Forest Service,* 97, 126–31. The seminal work delineating the watershed protection ideology is Schiff, *Fire and Water.* Also see Shands and Healy, *The Lands Nobody Wanted,* 14–15.

16. W. W. Ashe, "The Creation of the Eastern National Forests," 521; Ovid Butler, ed., "The National Forests of the East," *American Forests* 41 (Sept. 1935): 508; Willis C. Hawley, "Buying National Forests," *American Forests* 31 (May 1925): 293; Ise, *The United States Forest Policy,* 222; Kneipp, "Uncle Sam Buys Some Forests," 444; Alison T. Otis, William D. Honey, Thomas C. Hogg,

and Kimberly K. Lakin, *The Forest Service and the Civilian Conservation Corps: 1933–1942* (Washington, DC: USDA, Forest Service, 1986), 83.

17. Weeks Act of 1911, sec. 10.

18. Clark-McNary Act of 1924, Public Law 108-198, 108th Cong., 1st sess., 1924, secs. 1, 7, 8.

19. Ashe, "The Creation of the Eastern National Forests," 521–23; Butler, "National Forests of the East," 509; Hawley, "Buying National Forests," 293–95, 304; Kneipp, "Uncle Same Buys Some Forests," 444–46. The McNary-Woodruff Act provided $8 million for additional Forest Service acquisitions between 1928 and 1937.

20. Ashe, "The Creation of the Eastern National Forests," 521; "Message from the President of the United States Transmitting a Report of the Secretary of Agriculture in Relation to the Forests, Rivers, and Mountains of the Southern Appalachian Region" (Washington, DC: GPO, 1902), 190–91; Percy J. Paxton, "National Forests and Purchase Units of Region Eight," 2nd printing, Atlanta Region Eight Office, 1955, 3, 5.

21. Paxton, "National Forests and Purchase Units of Region Eight," 53.

22. Seyden, "Jefferson National Forest—Rich in History," 3.

23. National Forest Reservation Commission, *Progress of Purchase of Eastern National Forests* (Washington, DC: GPO, 1920), 16, 17; Paxton, "National Forests and Purchase Units of Region Eight," 3, 5, 49; *Coalfield Progress,* Mar. 28, 1935; Sept. 19, 1935; *Jefferson News Letter* 1 (May 25, 1936): 2, 3; *Smyth County News,* Sept. 5, 1935.

24. "New National Forests in the East," *American Forestry* 24 (June 1918): 346; Eller, *Miners, Millhands and Mountaineers,* 117–19; National Forest Reservation Commission, Progress of Purchase of Eastern National Forests, 16, 17; Natural Bridge National Forest Map (Washington, DC: GPO, 1930), 5; Otis et al., *The Forest Service and the Civilian Conservation Corps: 1933–42,* 83.

25. *Coalfield Progress,* Mar. 28, Sept. 19, 1935; Paxton, "National Forests and Purchase Units of Region Eight," 3; *Jefferson News Letter* 1 (May 25, 1936): 2, 3; *Smyth County News,* Sept. 5, 1935.

26. *Coalfield Progress,* June 24, July 1, 8, 1937.

27. M. L. Wilson, "Address at the Dedication of the Jefferson National Forest," July 1, 1937, TS, JNF Cultural Resources Division.

28. Although her geographical scope of what constitutes Appalachia is far too broad for some scholars, Wilma Dunaway's research on early landholdings is an important contribution to the literature. See Wilma A. Dunaway, "Speculators and Settler Capitalists: Unthinking the Mythology about Appalachian Landholding, 1790–1860," in Pudup et al., *Appalachia in the Making,* 50–75; Dunaway, *The First American Frontier,* chap. 3.

29. Kathryn Newfont, "Moving Mountains: Forest Politics and Commons Culture in Western North Carolina" (Ph.D. diss., Univ. of North Carolina–Chapel Hill, 2001), 29–31.

30. Eller, *Miners, Millhands and Mountaineers,* 54–57; Mastran and Lowerre, *Mountaineers and Rangers,* 3, 4. Absentee mineral interests and their local cohorts exploited subsurface mineral rights through the notorious "broad deed." See Dean Hill Rivkin, "Lawyering, Power, and Reform: The Legal Campaign to Abolish the Broad Form Mineral Deed," *Tennessee Law Review* 66 (1999): 467–98.

31. The land degradation is widely documented. For contemporary examples, see Bureau of Agricultural Economics, *Economic and Social Problems and Conditions of the Southern Appalachians* (Washington, DC: USDA, 1935), 2, 5; Otto, "Forest Fallowing among the Appalachian Mountain Folk: An Ethnohistorical Study," 15; Glenn, "Denudation and Erosion in the Southern Appalachian Region," 11–12, 22–23, 30; Ayres and Ashe, "The Southern Appalachian Forests," 19.

32. Perhaps the most egregious misunderstanding may be found in Barbara Rasmussen, *Absentee Landowning and Exploitation in West Virginia 1760–1920* (Lexington: Univ. Press of Kentucky, 1994), 138, 139, 142. Furthermore, as Ron Eller has wisely noted, even where outside interests historically exploited Appalachia they often found local collaborators. See Ronald D. Eller, *Uneven Ground: Appalachia since 1945* (Lexington: Univ. Press of Kentucky, 2008), 7–8.

33. A serious misunderstanding of this is found in Rasmussen, *Absentee Landowning and Exploitation,* 146, and in Donald Edward Davis, *Where There Are Mountains: An Environmental History of the Southern Appalachians* (Athens: Univ. of Georgia Press, 2000), 173. The latter represents a vague, unsubstantiated claim regarding the Cherokee National Forest, which, even if true, would only reflect a "fallacy of the lonely fact," as explicated by David Hackett Fischer, *Historians' Fallacies: Toward a Logic of Historical Thought* (New York: Harper & Row, 1970), 109–10.

34. Dana, *Forest and Range Policy,* 112; Worrell, *Principles of Forest Policy,* 82. In other cases in the southern Appalachians, impoverished people already surviving on the margins of a cash economy could not meet their tax payments. This factor did not become as significant in JNF acquisition as other Appalachian national forests, like the Cherokee. See Cherokee National Forest, "Final Environmental Impact Statement" (Atlanta: USDA, Forest Service, 1986).

35. The JNF bought around 37,000 acres from the Virginia Iron and Coal Company and the Hagan Estate for the Clinch Purchase Unit (soon to be the JNF's Clinch Ranger District) in 1935. Some of the larger NFRC purchases eventually contributing to the Wythe Ranger District included 8,695 acres from the Rustin Land, Mining and Manufacturing Company in Wythe, Grayson, and Carroll counties in 1923; and 15,857 acres in Pulaski and Wythe counties from the Bertha Mineral Company in 1939. All of these figures are compiled from the individual land files pertaining to these companies in the JNF Land Files. This corporate divestment tradition continued into more recent times, when in 1972 the Wythe District gained 45,800 acres in Bland County from the Consolidation Coal Company. Also see the *Roanoke Times,* Sept. 12, 1937.

36. NFRC, *1915 Annual Report,* 8. Settling title disputes was common procedure throughout the Southern Highlands. For further accounts, see "The Appalachian Forests," *American Forestry* 17 (July 1911): 381–83; Thomas D. Clark, *Greening of the South: The Recovery of Land and Forest* (Lexington: Univ. Press of Kentucky, 1984), 62–63; William L. Hall, "The Appalachian Work," *American Forestry* 18 (Mar. 1912): 192; Bret Wallach, "The Slighted Mountains of Upper East Tennessee," *Annals of the Association of American Geographers* 71, no. 3 (1981): 365–66.

37. This situation was broadly characteristic of all of southern Appalachia. See William L. Hall, "To Remake the Appalachians: A New Order in the Mountains that is Founded on Forestry—What the Government's Appalachian Forests Mean to the People in the Mountains and to the Millions Who Want Recreation," *World's Work* 28 (July 1914): 328, 330–32.

38. *New Castle Record,* May 28, 1938.

39. Virginia Coal and Iron Company, land acquisition abstracts, JNF Land Files.

40. Virginia Timber Corporation, land acquisition abstracts, JNF Land Files.

41. *Coalfield Progress,* July 22, 1937.

42. Ernie Karger, telephone interview, Denver, Mar. 12, 17, 1992.

43. Charles Sexton interview, Feb. 20, 1992.

44. William Campbell interview, Dec. 13, 1991.

45. For land acquisition for the Great Smoky Mountains National Park, see Margaret Lynn Brown, *The Wild East: A Biography of the Great Smoky Mountains* (Gainesville: Univ. Press of Florida, 2000), 96–99 (also 159–60); Durwood Dunn, *Cades Cove: The Life and Death of a Southern Appalachian Community 1818–1937* (Knoxville: Univ. of Tennessee Press, 1988), 243, 246, 248, 251; Daniel S. Pierce, *The Great Smokies: From Natural Habitat to National Park* (Knoxville: Univ. of Tennessee Press, 2000), chaps. 5 and 6. For an interesting retrospective assessment of Park Service policy, see T. Young, "False, Cheap and Degraded: When History, Economy and Environment Collided at Cades Cove, Great Smoky Mountains National Park," *Journal of Historical Geography* 32, no. 1 (2006): 169–89. Young documents the Park Service's post-displacement approach of preserving a manufactured "historic" presence by restoring log cabins, after razing more "modern" structures. At the cost of ecological health, the Park Service also continued hay cultivation with modern equipment in order to save costs and to maintain scenic views, combing cove pastoralism and forested slopes.

46. For the Shenandoah National Park land acquisitions, see Andrew H. Myers, "The Creation of Shenandoah National Park: Albemarle County Cultures in Conflict," *Magazine of Albemarle County History* 51 (1993): 52–89; Charles L. Perdue Jr. and Nancy Martin-Perdue, "Appalachian Fables and Facts: A Case Study of the Shenandoah National Park Removals," *Appalachian Journal* 7, nos. 1–2 (Autumn/Winter 1979–1980): 84; Charles L. Perdue Jr. and Nancy Martin-Perdue, "To Build a Wall around These Mountains": The Displaced People of Shenandoah," *Magazine of Albemarle County History* 49 (1991):

48–71; Gene Wilhelm Jr., "Shenandoah Resettlements," *Pioneer America* 14 (Mar. 1982): 15. For an account more sympathetic to Shenandoah Park advocates, see Dennis E. Simmons, "Conservation, Cooperation, and Controversy: The Establishment of the Shenandoah National Park, 1924–1936," *Virginia Magazine of History and Biography* 89 (1981): 387.

47. See Michael J. McDonald and John Muldowny, *TVA and the Dispossessed: The Resettlement of Population in the Norris Dam Area* (Knoxville: Univ. of Tennessee Press, 1982).

48. For an indication of the enduring legacy of displaced persons from these parks, see Carlos Santoa, "Shenandoah Park, Counties Aiming for Better Relations," *Richmond Times-Dispatch,* Dec. 7, 1992; Michael Ann Williams, *Great Smoky Mountains Folklife* (Jackson: Univ. Press of Mississippi, 1995), 143–69; and Katrina M. Powell, "Writing the Geography of the Blue Ridge Mountains: How Displacement Recorded the Land," *Biography* 25, no. 1 (2002): 73–94.

49. While this remains true enough on the surface, even Forest Service acquisitions during the Great Depression apparently incurred ill feelings. According to the Cherokee National Forest, "Final Environmental Impact Statement" (Atlanta: USDA, Forest Service, 1986), some of their local residents still begrudge the national forest for buying their family's land during 1930s and feel the federal government took advantage of them while in dire straits.

50. "Holston Working Circle Development Plan, 1941–50," pp. 12, 48–49, Albert Mustian Papers, Forest History Society, Durham, NC.

51. Ibid., 49.

52. Ibid., 13.

53. Before the advent of the Mount Rogers NRA in the 1960s, the JNF had condemned only occasional narrow corridors of land for rights-of-way and certainly had not displaced any local residents. Don Martindale interview, June 19, 1992.

54. Steen, *The U.S. Forest Service,* 79.

55. Campbell interview, Dec. 13, 1991. Campbell remembered several other examples of JNF tenancy in this interview. JNF tenancy is also indicated in Region 7 Inspection Report, Oct. 17, 1938, National Archives, Mid-Atlantic Branch, Philadelphia. For another example, see Hall, "Influences of the National Forests in the Southern Appalachians," 402–7.

56. David Conrad, "The Return of the Forests," unpublished history of Region 7 / Region 8, American Resources Group, 1988, 129. In 1938 the Region 7 inspection report listed "abject forest tenants in southwest Virginia coal field area." Such a designation proved unique among Region 7's southern Appalachia Forests, which included the George Washington, Monongahela, Cumberland, and Allegheny National Forests. See Region 7 Inspection Report, Oct. 17, 1938, National Archives, Mid-Atlantic Branch.

57. Mastran and Lowerre, *Mountaineers and Rangers,* 19.

58. USDA, Forest Service, *National Forest Manual*, 4-A.

59. Campbell interview, Nov. 14, 1991.

60. USDA, Forest Service, *National Forest Manual*, 4-A, 3-P.

61. C. B. Clark interview, Jan. 7, 1992.

62. *Smyth County News*, Nov. 4, 1937.

63. "Holston Working Circle Development Plan, 1941–1950," 7, 10, 11, 32.

64. *New Castle Record*, Apr. 9, 1938. In 1941 Van Alstine also began writing a "Ranger's Column" for the *New Castle Record*, his local newspaper, that reflected another example of typical Forest Service community relations efforts. Van Alstine tended to concentrate on fire prevention during fire season but also shared many of the district's and forest's other activities, involving silviculture, geology, wildlife, and recreation. He chastised timber trespassers, firebugs, and litterbugs, but also reported various community events such as Boy Scout activities and regularly mailed his column home whenever he and his family took vacation trips, such as their venture to Canada during the autumn of 1951.

65. Ernie Karger, pers. comm., Mar. 12, 17, 1992.

66. Probably no aspect differentiates the Forest Service of old from the modern agency more than firefighting policy, which naturally changed after a highly successful campaign against accidental fire. See chapter 9 for the change to endorsing "prescribed fire" and the place of fire in ecosystem management.

67. "A Summons to Save the Wilderness," in *Recreation*, comp. Phillip O. Foss (New York: Chelsea House, 1971), 394; Aldo Leopold, "The Wilderness and Its Place in Forest Recreation Policy," *Journal of Forestry* 19 (Nov. 1921): 718–21; USDA, Forest Service, *National Forest Manual*, 61-L; Worrell, *Principles of Forest Policy*, 30.

68. Giles County Historical Society, *Giles County, Virginia History-Families*, (Pearisburg, VA: Giles County Historical Society, 1983), 46; Henry M. Wilbur, "Facilities Support for Mountain Lake Biological Station," MS, 1992, Mountain Lake Biological Station, p. 10.

69. Henry M. Wilbur interview, Mar. 30, 1992.

6. The Depression Era

1. Franklin Roosevelt himself had practiced extensive conservation in his private life at his Hyde Park estate in New York, planting exhausted farmland and cut-over acres, establishing and maintaining various pine lots in the years following 1911, and experimenting in forestry both independently and with help from the New York State College of Forestry. Apparently he gained a love and appreciation of the natural outdoors as a child, when his father and others took him horseback riding, sailing, and for walks in the forest. His conservation ethic translated into political action during his career as New York governor and New York senator, assistant secretary to the Navy, and finally president. In fact, a cooperative Congress helped Roosevelt and others,

such as Secretary of Interior Harold Ickes and Soil Conservation Chief Dr. Hugh H. Bennett, herald in an entire new age of conservation. Unprecedented federal participation in coordinated funding and policy characterized this new age. See A. L. Riesch Owen, *Conservation under F.D.R.* (New York: Praeger, 1983), 4–6, 69, 157, 191, 193.

2. Gregg, "Uncovering the Subsistence Economy in the Twentieth-Century South: Blue Ridge Mountain Farms," 418–20, 424, 430–31; Hart, "Land Rotation in Appalachia," 148–66; Otto, "Forest Fallowing among the Appalachian Mountain Folk: An Ethnohistorical Study," 3–22; Otto, "Forest Fallowing in the Southern Appalachian Mountains: A Problem in Comparative Agricultural History," 51–63; Thomas, *Southern Appalachia, 1885–1915,* 30, 31–32. Secretary of Agriculture James Wilson probably did not appreciate land rotation and forest fallowing practice when he toured the area during the turn of the nineteenth and twentieth centuries. See James Wilson, "Report on the Forests and Forest Conditions of the Southern Appalachian Region" (Washington, DC: GPO, 1902), reproduced in Michael P. Branch and Daniel J. Philippon, eds., *The Height of Our Mountains: Nature Writing from Virginia's Blue Ridge Mountains and Shenandoah Valley* (Baltimore: Johns Hopkins Univ. Press, 1998), 205.

3. In regard to agriculture, such were the general conclusions for the entire Southern Appalachian Region, as reached by Glenn, "Denudation and Erosion in the Southern Appalachian Region" (esp. pp. 11–12, 22–23, 30). Also see Ayres and Ashe, "The Southern Appalachian Forests," 19.

4. Eller, *Miners, Millhands and Mountaineers,* xix, xiv, 226, 240; John F. Hart, "Loss and Abandonment of Cleared Farm Land in the Eastern United States," *Annals of the Association of American Geographers* 58 (Sept. 1968): 421–23, 434–35; David E. Whisnant, *Modernizing the Mountaineer* (Boone, NC: Appalachian Consortium Press, 1980), 226. For a series of contemporary publications addressing this matter, see Thomas Cooper, "What Is the Problem of Mountain Agriculture?" *Mountain Life and Work* 3 (July 1927): 13–15; L. C. Gray, "Economic Conditions and Tendencies in the Southern Appalachians as Indicated by the Cooperative Survey," *Mountain Life and Work* 9 (July 1933): 8; Charles D. Lewis, "Government Forests and the Mountain Problem," *Mountain Life and Work* 6 (Jan. 1931): 2–9; L. R. Neel, "Agriculture in the Southern Mountains," *Mountain Life and Work* 3 (Apr. 1927): 4–6, 10; W. D. Nicholls, "Families on Submarginal Land," *Mountain Life and Work* 10 (Apr. 1934): 26–28; USDA, *Economic and Social Problems and Conditions of the Southern Appalachians* (Washington, DC: GPO, Misc. Pub. No. 205, Jan. 1935), 5.

5. This type of interaction generally characterized national forests all over the southern Appalachian region. See NFRC, *1930 Annual Report,* 24, and M. A. Matton, "Appalachian Comeback," in *Trees, Yearbook of Agriculture 1949* (Washington, DC: GPO, 1949), 304–9.

6. Ayres and Ashe, "The Southern Appalachian Forests," 18; Pyne, *Fire in America,* 74–75, 143–45; Hall, "To Remake the Appalachians," 321, 334.

7. Seth G. Hobart, George W. Dean, and Edwin E. Rodger, *The History of the Virginia Division of Forestry: 1914–1981* (Charlottesville: Virginia Dept. of Forestry, 1981), 4, 5; *New Castle Record*, Mar. 1, 1941; Ralph R. Widner, ed., *Forests and Forestry in the American States: A Reference Anthology* (National Association of State Foresters, 1968), 298.

8. *Roanoke Times*, Mar. 31, 1940.

9. Ibid., Apr. 28, 1940.

10. Early major fires included two on the Glenwood Ranger District, in 1930 and 1942, that each burned a couple thousand acres of national forest land. The Holston Ranger District saw a few large fires in 1941, one that burned at least 5,000 acres. Both required 1,000–1,200 firefighters to suppress them. *Jefferson News Letter* 1 (May 25, 1936), 6; NFRC, *1936 Annual Report*, 30; *New Castle Record*, May 1, 1937; *Roanoke Times*, Apr. 14, 1940, Oct. 5, 1941; Lewis Smith, "Glenwood District Ranger, August 1931–September 1958," JNF Cultural Resources Division; *Smyth County News*, Apr. 29, 1937, May 8, 1941. When most Region 7 forests received sixty copies or fewer of the 1951 "Prevent Forest Fires" poster, the JNF received 325 (V. Logan to Ted Sharp, Records of the Forest Service, R7, National Archives, Mid-Atlantic Branch).

11. Karger, telephone interview, Mar. 17, 1992.

12. U.S. Forest Service, *Fire Handbook—Region Seven* (Washington, DC: GPO, 1931), 10–13.

13. U.S. Forest Service, *Fire Handbook—Region Seven*, 8.

14. Pyne, *Fire in America*, 172–73.

15. U.S. Forest Service, *Fire Handbook—Region Seven*, 105.

16. Hadacek interview, Dec. 16, 1991. As William Campbell recalled, during the 1930s the JNF paid fire fighters seventy-five cents an hour, while the State of Virginia paid only fifty cents an hour. Strangely, almost all the fires began on federal land (Campbell interview, Dec. 13, 1991). JNF Ranger Van Alstine commented, tongue-in-cheek, in 1952, "By some remarkable system I cannot understand, the State seldom has a large fire" (*New Castle Record*, Mar. 16, 1952).

17. Campbell interview, Dec. 13, 1991.

18. U.S. Forest Service, *Fire Handbook—Region Seven*, 100–101.

19. Campbell interview, Dec. 13, 1991.

20. J.H.W. Smith interview, Jan. 30, 1992. Occasionally, diligent rangers and the right circumstances netted the Forest Service with arson convictions. In May 1941, an all-night search on the Holston District ended in the capture of two young men who later confessed and pled guilty to burning fifty acres near Konnarock (*Smyth County News*, May 15, June 5, 1941).

21. Campbell interview, Dec. 13, 1991. A specific example occurred in November 1937 when a court fined Glenn Anders of Atkins twenty-five dollars after his private-land fire escaped onto JNF territory on Glade Mountain. *Smyth County News*, Nov. 11, 1937.

22. C. B. Clark interview, Jan. 7, 1992.

23. Karger interview, Mar. 17, 1992.

24. *Jefferson News Letter* 1 (May 25, 1936): 8, 10; *Smyth County News,* Apr. 29, 1937. At least part of the many numerous tabulations resulting from this survey may be found in the "Holston Working Circle Development Plan, 1941–1950."

25. Campbell interview, Dec. 13, 1991; *New Castle Record,* Apr. 9, 1938; USDA, Forest Service, *National Forest Manual,* 17-S; Ward Robens, "Historical Data, Glenwood Working Circle, Jefferson National Forest," Sept. 1949, Glenwood Ranger District, history files, Jefferson National Forest.

26. Conrad, "The Return of the Forests," 164; USDA, Forest Service, *National Forest Manual,* 135-S; Robens, "Historical Data, Glenwood Working Circle"; *Smyth County News,* Nov. 25, 1937, Feb. 3, 1938, Feb. 24, 1938.

27. Campbell interview, Dec. 13, 1991; *Smyth County News,* Apr. 22, 1937, Feb. 3, 1938; Steen, *The U.S. Forest Service,* 130, 216; Scott Weidensaul, *Mountains of the Heart: A Natural History of the Appalachians* (Golden, CO: Fulcrum Publishing, 1994), 174. Contemporary articles on the white pine blister rust include "The Pine Blister Disease," *American Forestry* 22 (Dec. 1916): 748–50; "White Pines Threatened by Destructive Disease," *American Forestry* 22 (Nov. 1916): 662–63; Haven Metcalf, "Summary of the White-Pine Blister Rust Situation," *Journal of Forestry* 16 (Jan. 1918): 85–89; Perley Spaulding, "The Blister Rust Disease of White Pine," *American Forestry* 22 (Feb. 1916): 97–98; Perley Spaulding, "The White-Pine Blister Rust Situation," *American Forestry* 22 (Mar. 1916): 137–38.

28. White Pine Blister Rust Control Act of 1940, Public Law 76-486, 76th Cong., 3rd sess., Apr. 26, 1940. This legislation became especially pertinent to the Jefferson, for the 1946 Blister Rust Control Survey listed a full 54,954 JNF acres as bearing white pines (Robens, "Historical Data, Glenwood Working Circle"). By 1941 the blister rust had arrived in Bland, Giles, Grayson, Pulaski, Smyth, and Wythe counties. By 1960 JNF timber management staff maintained control of the rust on 123,000 acres but reported that "butt rots are causing serious loss of timber volume of the Forest." The 1960s marked a turning point in blister rust incidence, though how much this reflected the JNF's eradication efforts or a natural cycle remains uncertain. By the early 1990s blister rust had all but disappeared from the JNF (John Hinrichs, JNF timber staff, pers. comm., May 1992; George Freeland, pers. comm., May 1992; National Archives, RG 95, 69A-2200, box 10; JNF, *1960 JNF Timber Management Plan,* 19; *Smyth County News,* Oct. 30, 1941).

29. The folk tradition claiming the fungus arrived on "Chinese chestnut" trees seems erroneous. "Chinese chestnut" is likely a nickname for the indigenous Chinquapin tree (*Castanea pumila*) which, unlike the American chestnut, naturally resisted the *Endothia parasitica* fungus.

30. For a biological history of the chestnut tree, as well as how the chestnut trees related to society and culture, see Susan Freinkel, *American Chestnut: The Life,*

Death, and Rebirth of a Perfect Tree (Berkeley: Univ. of California Press, 2007). Also see Stephen D. Solomon, "Chestnut Trees Return," *Scientific American Earth* 19, no. 1 (2009): 4. For a historical assessment of the chestnuts themselves, see Ralph H. Lutts, "Like Manna from God: The American Chestnut Trade in Southwestern Virginia." *Environmental History* 9 (July 2004): 497–525.

31. Robens, "Historical Data, Glenwood Working Circle."

32. The JNF's 1941 timber management plan called for the "continued removal of chestnut," and pulpwood and extractwood contractors cut around 200,000 cords on the Glenwood between 1930 and 1950. The Glenwood District had sold almost 75,000 million board feet of chestnut during its first thirty-eight years, but by 1951 rangers estimated only 4,400 million board feet remained on that district. Still, even though the JNF sold 11,510 million board feet of chestnut on the entire Forest between 1951 and 1954, this amounted to only 10 percent of the estimated whole, and a lack of roads rendered the rest inaccessible. See Ronald B. Craig, *Virginia Forest Resources and Industries* (Washington, DC: USDA Misc. Pub. no. 681, Apr. 1949), 1–2, 37; Robert W. Larson and Mackay B. Bryan, *Virginia's Timber,* Forest Survey Release No. 54 (Asheville, NC: USDA, Forest Service, Southeastern Forest Experiment Station, 1959), 3; Bill Morrison, "Writings on History of Forestry on the Jefferson," Nov. 1975, JNF Cultural Resources Division; National Archives, RG 95, 69A-2209, box 10; JNF, *1960 JNF Timber Management Plan,* 5; *New Castle Record,* Apr. 9, 1938; Smith, "Glenwood District Ranger, August 1931–September 1958"; Glenwood Working Circle, Jefferson National Forest, *Streamliner Timber Management Plan,* JNF Cultural Resources Division.

33. The literature documenting this is enormous. For a sampling, see Cavender, *Folk Medicine in Southern Appalachia,* 47–50, 197–201, and generally chapter 4; Hicks, *Appalachian Valley,* 20–21; Charles L. Perdue and Nancy Martin-Perdue, eds., *Talk about Trouble: A New Deal Portrait of Virginians in the Great Depression* (Chapel Hill: Univ. of North Carolina Press, 1996), 47; Rehder, *Appalachian Folkways,* 163–71, 226–29; Laurel Shackelford and Bill Weinberg, eds., *Our Appalachia: An Oral History* (New York: Hill & Wang, 1977), 92–93, 261; Thomas, *Southern Appalachia, 1885–1915,* 111, 116, 219–20, 224. For the economic importance of early flora gathering in the region, see Ina C. Yoakley, "Wild Plant Industry of the Southern Appalachians," *Economic Geography* 8 (July 1932): 311–17.

34. For accounts related to the local tradition of herb gathering in the Mount Rogers/Whitetop area, see William B. Gable and Edward H. Davis, eds., *An Oral History of Konnarock, Virginia* (Charlottesville: Virginia Foundation for the Humanities, 1997), 16, 18, 19, 25, 29, 30, 36, 37, 38, 39, 72, 73.

35. *New Castle Record,* Apr. 9, 1938.

36. Luther C. Hassinger, "The Lumber Industry of Southwest Virginia," *Historical Society of Washington County, Virginia Bulletin,* ser. 2, no. 4 (Spring 1967): 1–16; USDA, Forest Service, *National Forest Manual,* 17-S, 21-S; Robens, "Historical Data, Glenwood Working Circle."

37. Campbell interview, Dec. 13, 1991; Matton, "Appalachian Comeback," 307, 308; Robens, "Historical Data, Glenwood Working Circle."

38. Hermie Medley interview, Jan. 22, 1992. For other sources documenting aspects of the forest economy in southwestern Virginia 1929–30, see a series of county industrial surveys (such as *Industrial Survey of Giles County Virginia* or *Industrial Survey of Bland County Virginia*) compiled by R. L. Humbert et al. between Mar. 1929 and May 1930 and published by Virginia Tech Engineering Extension Division. They are all located in the Virginia Tech Newman Library stacks.

39. See the 1910 writings of Forest Service employee Treadwell Cleveland, "National Forests as Recreation Grounds," in Foss, *Recreation,* 366, 367, 368; Forest Service landscape architect Frank A. Waugh's 1918 study, "Recreation Uses on the National Forests," in Foss, *Recreation,* 377–78; for the historic roles of Waugh and Forest Service landscape architect Arthur H. Carhart, see Donald N. Baldwin, *The Quiet Revolution: Grass Roots of Today's Wilderness Preservation Movement* (Boulder, CO: Pruett Pub. Co., 1972).

40. Charles F. Wilkinson and H. Michael Anderson, *Land Resource Planning in the National Forests* (Washington, DC: Island Press, 1987), 316.

41. Baldwin, *Quiet Revolution,* 201; Mastran and Lowerre, *Mountaineers and Rangers,* 53–54; Dana and Fairfax, *Forest and Range Policy,* 131; Steen, *The U.S. Forest Service,* 209.

42. Arthur H. Carhart, "Recreation in the Forests," *American Forestry* 26 (May 1920): 268–72; Henry Graves, "A Crisis in National Recreation," *American Forestry* 26 (July 1920): 391–400; Swain, *Federal Conservation Policy 1921–1933,* 21. For the history of national parks, see Alfred Runte, *National Parks: The American Experience* (Lincoln: Univ. of Nebraska Press, 1979), and Richard West Sellars, *Preserving Nature in the National Parks* (New Haven: Yale Univ. Press, 1997).

43. USDA, Forest Service, *National Forest Manual,* 98-L.

44. *Roanoke Times,* June 9, 1940.

45. Mastran and Lowerre, *Mountaineers and Rangers,* 53–54; Steen, *The U.S. Forest Service,* 209. Parallel recreational development took place at state level, in territory neighboring the JNF. See Mack H. Sturgill, *Hungry Mother: History and Legends* (Marion, VA: Tucker Printing, 1986), 34, 43, 45–47, 86.

46. *Coalfield Progress,* Feb. 25, 1937, June 15, 1939.

47. Cowdrey, *This Land, This South,* 114–15; Henry Mosby and Charles Handley, *The Wild Turkey in Virginia* (Richmond: Virginia Commission of Game and Inland Fisheries, 1942), 3; *New Castle Record,* Sept. 10, 1938.

48. Ayres and Ashe, "The Southern Appalachian Forests," 26. The fieldwork for this study took place during 1900–1901.

49. For the broader picture involving wildlife rehabilitation for this period, see Cowdrey, *This Land, This South,* 142; Donald C. Swain, *Federal Conservation Policy 1921–1933* (Berkeley: Univ. of California Press, 1963), 43; Worster, *Nature's Economy,* 271.

50. A. R. Cochran memoirs, JNF Cultural Resources Division; Theodore Fearnow and I. T. Quinn, "Action on the Blue Ridge," in *The Yearbook of Agriculture, 1949* (Washington, DC: GPO, 1950), 586, 587.

51. A. R. Cochran, "Jefferson National Forest," *Virginia Wildlife* 10 (May 1949): 7. Leopold published his classic, *Game Management,* in 1933.

52. John B. Lewis, "Success Story: Wild Turkey," in *Restoring America's Wildlife 1937–1987: The First 50 Years of the Federal Aid in Wildlife Restoration (Pittman-Robertson Act)* (Washington, DC: U.S. Dept. of Interior, 1987), 38–39.

53. Pittman-Robertson Act of 1937, Public Law 75-415, 75th Cong., 1st sess., 1937, secs. 1, 2. For the Pittman-Robertson Act in a broader context, see Kier Sterling, "Zoological Research, Wildlife Management, and the Federal Government," in *Forest and Wildlife Science in America: A History,* ed. Harold K. Steen (Durham, NC: Forest History Society, 1999), 48.

54. Theodore Catton and Lisa Mighetto, *The Fish and Wildlife Job on the National Forests: A Century of Game and Fish Conservation, Habitat Protection, and Eco-system Management* (Washington, DC: USDA, Forest Service, 1998), 104.

55. K. P. Butterfield, "Forests and Game," *Virginia Wildlife* 7 (Nov. 1946): 10; Fearnow and Quinn, "Action on the Blue Ridge," 588; Terry Lewis, "Great Game Preserve Proposed in Virginia," *Virginia Forests* 43 (Spring 1988): 21.

56. By 1946, five such game managers worked on the JNF. Butterfield, "Forests and Game," 10–11; Cochran, "Jefferson National Forest," 24; USDA, Forest Service, *1949 Wildlife Management Handbook,* JF-2–1.

57. Fearnow and Quinn, "Action on the Blue Ridge," 588; Granger and Spillers, R7, JNF, General Integrating Inspection, Sept. 1948, National Archives, Mid-Atlantic Branch; *Roanoke Times,* June 14, 1936, Aug. 1, 1937; *Smyth County News,* Mar. 3, 1938.

58. *Coalfield Progress,* Apr. 2, 1936; *Jefferson News Letter* 1 (May 25, 1936): 5; Terry Lewis, "Great Game Preserve," 21; *New Castle Record,* Apr. 30, 1938, Oct. 14, 1939.

59. USDA, Forest Service, *Story of the Jefferson National Forest* (Washington, DC: GPO, 1964); Randall, "Jefferson's Forest," 62; Paul Shrauder interview, Dec. 10, 1991.

60. Clark, "First Thirty Years on the Clinch, 1935–1965"; Fearnow and Quinn, "Action on the Blue Ridge," 588; *New Castle Record,* Sept. 10, 1938, June 10, 1939; *Smyth County News,* Jan. 30, 1941; *Southwest Virginia Enterprise,* Feb. 13, 1940.

61. For general sources concerning illegal whiskey manufacture, see Jess Carr, *The Second Oldest Profession: An Informal History of Moonshining in America* (Englewood Cliffs, NJ: Prentice Hall, 1972); Joseph Earl Dabney, *Mountain Spirits: A Chronicle of Corn Whiskey from King James' Ulster Plantation to America's Appalachians and the Moonshine Life* (New York: Charles Scribner's Sons, 1974); Danny Fulks, "Moonshine Reflections," *Timeline* 7 (June–July 1990): 42–54; Wilbur R. Miller, *Revenuers and Moonshiners: Enforcing Federal Liquor Law in the Mountain South, 1865–1900* (Chapel Hill: Univ. of North

Carolina Press, 1991); Eliot Wigginton, ed., *The Foxfire Book* (Garden City, NY: Doubleday, 1972), 301–45. For accounts related to the Mount Rogers/Whitetop area, see Gable and Davis, *Oral History of Konnarock,* 15, 35, 36, 66. Also see Shackelford and Weinberg, *Our Appalachia,* 104–8, 247–50.

62. Campbell interview, Dec. 13, 1991.

63. Ibid.; C. B. Clark interview, Jan. 7, 1992.

64. J. N. Jefferson interview, Jan. 17, 1992.

65. Campbell interview, Dec. 13, 1991; C. B. Clark interview, Jan. 7, 1992. The informal policy toward moonshiners that the JNF pursued with its local constituencies was not unique among national forests in the Appalachian region. William C. Curnutt, JNF supervisor 1957–66, recalled a similar incident on the Monongahela National Forest in West Virginia. After discovering a still, he also left a note and requested that the device be moved after the owner had "run his batch," since, in this instance, the mash was obviously approaching maturity. Months later, upon accidentally meeting Curnutt, the moonshiner rewarded him with two quarts of moonshine for his kindness (*Roanoke Times,* Dec. 21, 1966).

66. Former JNF Deputy Supervisor Cecil Cordell recalled that he and others working on the Pisgah National Forest in North Carolina also contended with resident moonshiners in similar ways, though they tended to actually avoid whiskey stills. Many of the Pisgah employees themselves hailed from the region and knew where not to tread. But Forest Service involvement could run quite deep. Cordell recalled one well-known moonshiner who also had an art for gaining multiple national forest permits for gathering dogwood, then in high demand for textile mill shuttlecocks. One day the dogwood-gathering moonshiner stepped into the Pisgah ranger district office and said to young Cordell, "Tell the ranger the dogwood's ready to scale." Cordell asked him which permit he was operating under, but the moonshiner only repeated his message, and later Cecil learned that "the dogwood's ready to scale" was code indicating that a fresh batch of whiskey had just been run. Cordell's boss was a customer (Cecil Cordell interview, Dec. 18, 1991).

67. *Southwest Virginia Enterprise,* Apr. 1, 1949.

68. Kathryn Newfont briefly alludes to a similar dynamic regarding the national forest rangers in western North Carolina. See Newfont, "Moving Mountains," 45–47.

69. Dora Testerman interview, p. 14, Konnarock Oral History, archived at Emory and Henry College, available at http://library.ehc.edu/konnarock.html.

7. World War II and Postwar Transitions

1. *New Castle Record,* June 5, 1953, Feb. 1, 1962.

2. Ibid., Aug. 22, 1952. Increasingly during the late 1950s and early 1960s higher steel towers replaced older wooden ones, but only for a short period before aerial fire detection replaced lookout towers altogether. See C. B. Clark, "First Thirty Years on the Clinch, 1935–1965," JNF Cultural Resources Division;

Hobart et al., *History of the Virginia Division of Forestry*, 6, 16; *Roanoke Times*, Mar. 31, 1940.

3. Granger and Spillers, R7, JNF, GII, Sept. 1948, National Archives, Mid-Atlantic Branch; Pyne, *Fire in America*, 174.

4. Hobart et al., *History of the Virginia Division of Forestry*, 35.

5. Granger and Spillers, R7, JNF, GII, Sept., 1948, National Archives, Mid-Atlantic Branch.

6. G. H. Lentz, "War Activities of the Forest Service in Region Seven," Allegheny Section Meeting speech, Feb. 26, 1943; R. M. Evans, Regional Forester, Eastern Region (7), "A Three-Point Program for Better Forests," n.d., RG 95, Records of the Forest Service, R7, National Archives, Mid-Atlantic Branch; Steen, *The U.S. Forest Service*, 254.

7. C. B. Clark interview, Jan. 7, 1992. Ronald B. Craig, *Virginia Forest Resources and Industries* (Washington, DC: USDA Misc. Pub. no. 681, Apr. 1949), 1, 12, 30; *Southwest Virginia Enterprise*, Aug. 31, 1943, Nov. 5, 1943, Feb. 29, 1944. For an extensive listing of the numerous wood and paper products required by the war effort, see Richard G. Lillard, *The Great Forest* (New York: Alfred A. Knopf, 1947), 312–22.

8. "Jefferson National Forest Timber Management Program," statistics, JNF Timber Files; *Roanoke Times*, July 11, 1943. The labor shortage for acquiring and processing forest products was a national problem, with shortages of loggers throughout national forests and private woodlands. For the national picture, see Lucia, *The Big Woods*, 77, and Steen, *The U.S. Forest Service*, 250. For Virginia, see Craig, *Virginia Forest Resources and Industries*, 38–39.

9. *Coalfield Progress*, Apr. 19, 1951; Hadacek interview, Dec. 16, 1991; *New Castle Record*, Oct. 29, 1959; U.S. Dept. of Agriculture, Forest Service, *Facts about Region 7* (Upper Darby, PA: USDA, Forest Service, 1961–62), 29.

10. During August 1960, for example, New Castle District's Bruce Kibler and June Elmore became two of forty Region 7 fire personnel sent West for emergency duty (*New Castle Record*, Aug. 18, 1960). Firefighting information also derived from Hadacek interview, Dec. 16, 1991; Charles Saboites interview, Feb. 20, 1992.

11. Craig, *Virginia Forest Resources and Industries*, 25; *New Castle Record*, Aug. 14, 1953. In 1953 the JNF cut 7.9 million board feet. The postwar high was 38.8 million in 1969. For interim years, see appendix on annual timber-cut statistics.

12. Culhane, *Public Lands Politics*, 50–51.

13. David A. Clary, *Timber and the Forest Service* (Lawrence: Univ. Press of Kansas, 1986), 153.

14. Memo, Forester Gross (whose further identification remains unknown) to W.O. Forester Morriss, National Archives, RG 95, 69A-2209, box 10.

15. Campbell interview, Dec. 13, 1991; Morrison, "Writings on History of Forestry on the Jefferson"; Robens, "Historical Data, Glenwood Working Circle"; Glenwood Working Circle, *Streamliner Timber Management Plan*; USDA

Forest Service, *Story of the Jefferson National Forest* (Washington, DC: GPO, 1964). In 1945 portable circular saws, with an annual capacity of less than a million board feet, cut almost half of Virginia's total timber harvest of 955 million board feet. Only nine large operations, with annual capacities exceeding five million board feet, operated in Virginia by then, but cut only about 10 percent of the state total (Craig, *Virginia Forest Resources and Industries,* 1).

16. Granger and Spillers, R7, JNF, General Integrating Inspection, Sept. 1948, RG 95, National Archives, Mid-Atlantic Branch.

17. "U.S. Forest Service data," as cited by Marion Clawson, *The Federal Lands since 1956: Recent Trends in Use and Management* (Washington, DC: Resources for the Future, 1967), 60, table 4; Marion Clawson, *The Federal Lands Revisited* (Washington, DC: Resources for the Future, 1983), 99–101, 291, table A-11. Similarly, national park visitation more than tripled between 1955 and 1974, from around fourteen million to forty-six million. See Alfred Runte, *National Parks: The American Experience* (Lincoln: Univ. of Nebraska Press, 1979), 26.

18. *Coalfield Progress,* Dec. 25, 1958.

19. Paul Wayne Hirt, *A Conspiracy of Optimism: Management of the National Forests since World War Two* (Lincoln: Univ. of Nebraska Press, 1994), 152–57.

20. Watershed Protection and Flood Prevention Act of 1954, Public Law 83-566, 83rd Cong., 2nd sess., 1954.

21. Cochran, "Jefferson National Forest," 8; USDA, Forest Service, *Facts about Region 7,* 36, 43; Watershed Protection and Flood Prevention Act (PL 566), Aug. 4, 1954, *United States Statutes at Large* 1954, vol. 68, pt. 1, 666–68.

22. JNF 2520 Files, Glade Mountain Mining Restoration folder.

23. Ibid. Campbell interview, Dec. 13, 1991.

24. Hamilton K. Pyles to Murray H. Stevens, Jan. 15, 1960, JNF, Blacksburg Ranger District, 2820 files.

25. Don Blackburn interview, July 10, 1992; Campbell interview, Dec. 13, 1991.

26. BLM permit no. 045280, BLM mineral permit no. 045836, National Archives, RG 95, 62A-962, box 38; BLM prospecting permit no. 045170, National Archives, RG 95, 62A-962, box 36. Murray H. Stevens of the Appalachian Trail Conference also advocated Forest Service protection of this land for hikers' usage. See Murray H. Stevens to Richard E. McArdle, Jan. 4, 1960, JNF, Blacksburg Ranger District, 2820 files.

27. W. C. Curnutt to R-7 Forester, National Archives, RG 95, 70A-2049, box 5.

28. *Lynchburg News,* July 31, 1960; *Roanoke Times,* July 1, 3, 1960, Oct. 16, 1960.

29. *Bluefield Daily Telegraph,* Dec. 19, 1960, Jan. 14, 1961; *Giles Virginian-Leader,* Jan. 19, Feb. 25, Aug. 10, 1961; *Roanoke Times,* Oct. 16, Nov. 16, 19, Dec. 6, 20, 1960; *(Princeton-Bluefield) Sunset News-Observer,* Dec. 21, 1960; USDA, Forest Service, *Story of the Jefferson National Forest* (Washington, DC: GPO, 1964).

30. Clark, "First Thirty Years"; *Coalfield Progress,* Oct. 25, 1945; Fearnow and Quinn, "Action on the Blue Ridge," 589, 592; C. P. Mead, "Forests and Game," *Virginia Wildlife* 9 (Jan. 1948): 16; George B. P. Mullin, "The Growing Recre-

ational Use on the Jefferson National Forest," *Virginia Wildlife* 16 (Apr. 1955): 18; *New Castle Record*, June 22, 29, 1951, Jan. 9, 1953; Otis et al, *The Forest Service and the Civilian Conservation Corps*, 56; Reeves, "The History and Development of Wildlife Conservation in Virginia," 64–65. Some have doubted the numbers concerning bears, claiming bear hunters checked their game at more than one station to deliberately inflate population estimates, thus ensuring a bear-hunting season.

31. Burd McGinnes interview, Jan. 29, 1992.

32. Ibid; *New Castle Record*, Nov. 22, 1957. A sampling of these studies include the following: Joseph Peter Bachant, "The Distribution and Abundance of the Eastern Wild Turkey as Influenced by Forest Wildlife Management Practices" (master's thesis, 1963); Michael J. Blymyer, "Impact of Clearcutting on Indigenous Mammals of Southwest Virginia" (master's thesis, 1976); Robert L. Curtis Jr., "Climatic Factors Influencing Hunter Sightings of Deer on the Broad Run Research Area" (master's thesis, 1971); Robert Guy Hooper, "The Influence of Habitat Disturbance on Bird Populations" (master's thesis, 1967); J. S. Larson, "Wildlife Forage Clearings on Forest Lands: A Critical Appraisal and Research Needs," (Ph.D. diss., 1966); Earnest Davis Seneca, "Deer Distribution, Usage and Population on the Broad Run Management Area, Craig County, Virginia" (master's thesis, 1961); William Dickey Weekes, "A 15-Year Research Summary and Hunting Harvest Data Evaluation of the Broad Run Management Area" (master's thesis, 1974).

33. Clark, "First Thirty Years"; *Roanoke Times*, Nov. 1, 1964.

34. See U.S. Fish and Wildlife Service, Historical Hunting License Data, at http://wsfrprograms.fws.gov/Subpages/LicenseInfo/Hunting.htm.

35. Arthur Randolph Shields, "The Isolated Spruce and Spruce-Fir Forests of southwestern Virginia: A Biotic Study" (Ph.D. diss., Univ. of Tennessee, 1962), 27, 34–37, 159–65.

36. Earl Sludes, "A White Pine Provenance Study in the Southern Appalachians," USDA Forest Research Paper SE-2, Feb. 1963; Earl R. Sludes and Keith W. Dorman, "Performance in the Southern Appalachians of Eastern White Pine Seedlings from Different Provinces," USDA Forest Service Research Paper SE-90, Dec. 1971; Donald Beck, "Polymorphic Site Index Curves for White Pine in the Southern Appalachians," USDA Forest Research Paper SE-80, Apr. 1971; Charles E. McGee and Lino Della-Bianca, "Diameter Distributions in Natural Yellow Poplar Stands," USDA Forest Research Paper SE-25, Feb. 1967; Donald E. Beck and Lino Della-Bianca, "Yield of Unthinned Yellow Poplar," USDA Forest Research Paper SE-58, Mar. 1970; Donald E. Beck and Lino Della-Bianca, "Growth and Yield of Thinned Yellow-Poplar," USDA Forest Research Paper SE-101, Nov. 1972; and Donald E. Beck and Lino Della-Bianca, "Board-Foot Diameter Growth of Yellow-Poplar after Thinning," USDA Forest Research Paper SE-123, Jan. 1975.

37. Catherine Keever, "Present Composition of Some Stands of the Former Oak-Chestnut Forest in the Southern Blue Ridge Mountains," *Ecology* 34

(1953): 44–54; J. Frank McCormick and Robert B. Platt, "Recovery of an Appalachian Forest Following Chestnut Blight," *American Midland Naturalist* 104 (Oct. 1980): 264–73; Thomas C. Nelson, "Chestnut Replacement in the Southern Highlands," *Ecology* 36 (Apr. 1955): 352–53; Frank W. Woods and Royal E. Shanks, "Natural Replacement of Chestnut by Other Species in the Great Smoky Mountains National Park," *Ecology* 40 (July 1959): 349–61. USDA forest pathologists actually began experimenting with blight-resistant chestnut strains as early as 1909, when chestnut blight began to gain momentum. See Frederick H. Berry, "Chestnut Breeding in the U.S. Department of Agriculture," in *Proceedings of the American Chestnut Symposium* (Morgantown: West Virginia Univ. Books, 1978), 39–40.

38. *Roanoke Times,* June 12, 1964.

39. James Wilson to Gifford Pinchot, 1905, in Foss, *Recreation,* 364; Steen, *The U.S. Forest Service,* 75; Worrell, *Principles of Forest Policy,* 57. For a good overview of multiple use, see Hirt, *A Conspiracy of Optimism,* 171–92.

40. Multiple-Use and Sustained Yield Act of 1960, Public Law 86-517, 86th Cong., 2nd sess., 1960.

41. Ibid., secs. 1, 2.

42. Richard E. McArdle, "Multiple Use—Multiple Benefits," *Journal of Forestry* 51 (May 1953): 323–25, as quoted in Foss, *Recreation,* 401. And as historian Harold K. Steen wrote in 1976, "unanimity is an impossible and even undesirable goal in a democratic society, but there must be some middle ground that is acceptable to most" (Steen, *The U.S. Forest Service,* 323).

43. Frank Gregg, "Public Land Policy: Controversial Beginnings for the Third Century," in *Government and Environmental Politics: Essays on Historical Developments since World War Two,* ed. Michael J. Lacey (Baltimore: Johns Hopkins Univ. Press, 1989), 166–67; Mastran and Lowerre, *Mountaineers and Rangers,* 141; William E. Shands, Perry R. Hagenstein, and Marissa T. Roche, *National Forest Policy: From Conflict toward Consensus* (Washington, DC: Conservation Foundation, 1979), 5; Worrell, *Principles of Forest Policy,* 29.

44. Herbert Kaufman, *The Forest Ranger: A Study in Administrative Behavior* (Baltimore: Johns Hopkins Univ. Press, 1960), 68.

45. Multiple Use Sustained Yield Act, sec. 4(a).

46. Ibid., sec. 4(b).

47. Generally, see former Forest Service chief historian David Clary, *Timber and the Forest Service* (Lawrence: Univ. Press of Kansas, 1986); also, Shands et al., *National Forest Policy: From Conflict toward Consensus,* 5. Paul Hirt lists many sources related to the Forest Service Employees for Environmental Ethics, a group that arose within the agency during the late 1980s in protest of traditional timber emphasis. See Hirt, *A Conspiracy of Optimism,* 315–16. During the early 1990s, JNF Archaeologist Mike Barber felt this scenario continued, in which archaeology and cultural resources (as the minimal section of the already small recreation budget) in general were funded as a "support" contingency and not as true resources in themselves (various personal communications to the author and author observations, 1991–1993).

48. For more on the debate, see Helen M. Ingram and R. Kenneth Godwin, "Conservation and the Forces of Change," in *Public Policy and the Natural Environment* (Greenwich, CT: JAI Press, 1985), 4:168–69. Ingram and Godwin's argument suffers from some degree by using "conservation" to describe the advocacies of people like John Muir, who distinguished themselves from their contemporaries like Pinchot by advocating what we today call "preservation" (169–70), most readily ascribed to the U.S. Park Service today. For other opinions and observations, see Culhane, *Public Lands Politics*, 327; George R. Hall, "The Myth and Reality of Multiple Use Forestry," in *Politics, Policy, and Natural Resources*, ed. Dennis L. Thompson (New York: Free Press, 1972), 363–75, esp. 374; and, Walter A. Rosenbaum, *Environmental Politics and Policy* (Washington, DC: Congressional Quarterly Press, 1985), 262.

49. Ray Walters, pers. comm., Feb. 18, 1992.

50. Frome, *The Forest Service*, 84; Mastran and Lowerre, *Mountaineers and Rangers*, 97, 168; USDA, Forest Service, *National Forest Manual*, 4-A.

51. Frome, *The Forest Service*, 44. Several rangers and other JNF personnel expressed this kind of disillusionment and frustration to the author during 1991–93. Also evident were generational educational roots: certain personnel went to forestry school with the belief and desire that they would actually work in the forest. Later generations, educated during or after the bureaucratic transition, did not necessarily expect to be fieldworkers.

52. Campbell interview, Dec. 13, 1991; Frances Runyan interview, Dec. 9, 1991; and various personal communications from JNF personnel who arrived there in 1966 or shortly thereafter as a result of the administrative redesignation.

53. Conrad, "The Return of the Forests," 65–66, 320–25; Kaufman, *The Forest Ranger*, 9.

54. Charles Blankenship interview, Dec. 12, 1991; Clark, "First Thirty Years on the Clinch; Runyan interview, Dec. 9, 1991.

55. Kaufman, *The Forest Ranger*, presented the classic work depicting the Forest Service tradition of decentralization, cited heavily by Culhane, *Public Lands Politics*, in which Culhane effectively argues resistance of the Forest Service to capture by the timber industry, based on this earlier tradition. Traditional Forest Service emphasis on the timber resource to the detriment of nontimber resources has had a number of critics, especially since the 1960s. Ralph Nader was among the earlier critics [see Daniel R. Barney, *The Last Stand: Ralph Nader's Study Group Report on the National Forests* (New York: Grossman Publishers, 1974)]. David A. Clary, *Timber and the Forest Service* (Lawrence: Univ. Press of Kansas, 1986), generalizes from many citations from the Coconino National Forest, perhaps capturing the broader picture for western national forests. For an additional valuable study, see Hirt, *A Conspiracy of Optimism*.

56. Kaufman, *The Forest Ranger*, 83, 86. During the 1980s landscape architects began moving to individual districts for the first time after an initial concentration in the centralized supervisor's office. As Charles Blankenship noted, this perhaps represented a reverse trend back toward decentralization (Blankenship interview, Dec. 12, 1991).

57. Steen cites the figures of 90 percent foresters in 1958 down to 50 percent in 1973. See Steen, *The U.S. Forest Service*, 318.

58. Blankenship interview, Dec. 12, 1991.

59. See U.S. Congress, *Economics of Federal Timber Sales,* Hearings before the Subcommittee on Forests, Family Farms, and Energy of the Committee on Agriculture, House of Representatives, 99th Cong., 1st sess., 1985. The below-cost phenomenon is discussed at greater length later.

60. *Coalfield Progress,* Dec. 25, 1958.

61. Timber Management Program, JNF Timber Files; *Story of the Jefferson National Forest;* Mastran and Lowerre, *Mountaineers and Rangers,* 144. Nationally the Forest Service reached its peak of timber sales during the 1980s at 12 billion board feet. By 2005 sales were down to 3 billion board feet. See Hays, *The American People and the National Forests,* 137–38.

62. *Story of the Jefferson National Forest;* Thomas A. Hoots, "Some Facts about Timber Management and Roads on the Jefferson National Forest, Virginia," JNF Cultural Resources Division. Hoots wrote the mileage expanded from 328 miles in 1964 to over 1,000 miles by 1983. Forest Service road construction by itself has generated a great deal of controversy. The agency's argument centered around access to timber sales, current and future, and the added benefit of accommodating hikers, horseback riders, all-terrain vehicles, and other recreational uses. Critics charged the expense and maintenance of such roads did not justify their construction and compromised the forest's environmental quality in the process. See Hirt, *Conspiracy of Optimism,* 144–49.

63. Frome, *The Forest Service,* 28–29; Shands and Healy, *The Lands Nobody Wanted,* 142.

64. Frome, *The Forest Service,* 30.

65. USDA, Forest Service, Poverty Creek Unit Plan Environmental Statement, summary sheet, Jan. 19, 1972, JNF Cultural Resource Division.

66. Forest and Rangeland Renewable Resources Planning Act (PL 93–378), Aug. 17, 1974, *United States Statutes at Large* 1974, vol. 88, pt. 1, 476–80; National Forest Management Act (PL 94–588), *United States Statutes at Large,* 1976, vol. 90, pt. 2, 2949–2963; Dana, *Forest and Range Policy,* 323; Shands and Healy, *The Lands Nobody Wanted,* 136; Shands et al., *National Forest Policy,* vii, 2, 3.

67. A plethora of literature deals with these subjects alone, and the following are merely suggestive: W. J. Rorabaugh, *Berkeley at War: The 1960s* (New York: Oxford Univ. Press, 1989) offers a unique account of the birthplace where much of the 1960s political activism began and includes chapters on liberalism, Black Power, communism, and counterculture. Allen J. Matusow, *The Unraveling of America: A History of Liberalism in the 1960s* (New York: Harper & Row, 1984) offers a valuable account analyzing Keynesian economics and liberal politics, as well as a strong section on the counterculture. Fred Halsted, *Out Now: A Participant's Account of the American Movement against the Vietnam War* (New York: Monad Press, 1978), and Todd Gitlin, *The Sixties: Years of Hope, Days of Rage* (New York: Bantam Books, 1989), both offer works

that mix personal memoir within a larger sociopolitical, historical context. A good introductory work to the civil rights movement remains in James A. Geschwender, ed., *The Civil Rights Movement, Ghetto Uprisings, and Separatism* (Englewood Cliffs, NJ: Prentice Hall, 1971).

68. Following a protracted lawsuit, in which the FBI attempted to block publication, the re-release of Peter Matthiessen's *In the Spirit of Crazy Horse* (New York: Viking Press, 1991) offers a highly important, though quite disturbing, journalistic account of the violence surrounding the American Indian Movement, otherwise neglected by books dealing with activism in the 1960s and 1970s. The first Earth Day, now an annual event widely recognized, was not created in 1970 solely by countercultural elements as some then charged, and it generated no small amount of interest from mainstream society. A good indication of this lies in National Staff of Environmental Action, *Earth Day— The Beginning* (New York: Arno Press and New York Times, 1970), which offers a large number and variety of contributing essays by Walter Mondale, Eric Sevareid, Ralph Nader, Rene Dubos, Margaret Mead, and many others. Samuel P. Hays covers many aspects of both the modern environmental movement and recent environmental politics in *Beauty, Health, and Permanence: Environmental Politics in the United States, 1955–1985* (New York: Cambridge Univ. Press, 1987).

69. Harold Calhoun's comments in the "Mount Rogers NRA Lands Staff Comments Relative to Proposed Unit Plans," Aug. 6, 1973, Jefferson National Forest, Mount Rogers National Recreation Area Headquarters Land (2100 File).

8. The Mount Rogers National Recreation Area

1. See Charles B. Coal, *The Life and Adventures of Wilburn Waters: The Famous Hunter and Trapper of White Top Mountain* (1878; Richmond, VA: M. D. Hart, 1960).

2. In addition to his own research, Arthur Randolph Shields cited many of the earlier scientific studies conducted in the Mount Rogers / Whitetop area. See Arthur Randolph Shields, "The Isolated Spruce and Spruce-Fir Forests of Southwestern Virginia: A Biotic Study" (Ph.D. diss., Univ. of Tennessee, 1962). For more recent studies, see Kenneth Edward Diebel, "Isozyme Variation within the Fraser Fir Population on Mt. Rogers, Virginia" (Ph.D. diss., Virginia Polytechnic Institute and State Univ., 1989); Niki S. Nicholas, "Stand Structure, Growth, and Mortality in Southern Appalachian Spruce-Fir" (Ph.D. diss., Virginia Polytechnic Institute and State Univ., 1992), and Peter S. White, ed., *The Southern Appalachian Spruce-Fir Ecosystem: Its Biology and Threats* (Gatlinburg, TN: Uplands Field Research Laboratory, 1984). S. B. McTague Jr.'s specialized study, "Petrography and Petrology of 'Unakites' Located in the Mount Rogers Area, Southwestern Virginia" (master's thesis, Virginia Polytechnic Institute, 1967), 6–11, also contains a useful summary of previous geological studies in the Mount Rogers area.

3. David E. Whisnant, *All That Is Native and Fine: The Politics of Culture in an American Region* (Chapel Hill: Univ. of North Carolina Press, 1983), 181–252.

4. Patrick Jennings, interview by the author, Jan. 6, 1992.

5. USDA, Forest Service, *Facts about Region 7,* 16, 32; Jennings interview, Jan. 6, 1992.

6. Frome, *The Forest Service,* 28, 141; Hirt, *A Conspiracy of Optimism,* 202–3. Congress's establishment of the Bureau of Outdoor Recreation in the Department of the Interior in 1962 similarly reflected growing demands for national recreational needs.

7. Although the National Park Service administered their first National Recreation Area, the Coulee Dam NRA, as early as 1946, they clearly joined the post-ORRRC trend by designating their subsequent NRAs during and after the mid-1960s; nine between 1964 and 1968, and another seven during the 1970s. See Conservation Foundation, *National Parks for a New Generation: Visions, Realities, Prospects* (Washington, DC: Conservation Foundation, 1985), 313–31.

8. Comprised of the secretaries of Interior, Agriculture, Defense, Commerce, Health, Education, and Welfare, and the administrator of the Housing and Home Finance Agency.

9. Recreation Advisory Council, "Federal Executive Branch Policy Governing the Selection, Establishment, and Administration of National Recreation Areas," Mar. 26, 1963, in USDA, Forest Service, *Mount Rogers Final Management Plan* (Roanoke, VA: Jefferson National Forest, 1978), 117–18.

10. *Roanoke Times,* Feb. 18, 1962, Aug. 9, 1964; *Smyth County News,* Feb. 3, 1966.

11. 89th Cong., 1st sess., *Congressional Record* (Sept. 7, 1965): H 4112; Jennings interview, Jan. 6, 1992.

12. Charles A. Blankenship, "Mount Rogers National Recreation Plan," JNF Cultural Resources Division, 2, 6, 8, 10; Orville Freeman to Harold Cooley, Mar. 20, 1964, in H. Doc. 910, 89th Cong., 1st sess., 1964.

13. In addition to meeting with the Boards of Supervisors of the five counties in proximity to the NRA, a partial list of groups that planners and various others (including forest supervisors and the undersecretary of agriculture) visited included Troutdale School, Damascus Bank, Damascus High School, Konnarock School, Oak Hill Academy (near Mouth of Wilson), and the Mount Rogers Community Improvement Club (Whitetop). Blankenship interview, Dec. 12, 1991; Charles Blankenship, pers. comm., Dec. 2, 1992.

14. Blankenship, "Mount Rogers National Recreation Plan," 8; *Mount Rogers National Recreation Area, Jefferson National Forest, Virginia: A Review,* Forest Service publication, 1969, JNF Cultural Resources Division.

15. *Bristol Herald Courier,* Aug. 18, 1972.

16. Pam Daniels, *The Stewardship of the Land: A Selected Bibliography of Current Readings* (Monticello, IL: Council of Planning Librarians, Exchange Bibliography no. 452, Sept. 1973), 9; Samuel P. Hays, *Beauty, Health, and Permanence: Environmental Politics in the United States, 1955–1985* (New York: Cambridge Univ. Press, 1987), 166–67.

17. Appalachian Regional Commission et al., *Subdividing Rural America: Impacts of Recreational Lot and Second Home Development,* prepared by American Society of Planning Officials et al. (Washington, DC: GPO, 1976), 2, 11. This document contains much valuable information regarding the "recreational development land boom" of the late 1960s and early 1970s.

18. An indication of such support revealed itself in July 1973, when opposition had reached its early crescendo. During a meeting with Grayson County Board of Supervisors, the Forest Service frankly outlined its plan to acquire an additional 10,000 acres in Grayson County that would involve land occupied by sixty-five families. Supervisory Forester George Wolfel expected a higher percentage of condemnation than previous acquisition. At the same meeting, community activist Rev. William Gable vocally opposed this plan, but the board refused to endorse his petition and continued to support the Forest Service (Grayson County Board of Supervisors meeting description, July 12, 1973, Jefferson National Forest, Mount Rogers National Recreation Area Headquarters Land [2100 File]).

19. "Report to Accompany H.R. 10366," Senate Report No. 1182, Public Law 89-438, 89th Cong., 2nd sess., 1966.

20. The document that the JNF's interdisciplinary team produced was called the "Supplemental Landownership Adjustment Plan and Recommendations for Condemnation Policy Exceptions for the Mount Rogers NRA." "History of Mt. Rogers NRA Planning," chronology, Jefferson National Forest, Mount Rogers National Recreation Area Headquarters Land (2100 File).

21. Ben H. Bolen to Elbert Cox, Director of Commission of Outdoor Recreation, Sept. 14, 1967; Commonwealth of Virginia, Commission of Outdoor Recreation, Project Status Report, Nov. 21, 1967; and other various correspondences and documents in the Land Acquisition file for the Grayson Highlands State Park, Department of Conservation and Recreation, 203 Governor Street, Suite 302, Richmond, VA. Harold Calhoun remembers the Grayson Highlands acquisitions as particularly harsh, reminding him and others of the Shenandoah acquisitions of the 1920s and 1930s; letter to the author, Nov. 18, 1992.

22. NRA planners necessarily worked at a different level, but the resulting difference in perspective between the office and the field created a certain amount of tension among the NRA staff themselves. On August 6, 1973, amidst the growing controversy over condemnation, the JNF produced an interagency document entitled "Mount Rogers NRA Lands Staff Comments Relative to Proposed Unit Plans." Supervisor Penfold called this a "rather thought provoking document prepared by the Mount Rogers Lands Staff. They make some points which seem to me to be important and valid." But an anonymous member of the Unit Planning Team handwrote, sarcastically, "this appears to be a totally unnecessary 'report' as well as uncalled for. Perhaps we should call in all the District clerks, technicians and forest workers to review the work of the professional planners and let them decide on policy and procedure"

(Michael J. Penfold to Charles Blankenship and Mount Rogers Unit Planning Team, Sept. 6, 1973, "Mount Rogers NRA Land Staff Comments Relative to Proposed Unit Plans," Jefferson National Forest, Mount Rogers National Recreation Area Headquarters Land [2100 File]).

23. George A. Wolfel to George W. Kelly, Unit Planner, June 6, 1972, Jefferson National Forest, Mount Rogers National Recreation Area Headquarters Land (2100 File).

24. Harold K. Calhoun to the Record, Sept. 21, 1971, Jefferson National Forest, Mount Rogers National Recreation Area Headquarters Land (2100 File); Harold K. Calhoun to the Record, n.d., describing meeting with Ms. Atwood on Feb. 18, 1969, Jefferson National Forest, Mount Rogers National Recreation Area Headquarters Land (2100 File).

25. Uniform Relocation Assistance and Real Property Acquisition Policies Act of 1971, Public Law 91-646, 91st Cong., 1st sess., 1971, title 2, sec. 201, 205, 216.

26. The foregoing derived from Don Martindale, pers. comm., July 31 and Aug. 5, 1992; and various correspondence in Sessie Pruitt folder, JNF Land Files.

27. McDonald and Muldowny, *TVA and the Dispossessed,* 138.

28. All cases, listed under surnames, JNF Land Files.

29. In 1992 NRA acreage stood at between 115,000 and 116,000 acres (Jim Barker, NRA Realty specialist, pers. comm., Aug. 1992). All land acquired since 1975 has come from willing sellers.

30. Condemnation lawsuit numbers determined from Mount Rogers NRA / Wythe Ranger District Land Status Atlas, JNF Land Files. Subsequent information pertaining to specific land tracts derived from the Mount Rogers NRA Land Status Atlas and individual files concerning the forty-eight condemnation lawsuits, JNF Land Files.

31. James Brown and George Hillery, "The Great Migration 1940–1960," in *The Southern Appalachian Region: A Survey,* ed. T. Ford (Lexington: Univ. of Kentucky Press, 1962), 54–78; Jerome P. Pickard, "Counting Noses in Region and Nation: A Projection," *Appalachia* 13 (Mar.–Apr. 1980): 1–12; Clyde B. McCoy and Virginia McCoy Watkins, "Stereotypes of Appalachian Migrants," in Ergood and Kuhre, *Appalachia: Social Context Past and Present,* 100–105; Gail A. Leithauser and Andrew P. Levien, "New Growth Patterns Affect Billions in Investments," *Appalachia* 13 (July–Aug. 1980): 5–12; Jerome Pickard, "A Decade of Change for Appalachia," *Appalachia* 14 (July–Aug. 1981): 1–9; Jerome Pickard, "Appalachia's Decade of Change = A Decade of Inmigration," *Appalachia* 15 (Sept.–Oct. 1981): 24–28. Also see Shackelford and Weinberg, *Our Appalachia,* chap. 7; Phillip J. Obermiller, Thomas E. Wagner, and E. Bruce Tucker. eds., *Appalachian Odyssey: Historical Perspectives on the Great Migration* (Westport, CT: Praeger, 2000); Eller, *Uneven Ground,* 11–12, 20–27.

32. Gable interview, Jan. 8, 1992. William Gable did have some earlier connection to the area through his father, who was a Lutheran missionary in Konnarock during the 1930s. In any case, Samuel Hays recognized a national trend toward cooperation between outsiders and locals against development

during the modern environmental era. See Samuel Hays, *Beauty, Health, and Permanence: Environmental Politics in the United States, 1955–1985* (New York: Cambridge Univ. Press, 1987), 432.

33. Gable interview, Jan. 8, 1992; newsletter, Aug. 17, 1973, Jefferson National Forest, Mount Rogers National Recreation Area Headquarters Land (2100 File).

34. The Council of the Southern Mountains formed in 1913 as a nonprofit organization designed to help people living in the mountains empower themselves.

35. *Roanoke Times,* Feb. 2, 1967. As the *Wytheville Enterprise* reported on May 18, 1972, "plans call for all the things that go into making an area a top tourist attraction but without a neon light atmosphere."

36. Christopher James Klyza, "The Political Economy of Natural Resource Development: A Case Study of Mount Rogers National Recreation Area" (master's thesis, Duke Univ., 1983), 55, 80 (quote). One notable exception involved Albert C. Mock, one-time advocate of NRA development, who turned against the NRA when his own land, the largest condemned tract at more than 1,400 acres, fell under threat of acquisition. Low public turnout characterized some of the earlier meetings concerning the Unit Plan; such as the one in Marion on May 23 and the one in Independence, May 24, 1972. Those who did attend generally supported NRA development plans (See *Bristol Herald Courier,* May 24, 25, 1972).

37. Lionel Melancon to the *Record,* Mar. 23, 1973, Jefferson National Forest, Mount Rogers National Recreation Area Headquarters Land (2100 File). The *Richmond Times-Dispatch* article of May 13, 1973, "A Battle for Land in the Hills," documented something of the loose dichotomy between the people Gable represented and other citizens, more of the MRCDC persuasion.

38. *Bristol Virginia-Tennessean,* Aug. 12, 1973.

39. Newsletter mailed to box holders by William Gable, published by the Council of Southern Mountains, Aug. 17, 1973, Jefferson National Forest, Mount Rogers National Recreation Area Headquarters Land (2100 File). Also, Gable interview, Jan. 8, 1992; John C. Lowe interview, May 5, 1992.

40. Gable interview, Jan. 8, 1992; Lowe interview, May 5, 1992. Gable provided a free list of comparable land prices and appraisals for anyone wishing to contest Forest Service purchase offers (see notice of such in the *Plow,* Apr. 15, 1978).

41. This procedure as outlined to the author, July–August 1992, by JNF Realty specialist Don Martindale, NRA lands staff.

42. Uniform Relocation Assistance and Real Property Acquisition Policies Act, title 3, sec. 301. Nevertheless, during the August 1973 regional meeting NRA Lands Staff indicated the financial burden they had incurred, with prices "tripling" in value since the outset of NRA legislation (Regional Forester's Staff Meeting minutes, Atlanta, Aug. 16, 1973, Jefferson National Forest, Mount Rogers National Recreation Area Headquarters Land [2100 File]).

43. Don Martindale interview, June 19, 1992.

44. By the end of 1973, the NRA had filed a total of twenty-nine condemnation lawsuits, many of which continued through litigation during the following year. August 1973 R-8 Staff meeting minutes, Jefferson National Forest, Mount Rogers National Recreation Area Headquarters Land (2100 File).

45. Owners named in lawsuits concerning these tracts were, respectively, Annette B. Mock et al.; George E. West et al.; Osborne, Heirs, et al.; and Wythe M. Hull Jr. et al. (JNF Land Files).

46. Ballard, "Where Did Hillbillies Come From?" 138.

47. Mandel Sherman and Thomas R. Henry, *Hollow Folk* (New York: Thomas Y. Crowell Co., 1933).

48. Dunn, *Cades Cove,* 255–56.

49. The Citizens for Southwest Virginia, a group that arose to counter NRA development plans, later concurred with Gable's perspective. See Citizens for Southwest Virginia, "Response of the Citizens for Southwest to the Draft Environmental Impact Statement for the Mt. Rogers National Recreation Area and Mt. Rogers Scenic Highway," Aug. 1, 1978, Virginia State Library, Richmond, 12.

50. This exchange took place on Sept. 13, 1973, when Michael Penfold addressed the Mount Rogers Planning District Commission (transcript in Jefferson National Forest, Mount Rogers National Recreation Area Headquarters Land [2100 File]). Penfold was later quoted in the *Bristol Herald Courier,* Sept. 14, 1973; Gable, in the *Bristol Virginia-Tennessean,* Sept. 25, 1973 and the *Washington County News,* Feb. 21, 1974. The latter is cited in Edwin H. Rhyne, "Recreation, Continuity, and Change: The Impact of the Socio-Cultural Environment of the Development of a NRA in the Vicinity of Mount Rogers, Virginia," Oct. 1974, pp. 16, 18, Jefferson National Forest, Mount Rogers National Recreation Area Headquarters Land (2100 File).

51. See the comments of Doris Spence Stevenson in the *Plow,* Apr. 1, 1978.

52. Rhyne, "Recreation, Continuity, and Change."

53. For another example in Virginia, see Ken Ringle, "Land Rush Changes Massanutten Life," *Washington Post,* Sept. 4, 1973. The Appalachia Regional Commission study provided an overview of this situation in *Subdividing Rural America* (esp. pp. 10, 81, 88). Hal Rothman describes a similar, historic example surrounding the Pajarito Plateau region of northern New Mexico during turn of the century. See Hal K. Rothman, *On Rims and Ridges: The Los Alamos Area since 1880* (Lincoln: Univ. of Nebraska Press, 1992), 82, 276.

54. David A. Clary, *Timber and the Forest Service* (Lawrence: Univ. Press of Kansas, 1986), 100–101, 102, 152; Marion Clawson, *Forests for Whom and for What?* (Baltimore: Resources for the Future, by the Johns Hopkins Univ. Press, 1975), 104–5, 166; Hirt, *A Conspiracy of Optimism,* 152–59; Shands et al., *National Forest Policy,* 5.

55. Bristol *Herald Courier* Jan. 27, 1974; *Roanoke Times,* Jan. 13, 1974.

56. Hirt, *A Conspiracy of Optimism,* 268–69. The Mount Rogers NRA and other Forest Service recreation areas tended to remain underfunded, as reflected in

the U.S. General Accounting Office's report, "National Forests: Special Recreation Areas Not Meeting Established Objectives," Report to the Chairman, Subcommittee on National Parks and Public Lands, Committee on Interior and Insular Affairs, House of Representatives, GAO/RCED-90–27, 1990.

57. George Wolfel, JNF supervisory forester 1969–1977, to the author, Oct. 1992.

58. Shands and Healy, *The Lands Nobody Wanted*, 219.

59. Congress addressed this problem by enacting the Payment in Lieu of Taxes Act of 1976, Public Law 94-565, 94th Cong., 2nd sess., 1976.

60. Appalachian Regional Development Act of 1965, Public Law 89-4, 89th Cong., 3rd sess., 1965.

61. See the "Letter from Secretary of Agriculture Orville L. Freeman," concerning the Mount Rogers NRA, in Orville Freeman to Harold Cooley, Mar. 20, 1964, in H. Doc. 910, 89th Cong., 1st sess.

62. "Presentation to the Mount Rogers Planning District Commission, Sept. 13, 1973," TS, Jefferson National Forest, Mount Rogers National Recreation Area Headquarters Land (2100 File).

63. See Andy Leon Harney, "Appalachian Stoneware for America's Dinner Tables: The Iron Mountain Story," *Appalachia* 6 (Oct.–Nov. 1972): 49–56.

64. Charles Blankenship, pers. comm., Dec. 14, 1992.

65. Gable interview, Jan. 8, 1992.

66. To trace the evolution among those studying recreation-oriented economics, see the following works: Harry Clement, *Your Community Can Profit from the Tourist Business,* Office of Area Development, U.S. Dept. of Commerce, 1957; Outdoor Recreation Resources Review Commission, *Economic Studies of Outdoor Recreation,* report no. 24 (Washington, DC: GPO, 1962); Marion Clawson and Jack L. Knetsch, *Economics of Outdoor Recreation* (Baltimore: Johns Hopkins Univ. Press, 1966); and Jack L. Knetsch, "Outdoor Recreation as a Vector for Economic Development and Other Social Problems," and Glen D. Fulcher, "Wise Use of Economics and Economists in Public Land Natural Resource Management," in *Outdoor Recreation: Advances in Application of Economics,* General Technical Report WO-2, ed. Jay M. Hughes and R. Duane Lloyd (Washington, DC: USDA, Forest Service, Mar. 1977). Additional sources include Gottfried, "Observations on Recreation-Led Growth in Appalachia," 44–50; Leland Leon Nicholls, "A Geographical Analysis of Selected Ski Resorts in the South" (Ph.D. diss., Univ. of Tennessee, 1972), and *Subdividing Rural America,* 2, 61, 70–76.

67. Mastran and Lowerre, *Mountaineers and Rangers,* 161, 162.

68. Hays, *Beauty, Health, and Permanence,* 431.

69. Cowdrey, *This Land, This South,* 182–84; Hays, *Beauty, Health, and Permanence,* 49; Jeffrey K. Stine, "Environmental Politics in the American South: The Fight Over the Tennessee-Tombigbee Waterway," *Environmental History Review* 15 (Spring 1991): 1–24.

70. For a chronological and topical account of the Blue Ridge Reservoir event, see Thomas J. Schoenbaum, *The New River Controversy* (Winston-Salem, NC:

J. F. Blair, 1979). For an anthropological study of the Ashe County, North Carolina, people affected by this event, see Stephen W. Foster, *The Past Is Another Country: Representation, Historical Consciousness, and Resistance in the Blue Ridge* (Berkeley: Univ. of California Press, 1988), especially pp. 147–60, 167–76, 185, 193. By 1977 development opponents had also defeated Appalachian Power's Brumley Gap pump-storage proposal in southwestern Virginia, further reflecting the antidevelopment mood of the area. See Richard Cartwright Austin, "The Battle for Brumley Gap," *Sierra* 69 (Jan.–Feb. 1984): 120–24.

71. Neil Walp "The Market for Recreation in the Appalachian Highlands," *Appalachia* 4 (Nov.–Dec. 1970): 32; URS Research Company, *Appalachian Research Report No. 14: Recreation Potential in the Appalachian Highlands: A Market Analysis* (reprint, Washington, DC: Appalachian Regional Commission, Mar. 1971), 147.

72. Charles Blankenship, pers. comm., Dec. 2, 1992. The NRA's 1977 Draft Environmental Impact Statement noted that the Pine Mountain ski area was physically but not economically feasible, and zoning would be required for any further consideration (p. 28).

73. Charles Blankenship, "Why Not a Small, Quiet NRA?" Jefferson National Forest, Mount Rogers National Recreation Area Headquarters Land (2100 File).

74. Regional Forester's Staff Meeting minutes, Atlanta, Aug. 16, 1973, 2100 File. Beginning in 1973, NEPA required that unit plans replace multiple use plans. See Wilkinson and Anderson, *Land Resource Planning in the National Forests*, 33–34.

75. Regional Forester's Staff Meeting minutes, Atlanta, Aug. 16, 1973, Jefferson National Forest, Mount Rogers National Recreation Area Headquarters Land (2100 File).

76. "Mount Rogers NRA Lands Staff Comments Relative to the Proposed Unit Plans," Aug. 6, 1973, Jefferson National Forest, Mount Rogers National Recreation Area Headquarters Land (2100 File).

77. "History of Mt. Rogers NRA Planning," chronology, Jefferson National Forest, Mount Rogers National Recreation Area Headquarters Land (2100 File).

78. Michael J. Penfold to Regional Forester, R-8, Mar. 22, 1974, Jefferson National Forest, Mount Rogers National Recreation Area Headquarters Land (2100 File).

79. In June 1978 the National Parks and Conservation Association protested further Mount Rogers NRA development, concentrating on what had then become the major areas of debate: the scenic highway, a proposed ski area, and any other developments emphasizing artificial rather than natural features. See National Parks and Conservation Association, "Assault on Mount Rogers National Recreation Area," *National Parks and Conservation Magazine* 52 (June 1978): 24–25.

80. USDA, Mount Rogers NRA Final Management Plan and Environmental Impact Statement," vi–vii.

81. Citizens for Southwest Virginia, *Newsletter,* May 1978; *Plow,* Aug. 4, Oct. 1, 1978, Mar. 3, 1979. An actual copy of the petition remains missing, though Forest Service personnel, especially Blankenship, Melancon, and Kinman, insist that the wording of the petition was so inflammatory (and misleading or false) that almost anyone who did not question the validity of the petition's premise would have signed it. For the perspective of this organization, see Citizens for Southwest Virginia, "Response of the Citizens for Southwest Virginia to the Draft Environmental Impact Statement for the Mt. Rogers National Recreation Area and Mt. Rogers Scenic Highway," *Newsletter,* Aug. 1, 1978.

82. Citizens of Southwest Virginia, "Response of the Citizens."

83. *Washington County News,* Mar. 20, 1980.

84. Charles Blankenship, pers. comm., Dec. 2, 1992. The Forest Service created a separate plan for the Mount Rogers NRA in 1978 but included the NRA within the JNF's 1985 Forest Plan, perhaps further reflecting its lowered status.

85. *United States of America v. 615.10 Acres of Land . . . in Grayson, et al, Counties,* 327 F. Supp 697.

86. For all the contentiousness during the 1970s surrounding the Mount Rogers NRA, on a macro scale among federal agencies it may have been the Forest Service's own tradition of multiple use and interaction with local communities that helped invite this level of conflict. For a short study contrasting the Mount Rogers NRA with the Big South Fork NRA in Tennessee and Kentucky (managed by the Army Corps of Engineers), see Benita J. Howell, "National Recreation Areas in Appalachia: Citizen Participation in Planning and Management," in *Culture: The Missing Element in Conservation and Development,* ed. R. J. Hoage and Katy Moran (Dubuque, IA: Kendall Hunt Publishing Co., 2000), 51–64.

87. See the *Roanoke Times and World News,* Aug. 23, 1992, and Jan. 10, 1993, for extensive coverage of this story.

88. L. Sue Greer, "Rationalization, Power and the Forest Service: A Case Study of Conflict over the Mount Rogers National Recreation Area, Virginia" (Ph.D. diss., Univ. of Kentucky, 1984) reflects exceptionally poor scholarship. The dissertation's factual inaccuracy's range from failing to identify the Unaka National Forest correctly (19, 59) to falsely asserting, with no supporting evidence, that an industrial economy was developing in the region (193), to inaccurately identifying the late date of 1978 as the beginning of activism against the NRA (36, 135). Numerous other mistakes, trivial and profound, abound. Greer uncritically accepted the *Plow's* seriously inaccurate reporting of "forty-nine" displaced families (26, 217). Greer focused on the Draft Environmental Impact Statement as if it were a done-deal policy plan rather than a document legally receptive to public feedback (26–31). A long quotation on page 35 receives no attribution. Discussion of tax issues ignored the Payment in Lieu of Taxes Act (41). Evaluation of outsiders' role in the controversy appears

inconsistent (137, 194). Greer makes vague, unsupported, and false asser-
tions—JNF acquisitions of the 1930s involved property condemnation (259–
60) and "a significant proportion of the parcels acquired in the 1960s were
tenanted" (168). "Straight condemnation process" (272) remains undefined.
"Mrs. Sherman Richardson" is cited as one of the "dispossessed" (270) but
does not appear anywhere among all the NRA condemnation case records.

89. Gable and Davis, *An Oral History of Konnarock, Virginia,* 2, 72, 76. For a similar situation in western North Carolina during this same time period, see
Hicks, *Appalachian Valley,* 52–53.

9. From Commodity Interests to Ecological Forestry

1. Samuel P. Hays, *Wars in the Woods: The Rise of Ecological Forestry in America*
(Pittsburgh: Univ. of Pittsburgh Press, 2007), 82, 83, 85, 86, 187, 189; Paul
Mohai et al., *Change in the USDA Forest Service: Are We Heading in the Right
Direction?* USDA Forest Service General Technical Report NC-172 (Washing-
ton, DC: GPO, 1994), 18.

2. Two examples can be found in David A. Clary, *Timber and the Forest Service*
(Lawrence: Univ. Press of Kansas, 1986) and Hirt, *A Conspiracy of Optimism.*
More recently, Samuel P. Hays contends that this deep tradition continues.
See Hays, *Wars in the Woods,* 189.

3. Blankenship interview, Dec. 12, 1991; Hays, *Wars in the Woods,* 9.

4. Cochran, "Jefferson National Forest," 8; Timber Management Program statistics, JNF Timber Files; JNF silviculturist George Freeland, pers. comm., June
1992. Thomas Cox describes the broader industrial and economic changes
affecting the wood products industries after World War II in Thomas R. Cox,
Robert S. Maxwell, Phillip Drennon Thomas, and Joseph J. Malone, *This Well-
Wooded Land: Americans and Their Forests from Colonial Times to the Present*
(Lincoln: Univ. of Nebraska Press, 1985), 244–45.

5. Blankenship interview, Dec. 12, 1991; Morrison interview, Jan. 16, 1992.

6. For examples of the Appalachian Forest Management Group's point of view,
see *Roanoke Times,* Mar. 15, 1991, and Feb. 7, 1994.

7. Dana and Fairfax, *Forest and Range Policy,* 208, 225, 227, 229.

8. Shands et al., *National Forest Policy,* 6.

9. Ibid.; Shands and Healy, *The Lands Nobody Wanted,* 28; Steen, *The U.S. Forest
Service,* 315–17; USDA, Forest Service, *Jefferson National Forest Land Resource
Management Plan* (Roanoke, VA: Land Management Planning, 1985), on file
with JNF Timber Division, Roanoke.

10. Hadacek interview, Dec. 16, 1991. Lest it be forgotten, many traditional
Forest Service foresters opposed clear-cutting from the outset. As Reginald
Kinman recalled, he and JNF supervisor William C. Curnutt (1957–66) were
driving through the Blacksburg Ranger District one day and came upon
one of the early clear-cuts. After stopping and gazing upon the scene for a
moment, Curnutt remarked, "Well any damned high school dropout can do
that" (Reginald Kinman interview, Feb. 4, 1992).

11. Writers sometimes misuse the term, using "clear-cutting" where they should use something like "clean-cutting" or "timber stripping." For examples, see Daniel B. Botkin, *Discordant Harmonies: A New Ecology for the Twenty-First Century* (New York: Oxford Univ. Press, 1990), 164–65; Hirt, *A Conspiracy of Optimism*, xxvii; Randal O'Toole, *Reforming the Forest Service* (Washington, DC: Island Press, 1988), 157.

12. The description of the forty-two-acre clear-cut describes exactly a tract of JNF land on the Blacksburg District that the author "cleaned up" through a purchase order contract, circa 1979, after a professional logger harvested what he deemed marketable timber. See the preface for a more detailed description.

13. See the description of Monongahela cutting practices in the June 8, 1973, testimony of Dr. Leon Minckler, *Congressional Record*, 93rd Cong., 1st sess., 1973, vol. 119, pt. 15, pp. 18905–7.

14. Steen, *The U.S. Forest Service*, 314–15.

15. This figure was certainly the lowest since JNF timber statistics recording began in 1946. See Appendix C, "Jefferson National Forest Timber Management Program," statistics, JNF Timber Files. While focus in the federal court centered on the Monongahela, it should be noted that observers of the time also visited the JNF and offered different, if less vociferous, criticism. In 1973 Leon Minckler, a Syracuse University adjunct professor of silviculture, criticized a 169-acre clear-cut on the JNF on economic grounds, citing low value timber and high administration costs. See testimony of Dr. Leon Minckler, *Congressional Record*, 93rd Cong., 1st sess., 1973, vol. 119, pt. 15, pp. 18905–7.

16. National Forest Management Act of 1976, Public Law 94-588, 94th Cong., 2nd sess., 1976.

17. For a participant's views in the NFMA legislation, see the comments of Arnold W. Bolle, dean emeritus, University of Montana, School of Forestry, in Wilkinson and Anderson, *Land Resource Planning in the National Forests*, 1–6.

18. Hoots, "Some Facts about Timber Management." In this same publication, Hoots explained the JNF's continuing even-age timber management policy. He described the forest's timber-cutting procedure, which entailed a 80–100-year cycle involving four stages of precommercial thinning, noncommercial thinning, commercial thinning, and a final clear-cut, which would begin the process all over again. But he indicated that this procedure was "such a gradual and scattered process . . . that it will essentially happen unnoticed."

19. Shands et al., *National Forest Policy*, 2, 7–9, 12, 13, 16, 20. In 1992, according to JNF timber staff officer Gretchen Merrill, the tangible results of the NFMA legislation had only begun making itself felt in recent years.

20. Steen, *The U.S. Forest Service*, 316.

21. Ibid. The criticism of selective cutting or all age management as being synonymous with creaming only the most marketable trees was a very common comment among JNF foresters during the author's JNF employment, 1991–93.

22. The author heard several such comments from various silviculturists, forest pathologists, and other foresters during his 1991–93 employment with the JNF.

23. John R. Erickson, "Changing Resource Quality: Its Impact in Harvesting and Transportation," in *Impacts of the Changing Quality of Timber Resources,* ed. Richard L. Porterfield and John B. Crist (Madison, WI: Forest Products Research Society, 1978), 16–17.

24. *Coalfield Progress,* Aug. 3, 1982; *Cumberland Times,* Feb. 29, July 18, 1984; *Smyth County News,* Aug. 2, 1990.

25. The later stages of the controversy originated with two 1984 reports, one from General Accounting Office and another from the Congressional Research Service, which more widely broadcast the below-cost economic scenario. Economists and other specialists, however, recognized the problem during the 1960s and 1970s. See Marion Clawson, *Forests for Whom and for What?* 104; Hirt, *A Conspiracy of Optimism,* 233–35; J. Greg Jones and Ervin G. Shuster, "An Analysis of the Appropriateness of 'Below Cost' Timber Sales on National Forests" (Feb. 1985), in U.S. Congress, *Economics of Federal Timber Sales,* 45. Also see "The Feds Can't See Their Losses in the Trees," *Wall Street Journal,* Nov. 14, 1984 (reproduced in *Economics of Federal Timber Sales,* p. 303); William E. Shands and Thomas E. Waddell, *Below-Cost Timber Sales in the Broad Context of National Forest Management* (Washington, DC: Conservation Foundation, 1988).

26. Clawson, *Forests for Whom and for What?* 101–3, 138.

27. Ibid., 104.

28. For an overview of the Sagebrush Rebellion, see C. Brant Short, *Ronald Reagan and the Public Lands: America's Conservation Debate, 1979–1984* (College Station: Texas A&M Univ. Press, 1989). For a sampling of Sagebrush Rebellion literature, see John Baden and Richard Stroup, eds., *Bureaucracy vs. Environment* (Ann Arbor: Univ. of Michigan Press, 1981); John Baden and Richard Stroup, *Natural Resources: Bureaucratic Myths and Environmental Management* (San Francisco: Pacific Institute for Public Policy Research, 1983); Sterling Brubaker, ed., *Rethinking the Federal Lands* (Washington, DC: Resources for the Future, 1984); Robert T. Deacon and M. Bruce Johnson, eds., *Forestlands: Public and Private* (San Francisco: Pacific Institute for Public Policy Research, 1985); Robert H. Nelson, "The Future of Federal Forest Management: Options for Use of Market Methods," in *Federal Lands Policy,* ed. Phillip O. Foss (Westport, CT: Greenwood Press, 1987): 159–76. Perhaps the culmination of the private / public forest debate and Sagebrush Rebellion economic analysis is found in O'Toole, *Reforming the Forest Service* .

29. See *Economics of Federal Timber Sales,* Hearings before the Subcommittee on Forests, Family Farms, and Energy of the Committee on Agriculture, House of Representatives, 99th Cong., 1st sess., 1985.

30. During fiscal 1990 only 47 of 122 national forests escaped the below-cost category. All Southern Appalachian national forests were below cost. (See *Roanoke Times and World News,* Feb. 13, 1991).

31. In 1886, before industrial logging arrived in Appalachia, J. D. Imboden reported only small, poor-growth trees on sandstone ridges and mountain areas of high slate and shale content, such as those commonly found in Craig and other southwestern Virginia counties. J. D. Imboden, "Internal Commerce of the United States: Virginia," in *House Executive Documents,* 49th Cong., 2nd sess., 1886, 168. For a lengthy description of the regions "shale barrens," see Brooks, *The Appalachians,* chap. 10.

32. Shands et al., *National Forest Policy,* 2.

33. Hays, *The American People and the National Forests,* 40–41, 47–48, 98.

34. "Statement of Stephen A. Bennett," in U.S. Congress, *Economics of Federal Timber Sales,* 650; "Statement of H. E. Matics, Manager, Wood Department, Westvaco Corp. [Covington]," in U.S. Congress, *Economics of Federal Timber Sales,* 551.

35. John Hinrichs, JNF sales forester, pers. comm., Mar. 1, 1993; Hank Sloan, pers. comm., Feb. 23, 1993. Bennett's resource base likely came from all three nearby national forests: the JNF, George Washington, and Monongahela. See "Statement of Stephen A. Bennett, Bennett Logging & Lumber Co., on Behalf of the Virginia Forestry Association," in U.S. Congress, *Economics of Federal Timber Sales,* 514.

36. For examples in the congressional testimony, see U.S. Congress, "Statement of Stephen A. Bennett," 515–16, 650, and "Statement of H. E. Matics," 551–53. For examples in the region's largest daily newspaper, see the *Roanoke Times,* Feb. 23, Mar. 15, Apr. 13, 1991, Mar. 30, 1992, Aug. 28, 1993, Feb. 7, Mar. 10, 1994.

37. For an example in the congressional testimony, see "Statement of Edward E. Clark, Jr., President, Virginia Wilderness Committee," in U.S. Congress, *Economics of Federal Timber Sales,* 252. For newspaper examples, see *Roanoke Times,* Mar. 28, 1991, Jan. 29, 1992, Feb. 5, May 26, 1994.

38. Timber statistic courtesy of Jim Sitton, George Washington–Jefferson National Forest timber staff officer, July 30, 1997; USDA, Forest Service, GW-JNF budget documents, fiscal 1990–97. Much general information and observation gained through Jim Loesel interview, April 22, 1997, and in subsequent communications (JNF Budget Documents, courtesy of and located with Jim Loesel, Citizens Task Force, Roanoke, VA). For more detailed timber and budget statistics, see appendices.

39. As early as 1922, E. H. Frothingham objectively observed the effects of fire in the Appalachian hardwood forest. Frothingham noted the fire resistant conifer species, pitch, shortleaf, and table mountain pines, and documented the plant succession in recently burned areas, which grew populated with chinquapins, mountain laurel, rhododendron, huckleberry bushes, and "pink" locust. Frothingham's research on the effects of fire was among the earliest, and he knew that far more was needed before fire could be employed "effectively as a silvicultural agent." But his very recognition of fire's potential as a natural (or at least positive) environmental agent was, in itself, well before

the trend toward prescribed burning. See Frothingham, "Forest Research in the Southern Appalachians," 124. For a somewhat later study see, Kenneth H. Garren, "Effects of Fire on Vegetation of the Southeastern United States," *Botanical Review* 9 (1943): 617–54.

40. JNF officials reiterated the fundamental difference of an Appalachian hardwood forest, or its small wilderness areas, in relation to the agency "let burn" policy. See, *Roanoke Times,* Nov. 19, 1972.

41. Kaufman, *The Forest Ranger,* 82; Schiff, *Fire and Water,* 53, 62, 64, 67, 83, 111.

42. Spurr and Barnes, *Forest Ecology,* 434.

43. David H. Van Lear and Thomas A. Waldrop, "Effects of Fire on Natural Regeneration in the Appalachian Mountains," in *Guidelines for Regenerating Appalachian Hardwood Stands: Workshop Proceedings,* ed. Arlyn W. Perkey, H. Clay Smith; W. E. Kidd. (Morgantown: West Virginia Univ. Books, 1988), 56–70.

44. *Roanoke Times and World News,* Dec. 27, 1993, Apr. 22, 1994.

45. Bob Boardwine, pers. comm., Jan. 28, 1992.

46. The Surface Mining Control and Reclamation Act of 1977, Public Law 95-87, 95th Cong., 1st sess., 1977, secs. 101, 102, recognized the need for "reasonable regulation" of non-coal mining operations and took the first provisional step of gathering more information for making such regulation possible.

47. For the most detailed study of this subject, see Russell H. England, "Some Effects of Abandoned Manganese Strip Mines in Smyth County, Virginia on Stream Ecology" (master's thesis, Virginia Polytechnic Institute, 1968), esp. 32, 33, 72, 73, and abstract.

48. Blackburn interview, July 10, 1992.

49. JNF staff officer William D. Blackburn to R-8 Watershed and Minerals Director John Corliss, July 23, 1980, JNF 2520 Files, Georges Branch Plan.

50. J. Lee Bardwell to Supervisor Hoots, May 13, 1983, JNF 2520 Files, Georges Branch Plan.

51. Morrison interview, Jan. 16, 1992.

52. *Smyth County News,* Aug. 29, 1985.

53. By summer of 1986 the TVA assessed the project's achievement and gave JNF supervisor Thomas Hoots an "Award for Excellence in Reclamation." In February 1990 the American Fisheries Society gave the JNF an Award for Riparian Excellence. JNF 2520 Files, Georges Branch Plan; *Inside TVA* 11 (Feb. 13, 1990), 3.

54. Blackburn interview, July 10, 1992.

55. Ibid.

56. Minerals Leasing Act of 1920, Public Law 66-146, 66th Cong., 2nd sess., 1920.

57. Acquired Lands Act of 1947, Public Law 80-382, 80th Cong., 1st sess., 1947, sec. 2.

58. John Sieker to R-7, Jefferson, General, July 25, 1958, National Archives, RG 95, 62A-962, box 36. The 1947 Materials Disposal Act gave the JNF and other national forests sole authority to sell common minerals such as gravel. See Materials Disposal Act of 1947, Public Law 80-291, 80th Cong., 1st sess., 1947, and Wilkinson and Anderson, *Land Resource Planning in the National Forests*, 249–50, 262, 264.

59. Blackburn interview, July 10, 1992.

60. As William Shands wrote in 1977, "the Forest Service has considerably more power in cases in which it owns the mineral rights as well as the surface. It can impose conditions on how minerals may be extracted, or it may bar mining entirely." See Shands and Healy, *The Lands Nobody Wanted*, 60.

61. Blackburn interview, July 10, 1992.

62. The broad form mineral deed acquired a nefarious reputation for deception in which pre-strip mining era landowners inadvertently compromised their subsurface mineral interests and, especially in the strip-mining era and afterward, suffered unforeseen environmental damages to their land. See Rivkin, "Lawyering, Power, and Reform," 467–98.

63. *Roanoke Times and World News*, Feb. 17, 1992.

64. A cursory survey of the telephone directories in 1997 revealed the following numbers of lawyer *offices* in three small towns—all with populations of less than 5,000—in one small region of Wise and Lee counties, Virginia: Big Stone Gap, ten; Wise, seventeen; Norton, thirty-two.

65. Blackburn interview, July 10, 1992.

66. Surface Mining Control and Reclamation Act of 1977, sec. 522(e)2. This section of the law states that strip-mining can take place, but only as long as "there are no significant recreational, timber, economic, or other values which may be incompatible with such surface mining operations," which has resulted in no surface coal mining on the JNF's Clinch District.

67. Blackburn interview, July 10, 1992.

68. Ibid.; *Coalfield Progress*, May 17, 1989; Jefferson National Forest, *General Report to the Public for 1991*, 21; USDA, Forest Service, "Jefferson National Forest, Virginia, Facts 1981" (Roanoke, VA: Jefferson National Forest), 3.

69. Blackburn interview, July 10, 1992.

70. The Federal Onshore Oil and Gas Leasing Act was included in Title V: Energy and Environmental Programs, Omnibus Budget Reconciliation Act of 1987, Public Law 100-203, 100th Cong., 1st sess., 1987, sec. 5101.

71. Ibid., secs. 5102(a), 5102(B), 5102(g), 5102(h).

72. Morrison interview, Jan. 16, 1992. For an example of hydrological study on the JNF, see Schaeffer, "Effects of Land Use on Oxygen Uptake by Microorganisms."

73. Swank and Crossley, *Forest Hydrology and Ecology at Coweeta*, vi, 17–31.

74. Frome, *The Forest Service*, 157; Shrauder interview, Dec. 10, 1991; USDA, Forest Service, *Story of the Jefferson National Forest.*

75. Shrauder interview, Dec. 10, 1991.

76. USDA, Forest Service, *Wildlife Habitat Management Handbook,* FSH 2609.23R (Washington, DC: GPO, 1971), 20.2-1–20.3-1.

77. Samuel P. Hays, "Three Decades of Environmental Politics: The Historical Context," in Lacey, *Government and Environmental Politics,* 24.

78. Land and Water Conservation Fund Act of 1964, Public Law 88-578, 88th Cong., 2nd sess., 1964, sec. 6(a)(1).

79. Endangered Species Act of 1973, Public Law 93-205, 93rd Cong., 1st sess., 1973.

80. Bonnie Christensen, "From Divine Nature to Umbrella Species: The Development of Wildlife Science in the United States," in Steen, *Forest and Wildlife Science in America,* 221.

81. Shrauder interview, Dec. 10, 1991; USDA, Forest Service, "Jefferson National Forest General Report to the Public for 1991," 12.

82. Ronnie E. Brenneman, "Impact of Deer on Forest Regeneration," and Edwin D. Michael, "Effects of White-Tailed Deer on Appalachian Hardwood Regeneration," in Perkey et al., *Guidelines for Regenerating Appalachian Hardwood Stands.*

83. Maurice Brooks, *The Appalachians* (Boston: Houghton Mifflin Co., 1965), vii, 299–300; Frome, *The Forest Service,* 157, 162–63; USDA, Forest Service, *The Fairest One of All* (Milwaukee: USDA, Forest Service, 1973), 3–4.

84. Hewlette Crawford, "Deer Range Potential in Selective and Clearcut Oak-Pine Stands in Southwest Virginia," USDA Forest Research Paper SE-134, June 1975, 11.

85. Frome, *The Forest Service,* 157, 165–66; Wilkinson and Anderson, *Land Resource Planning in the National Forests,* 309–10.

86. *Coalfield Progress,* Dec. 20, 1988; Donald W. Linzey, ed., *Proceedings of the Symposium on Endangered and Threatened Plants and Animals of Virginia* (Blacksburg: Virginia Polytechnic Institute and State Univ., 1979), 626–27; Sylvia Whitworth, JNF fish biologist, pers. comm., Apr. 6, 1992.

87. Donald W. Linzey, ed., *Proceedings of the Symposium on Endangered and Threatened Plants and Animals of Virginia* (Blacksburg: Virginia Polytechnic Institute and State Univ., 1979), 626–30.

88. Weidensaul, *Mountains of the Heart,* 71–73; Richard H. Yahner, *Eastern Deciduous Forest: Ecology and Wildlife Conservation* (Minneapolis: Univ. of Minnesota Press, 1995), 108–23.

89. As Albert Cowdrey points out, southeastern national forests have greater silvicultural biomass and genetic variety, plus wildlife habitat potential for bear and beaver not found in the area's predominant commercial forests. See Cowdrey, *This Land, This South,* 178–79.

90. Wilderness Act of 1964, Public Law 88-577, 88th Cong., 2nd sess., 1964.

91. Eastern Wilderness Act of 1975, Public Law 93-622, 93rd Cong., 2nd sess., 1975.

92. Blankenship interview, Dec. 12, 1991; Frome, *The Forest Service,* 187; Dennis M. Roth, *The Wilderness Movement and the National Forests: 1964–1980* (Washington, DC: USDA, Forest Service, 1984), 9, 41.

93. Eastern Wilderness Act of 1975, sec. 2(b), 6.

94. Ibid., sec. 4(9).

95. See appendix C for a list of JNF wilderness areas and their respective acreages.

96. Morrison interview, Jan. 16, 1992.

97. *Roanoke Times and World News,* Dec. 21, 1986.

98. Phil Gersmehl concluded that these balds were not natural at all, but rather caused by early cattle herders who believed higher meadows would be healthier and that the surrounding forest would act as natural fencing. See Gersmehl, "Factors Leading to Mountaintop Grazing in the Southern Appalachians," 67–72.

99. For a lengthy, earlier analysis, see Jon R. Luoma, "Forests Are Dying but Is Acid Rain Really to Blame?" *Audubon* 89 (Mar. 1987): 36–51. Orie Loucks, "In Changing Forests: A Search for Answers," in *An Appalachian Tragedy: Air Pollution and Tree Death in the Eastern Forests of North America,* ed. Harvard Ayers, Jenny Hager, and Charles E. Little (San Francisco: Sierra Club Books, 1998), 85, 90, 92, 95. In addition to Loucks's article, this work contains a substantial bibliography of recent scientific publications related to the upland Appalachians. Also see Anita Kristine Rose, *Virginia's Forests, 2001,* Resource Bulletin SRS-120 (Asheville, NC: USDA FS Southern Research Station, 2007), 40–43.

100. All of the 1992 information was gathered from the Associated Press article published in the *Bristol Herald Courier,* Feb. 27, 1992. For a later account of the acid rain problem in the Southern Appalachian region, see *Roanoke Times and World News,* Sept. 2, 1993.

101. For individual surveys, see Thomas C. Evans, G. E. Morrill, and John Carow, *Virginia: Preliminary Tables of Forest Area, Timber Volumes and Utilization* (Asheville, NC: USDA, Forest Service, Appalachian Forest Experiment Station, 1941); Mackay B. Bryan, *Forest Statistics for the Mountain Region of Virginia, 1957,* Forest Survey Release No. 52 (Asheville, NC: USDA Forest Service, Southeast Forest Experiment Station, 1958); Robert W. Larson and Mackay B. Bryan, *Virginia's Timber,* Forest Survey Release No. 54 (Asheville, NC: USDA, Forest Service, Southeastern Forest Experiment Station, 1959); Herbert A. Knight and Joe P. McClure, *Virginia's Timber, 1966,* U.S. Forest Service Resource Bulletin SE8 (Asheville, NC: USDA, Forest Service, Southeastern Forest Experiment Station, 1967).

102. See Joe P. McClure, Noel D. Cost, and Herbert A Knight, *Multiresource Inventories—A New Concept for Forest Survey,* USDA Forest Service Research Paper SE-191, Apr. 1979.

103. Robert T. Colona and Louise Givens-Reynolds, "The JNF: A Stressed and Aging Condition," JNF Timber files; Steven W. Oak, Cindy M. Huber, and Raymond M. Sheffield, *Incidence and Impact of Oak Decline in Western Virginia, 1986,* USDA Resource Bulletin SE-123, Nov. 1991; USDA, Forest

Service, "Jefferson National Forest General Report to the Public for 1991," 13. For a more recent analysis, see Loucks, "In Changing Forests," 94–95.

104. *Roanoke Times and World News,* Apr. 27, 1995; Weidensaul, *Mountains of the Heart,* 208–10; Yahner, *Eastern Deciduous Forest,* 53–55.

105. Loesel interview, Apr. 22, 1997.

106. Mohai et al., *Change in the USDA Forest Service,* 1, 7, 8, 9, 11, 13.

107. Fred Huber, pers. comm., July 1992; George Freeland, pers. comm., Feb. 1992. Today RNAs provide various natural ecosystems as something of a norm against which similar, modified areas might be gauged. Only by the early 1990s, however, did the Forest Service request the JNF to designate such areas. The main obstacle lay in proper quality sites, with the Blacksburg District's unique fern stand at the foot of Pott's Mountain presenting too small an area and various spruce/fir stands around the Mount Rogers NRA complicated by the area's previous recreation designation.

108. Giles County Historical Society, *Giles County,* 46. In the early 1960s the Mountain Lake area and its biological research received much attention from Maurice Brooks, then researching for his natural history book, an important precedent in its field. See, Brooks, *The Appalachians,* 129, 200–204, 207, 225, 307.

109. Henry Wilbur interview, Mar. 30, 1992.

110. Wilbur, "Facilities Support for Mountain Lake Biological Station," 10; *Mountain Lake Biological Station,* 1992 UVA course brochure.

111. Wilbur interview, Mar. 30, 1992.

112. Wilbur, "Facilities Support for Mountain Lake Biological Station," 11; Wilbur interview, Mar. 30, 1992.

113. Aside from Mountain Lake and the Mount Rogers / Whitetop area, the JNF has almost unlimited potential to local graduate students as a "living laboratory" for a variety of experiments involving "pure" science. Two such studies include Paul C. Edmunds, "Seasonal Fluctuations of Rotifer Populations Related to Selected Biological, Chemical and Physical Parameters in a Small Mountain Pond, Jefferson National Forest, Virginia" (master's thesis, Virginia Polytechnic Institute and State Univ., 1974). In this work Edmunds used the familiar Pandapas Pond as a relatively undisturbed environment for his 1972–73 research. Harold S. Adams's 1974 dissertation on plant ecology was also one of pure science, and it was especially helpful within the context of other Southern Appalachian ecosystem studies. The JNF provided him with "relatively undisturbed vegetation [which he] considered suitable for this study." See Harold S. Adams, "Analysis of Vegetation on the South Slopes of Peters Mountain, Virginia" (Ph.D. diss., Virginia Polytechnic Institute and State Univ., 1974).

114. See Diebel, "Isozyme Variation within the Fraser Fir Population"; Richard D. Reinhardt, "Comparative Study of Composition and Distribution Patterns of Subalpine Forests in the Balsam Mountains of Southwest Virginia and the Great Smoky Mountains," in White, *The Southern Appalachian Spruce-Fir Eco-*

system, 87–99. Also see Robert H. Mohlenbrock, "Mount Rogers, Virginia," *Natural History* 99 (Dec. 1990): 72–74; Nicholas, "Stand Structure, Growth"; Joanna Samuels, "The Adelgid Strikes Again," *Audubon* 98 (Mar.–Apr. 1996): 24.

115. The Forest Service uses this term to describe its management approach. For a broader and more scientific view of the phrase's definition, see J. P. Kimmins, *Forest Ecology: A Foundation for Sustainable Forest Management and Environmental Ethics in Forestry* (Upper Saddle River, NJ: Prentice Hall, 2004), 517–30, 534–36.

116. Boyce and Oliver, "History of Research in Forest Ecology and Silviculture," in Steen, *Forest and Wildlife Science in America,* 421–22, 426, 431–32; Kimmins, *Forest Ecology,* 29, 31–34, 533, 610; Donald Worster, "The Ecology of Order and Chaos," in *Out of the Woods: Essays in Environmental History,* ed. Char Miller and Hal Rothman (Pittsburgh: Univ. of Pittsburgh Press, 1997), 3–17. Also, in general, see Daniel B. Botkin, *Discordant Harmonies: A New Ecology for the Twenty-First Century* (New York: Oxford Univ. Press, 1990).

117. Richard Freeman, "The Ecofactory: The United States Forest Service and the Political Construction of Ecosystem Management," *Environmental History* 7 (Oct. 2002): 632–58; Hays, *Wars in the Woods,* 84–85. For all the shortcomings of ecosystem management, Donald Edward Davis ignores it altogether (along with southern Appalachian national forest wilderness areas). Instead, Davis perpetuates the obsolete myth of these national forests as timber plantations. See Donald Edward Davis, *Homeplace Geography: Essays for Appalachia* (Macon, GA: Mercer Univ. Press, 2006), chap. 12.

118. GW-JNF budget statistics, fiscal 1990–92, 1994–97. See appendix for detailed figures.

119. Loesel interview, Apr. 22, 1997; *Roanoke Times and World News,* Oct. 15, 1993; *Roanoke Times,* Mar. 6, 1994.

120. Merrill R. Kaufmann, *An Ecological Basis for Ecosystem Management,* General Technical Report RM-246 (Fort Collins, CO: USDA Forest Service, Rocky Mountain Forest and Range Experiment Station, 1994), 2–4; Mohai et al., *Change in the USDA Forest Service,* 1, 7, 8, 9, 11, 13, 17.

121. Boyce and Oliver, "History of Research in Forest Ecology and Silviculture," 421, 422, 426, 431–32, 436–37.

10. Cultural Resources

1. For some indications of the type of work generated by Forest Service cultural resource management, see Theodore J. Karamanski, "Logging, History, and the National Forests: A Case Study of Cultural Resource Management," *Public Historian* 7, no. 2 (1985): 27–40, and Alan S. Newell, "Identification and Interpretation: Managing Cultural Resources in the U.S. Forest Service," *Public Historian* 9, no. 2 (1987): 147–53.

2. Historic Sites Act of 1935, Public Law 74-292, 74th Cong., 1st sess., 1935.

3. National Historic Preservation Act of 1966, Public Law 89-665, 89th Cong., 2nd sess., 1966.

4. Archaeological and Historic Preservation Act of 1974, Public Law 93-291, 93rd Cong., 2nd sess., 1974.

5. Ibid.; Dana, *Forest and Range Policy,* 217; Frome, *The Forest Service,* 85.

6. The Archaeological Resources Protection Act of 1979, Public Law 96-95, 96th Cong., 1st sess., 1979, sec. 7.

7. During July 1982, JNF archaeologist Mike Barber led an "emergency dig" on eight rock shelters on the Clinch District (which remains the JNF's most troubled area in terms of archaeological destruction). Looting continued to pose a serious problem on the Clinch, and in 1985 the JNF offered a $500 reward for reporting illegal pot hunters. Barber, Clinch Ranger Saboites, and others posted signs warning against looting and reiterated the legal consequences of fines and jail time—all to little avail in the long run. See, *Coalfield Progress,* July 27, 1982, Aug. 1, 1985.

8. This account was the basis for the author's article "Green Cove Station: An Appalachian Train Depot and Its Community," *Virginia Cavalcade* 42 (Autumn 1992): 52–61. The information on page 53 of the *Virginia Cavalcade* article was added by that periodical's editorial staff, derived from a *Galax Gazette* newspaper article that the author purposely excluded, as no one in Green Cove could verify its veracity, nor the existence of its main character, Anne Gentry, as a former resident of the small community.

9. Norfolk & Western finished buying the Virginia-Carolina Railroad in 1916 and called the line between Abingdon, Virginia, and Elkland, North Carolina, the Virginia-Carolina Branch. By 1919 N&W had renamed this line the Abingdon Branch. Popularly, at least around the Damascus and Green Cove area, the rail line was sometimes called the "Peavine" (not to be confused with eastern Tennessee railroads of the same name). See Charles H. Faulkner, "Industrial Archaeology of the 'Peavine Railroad': An Archaeological and Historical Study of an Abandoned Railroad in East Tennessee," *Tennessee Historical Quarterly* 44 (Spring 1985): 47. The train itself, of course, gained the well-known nickname of "Virginia Creeper," attributed variously to the train's slow pace over the mountainous grades or the vine of the same name growing in the region. Don Piedmont, N&S Director of Public Relations, provided the cost of the Green Cove station, pers. comm., Sept. 16, 1991; Adele Edwards, interview by the author, Sept. 17, 1991, provided the completion date. The completion date was verified by Virginia-Carolina Railway Company Annual Report, June 30, 1915, box 2.51, N&W Archives (Virginia Tech Special Collections, Blacksburg, VA). For a reference to the first train that reached North Carolina, on July 14, 1914, see Robert S. Jones, "Days of Iron and Steam," *Plow,* May 1976, 16.

10. Blanch Clark interview, Oct. 7, 1991.

11. Lewis Preston Summers, *History of Southwest Virginia, 1746–1786, Washington County, 1777–1870* (1903; Baltimore: Genealogical Publishing Co., 1966), 25.

12. William H. Nichols described this region in his study as twenty Virginia and Tennessee counties lying "within the watershed of the Tennessee River and . . .

drained by some of its major tributaries, the Holston and French Broad Rivers . . . and the Powell and Clinch Rivers." See William H. Nichols, "Economic Development in Upper East Tennessee Valley," *Journal of Political Economy* 64 (Aug. 1956): 278–79.

13. Nichols, "Economic Development in Upper Tennessee Valley," 289, 289n14, 289n15.

14. Summers, *History of Southwest Virginia,* 557. Determining true population fluctuation requires a knowledge of boundaries of the county and its districts, and any changes in those boundaries that may have occurred during the time in question. The present boundaries of Washington County were established by 1840 and certainly did not change between 1890 and 1910. As far as the Holston District, the major concern here, the Reassessment of Lands books for Washington County in 1890, 1900, and 1910 reveal, if anything, a slight *drop* in acreage for the Holston District—but certainly no expansion. Between 1890 and 1900, the county outside the Holston District grew 7.5 percent, while the district itself grew 45 percent. During the next ten years, the county outside the district grew 6.5 percent, while the district grew 60 percent. Thus, the remarkable population increase between 1890–1910 certainly reflects only in-migration and not growth of the district boundaries. See Reassessment of Lands books for 1890, 1900, 1910 in the Washington County Courthouse, Abingdon, VA; William Thorndale and William Dollarhide, *Map Guide to the U.S. Federal Censuses, 1790–1920* (Baltimore: Genealogical Pub. Co., 1987), 349–59.

15. Washington County Census, Holston Magisterial District, 1880.

16. Other districts lost population or gained only a relatively small number, the next highest gain being 19 percent in the Saltville District (including part of the town of Saltville). See Summers, *History of Southwest Virginia,* 851, for a population table.

17. Washington County Census, Holston Magisterial District, 1900.

18. Washington County Census, Holston District, 1910.

19. Martin Hassinger interview, July 24, 1979, interviewed by Mary Kegley and Lionel Melancon in Bristol, VA.

20. Deed Book 78, p. 279, Washington County Courthouse, Abingdon, VA.

21. *Bristol Herald Courier,* Sunday, June 7, 1987; Summers, *History of Southwest Virginia,* 76; Walter H. Hendricks, "Daniel Boone as a Virginian," *Historical Society of Washington County, Virginia, Bulletin* 24 (1987): 3, 4, 6, 8.

22. Goodridge Wilson, "The Southwest Corner," *Roanoke Times,* Oct. 23, 1963; Historical Society of Washington County, ser. 2, no. 4 (Spring 1967): 9; O. Winston Link and Tim Hensley, *Steam Steel and Stars: America's Last Steam Railroad* (New York: Harry N. Abrams, 1987), 128–29.

23. Eller, *Miners, Millhands and Mountaineers,* 87, 93, 98. Martin Hassinger recalled the biggest tree their company ever cut as being a red oak tree approximately eight to nine feet in diameter, located somewhere in the Green Cove Valley (pers. comm. to author, June 17, 1992). Hassinger showed the author

a photograph of this black oak, which was eight foot in diameter at breast height, as well as an eleven-foot-diameter tulip poplar. The enormously rich Hassinger photographic collection now resides at the Archives of Appalachia, East Tennessee State Univ., Johnson City, Tennessee.

24. Luther C. Hassinger, "The Lumber Industry in Southwest Virginia," *Historical Society of Washington County, Virginia, Bulletin,* ser. 2, no. 4 (Spring 1967): 9; *Plow,* July 1976, 15.

25. "Rugged Men Laid the Foundation," *Bristol Herald Courier/Virginia-Tennessean,* Apr. 26, 1990.

26. Lewis, "An Abingdon Branch Accolade," 23.

27. Martin Hassinger, "The Hassinger Lumber Company," *Plow,* July 1976, 13, 14; Claude H. Weaver, "Yet Another Tale of Railroading Days," *Plow,* June 1, 1978, 12.

28. Edwards interview, Sept. 17, 1991; Lewis, "An Abingdon Branch Accolade," 23; Appalachian Oral History Project, transcript no. 172, interview with Eleanor Buchanan, Adele Edwards, and Emma Wilson, June 30, 1975, 3, Appalachian Oral History Project archives, Emory and Henry College, Emory, Virginia.

29. Robert S. Jones, "Days of Iron and Steam," *Plow,* May 1976, 16; Pearl Dinkens Hensley (with Paige Proffitt), "Through Whitetop on the Memory Train," *Plow,* Nov. 18, 1977, 19; Blanch Clark interview, Oct. 7, 1991; Appalachian Oral History Project, transcript no. 172, 2, 11.

30. Blanch Clark interview, Oct. 7, 1991.

31. Hassinger interview, July 24, 1979.

32. Appalachian Oral History Project, transcript no. 172, 1–2.

33. Megaera Harris, research administrator/historian, Office of the Postmaster General, Oct. 9, 1991.

34. Clerk's Order Books, nos. 2–8, Washington County Courthouse, Abingdon, VA. Edwards interview, Sept. 17, 1991.

35. Edwards interview, Sept. 17, 1991.

36. Virginia-Carolina Railroad Company Annual Report, June 30, 1915, box 2.51, N&W Archives, Special Collections, Virginia Polytechnic Institute and State Univ.

37. Edwards interview, Sept. 17, 1991.

38. Ibid..

39. Box 2.51, N&W archives; 1932–33 Abingdon Branch Timetables, N&W archives, Special Collections, Virginia Polytechnic Institute and State Univ.

40. Edwards interview, Sept. 17, 1991.

41. Blanch Clark interview, Oct. 7, 1991.

42. Mike Testerman, "A Plan to Save the West Jefferson Line," *Plow,* Sept. 1976, 9–10.

43. Mt. Rogers Headquarters, land files, Map V-10-VA 215, sheet no. 16; Land Lease Agreement H-954.

44. Lewis, "An Abingdon Branch Accolade," 27.

45. Blanch Clark interview, Oct. 7, 1991.

46. Lewis, "An Abingdon Branch Accolade," 27.

47. Any official document verifying the official discontinuation of Buchanan's Western Union telegraph operations has remained elusive. Two people, however, who personally remember the station, have claimed that the end of the telegraph coincided exactly with the end of the post office (Herman Lucille Blevins, pers. comm., Oct. 6, 1991; Blanch Clark, pers. comm., Oct. 7, 1991).

48. Harris, pers. comm., Oct. 17, 1991.

49. Blevins interview, Oct. 15, 1991; Blanch Clark interview, Oct. 7, 1991.

50. Blanch Clark interview, Oct. 7, 1991.

51. W. M. Buchanan's final notary public registration is listed in Clerk's Order Book no. 8, Washington County Courthouse, Abingdon, VA.

52. Journalists writing for the *Plow* visited the store as late as May 1976 and reported it open.

53. Carolyn Rosenbaum, "Explore the Tracks of Yesterday," *Bristol Herald Courier/Virginia-Tennessean,* June 7, 1987.

54. For an example in east Tennessee similar to the Virginia Creeper that served Green Cove, see Faulkner, "Industrial Archaeology," 51–52.

11. Sacred Land

For the passage from "Brier Sermon—'You Must Be Born Again,'" see Jim Wayne Miller, *The Brier Poems* (Frankford, KY: Gnomon Press,1997), 68–69. The poem originally appreared in Miller, *The Mountains Have Come Closer* (Boone, NC: Appalachian Consortiuum Press, 1980), 52–64 (pp. 59–60 for excerpt).

1. In general see Edwin Bernbaum, *Sacred Mountains of the World* (San Francisco: Sierra Club Books, 1990), 206–17; Mircea Eliade, *Patterns in Comparative Religion,* trans. Rosemary Sheed (New York: Meridian Books, 1966), 266–68, 298–326; Mircea Eliade, *The Sacred and the Profane,* trans. Willard R. Trask (1957; New York: Harper & Row, 1959), 38–41.

2. Joseph Campbell, *Transformations of Myth through Time* (New York: Harper & Row, 1990), 42, 46. For a variation upon this theme, directly relevant to modern environmentalism, see Bernbaum, *Sacred Mountains,* 256.

3. Bernbaum, *Sacred Mountains,* chap. 12; Mircea Eliade, *The Myth of the Final Return,* trans. Willard R. Trask (New York: Pantheon Books, 1954), 12; Eliade, *Patterns in Comparative Religion,* 99–101; H.W.F. Saggs, *Civilization before Greece and Rome* (New Haven: Yale Univ. Press, 1989), 269, 282, 283, 285, 294.

4. Deborah V. McCauley, *Appalachian Mountain Religion: A History* (Urbana: Univ. of Illinois Press, 1995), 53.

5. Catherine L. Albanese, *America: Religions and Religion* (Belmont, CA: Wadsworth Pub. Co., 1981), 222, 226, 227, 235, 242–43; McCauley, *Appalachian Mountain Religion,* 7.

6. Albanese, *America: Religion and Religions,* 235–36; Gerald Milnes, *Signs, Cures and Witchery: German Appalachian Folklore* (Knoxville: Univ. of Tennessee Press, 2007), 24, 34; Thomas, *Southern Appalachia, 1885–1915,* 247; Rehder, *Appalachian Folkways,* 154–55; Wigginton, *The Foxfire Book,* 212–27.

7. Albanese, *America: Religion and Religions,* 235; Wigginton, *Foxfire Book,* 212.

8. Eliot Wigginton and Margie Bennett, eds. *Foxfire 9* (Garden City, NY: Doubleday, 1986), 321–28.

9. Eliade, *Patterns in Comparative Religion,* 269.

10. Ibid., 324.

11. In general see Nicole Boivin and Mary Ann Owoc, eds., *Soils, Stones and Symbols: Cultural Perceptions of the Mineral World* (London: ULC, 2004). Most of these examples stem back to prehistory, including Stonehenge and the Hopewellian mounds. For a study concerning traditional and contemporary Indonesia, see O. W. Hampton, *Culture of Stone: Sacred and Profane Use of Stone among the Dani* (College Station: Texas A&M Univ. Press, 1999).

12. Lee Smith, *Oral History* (New York: G. P. Putnam's Sons, 1983), 34, 88. The buckeye in the pocket was also believed to stave off rheumatism. See Shackelford and Weinberg, *Our Appalachia,* 79.

13. Wolfe, *Children of the Heav'nly King,* 28–29.

14. Ibid,, 29.

15. Ibid., 27–28.

16. See http://www.trailofthelonesomepine.org/.

17. Warren I. Titus, *John Fox, Jr.* (New York: Twayne, 1971), 96.

18. Ibid.

19. In addition to Eliade, *Patterns in Comparative Religion,* 298–326, see J. H. Philpot, *The Sacred Tree* (New York: Macmillan, 1897) and Alexander Porteous, *The Forest in Folklore and Mythology* (New York: Macmillan, 1928).

20. Albanese, *America: Religions and Religion,* 236. For examples of southwestern Virginia stories supporting Albanese's point, see Chase, *The Jack Tales,* 21–30, 40–46, 135–50, 151–60, 186–87, 189, 191, 198–99. Also see "How Jack Got a New Shirt" at http://www.ferrum.edu/applit/texts/JackShirt.htm.

21. Smith, *Oral History,* 41.

22. As told to Robert Coles in Robert Coles, *Migrants, Sharecroppers, Mountaineers* (Boston: Little, Brown, 1971), 205–6.

23. Charles K. Wolfe, *Children of the Heav'nly King: Religious Expression in the Central Blue Ridge* (Washington, DC: Library of Congress Recording Laboratory, ca. 1982), 42–43. This is a vinyl sound recording, but the accompanying liner notes represent a full transcription; page numbers regarding this work refer to this transcription.

24. Ibid., 42–43, also 6, 7, 38, 40. Also see John C. Campbell, *The Southern Highlander and His Homeland* (New York: Russell Sage Foundation, 1921), 177, 180, 193; McCauley, *Appalachian Mountain Religion,* 11, 15, 133, 159, 274, 335;

Emma Bell Miles, *The Spirit of the Mountains* (Knoxville: Univ. of Tennessee Press, 1975), 142–43 and generally chap. 7.

25. Gable and Davis, *An Oral History of Konnarock, Virginia*, 56–57. It might be noted that in the mountains such "spigot water" very often came gravity-fed from natural springs. The only other option in rural areas is well water, which overwhelmingly predominated in the flatlands.

26. Albanese, *America: Religion and Religions*, 241, 243.

27. McCauley, *Appalachian Mountain Religion*, 336. Oddly, in 1995 McCauley asserted that the mountains as "a subliminal force of extraordinary power on mountain people" had not yet been explored in the field of Appalachian mountain religion when, in fact, Albanese (whom McCauley cites) had mapped out a compelling conceptual framework in 1981. See Albanese, *America: Religion and Religions*, chap. 10; McCauley, *Appalachian Mountain Religion*, 336.

28. Quoted in Charles W. Conn, *Like a Mighty Army: A History of the Church of God, 1886–1976*, rev. ed. (Cleveland, TN: Pathway Press, 1977), 79.

29. Shackelford and Weinberg, *Our Appalachia*, 294–95.

30. Miles, *The Spirit of the Mountains*, 17, 18, 19.

31. Ibid., 140.

32. Loyal Jones, "Appalachian Values," in *Voices from the Hills: Selected Readings of Southern Appalachia*, ed. Robert J. Higgs and Ambrose N. Manning (New York: Frederick Ungar, 1975), 508 .

33. Coles, *Migrants, Sharecroppers, Mountaineers*, 208, 219.

34. Ibid., 202.

35. Shackelford and Weinberg, *Our Appalachia*, 75.

36. Edward O. Guerrant, *The Galax Gatherers* (1910; Knoxville: Univ. of Tennessee Press, 2005), 5, 21, 22, 74, 142. Deborah V. McCauley has been one of Guerrant's most strident critics, while David Whisnant saw him more sympathetically and accurately as a Kentucky resident who functioned as a liaison between the deep mountains and the flatlands. See Mark Huddle's preface in the above work, "Home Missions Revisited," xxiv–vii

37. Guerrant, *Galax Gatherers*, 116.

38. Albert J. Fritsch, *Appalachia: A Meditation* (Chicago: Loyola Univ. Press, 1986).

39. Ibid., 18.

40. Charles H. Lippy, "Popular Religiosity in Central Appalachia," in *Christianity in Appalachia: Profiles in Regional Pluralism*, ed. Bill J. Leonard (Knoxville: Univ. of Tennessee Press, 1999), 42, 46–47, 49.

41. James Still, *River of Earth* (1940; Lexington: Univ. Press of Kentucky, 1978), vii, 76.

42. See, for example, Shunryu Suzuki, *Zen Mind, Beginner's Mind* (New York: Weatherhill, 1970), 102–4.

43. Coles, *Migrants, Sharecroppers, Mountaineers*, 203.

44. Cathryn McCue, "Family Honor at Stake in Mountain," *Roanoke Times and World News,* July 10, 1994.

45. For a detailed treatment of this subject, see Will Sarvis, "Land and Home in the American Mind," *Journal of Natural Resources and Environmental Law* 22, no. 2 (2008–9): 107–37.

46. Various informants have expressed this exact phrase to the author in lands as divergent as the mountains of Virginia, the drained delta flatlands of far southeastern (e.g., "Bootheel") Missouri, the mountains, valleys, and ocean shore of the Pacific Northwest, and the rainforests of southeastern Alaska.

47. McCauley, *Appalachian Mountain Religion,* 61.

48. For West Virginia examples, see Patrick W. Gainer, *Witches, Ghosts and Signs: Folklore of the Southern Appalachians* (Grantsville, WV: Seneca Books, 1975), 100–109 and Milnes, *Signs, Cures and Witchery,* 92–93 and, in general, chap. 12. The Konnarock area examples mentioned before appear in *Oral History of Konnarock,* 16, 18, 19, 25, 29, 30, 36, 37, 38, 39, 72, 73. Also see Rehder, *Appalachian Folkways,* 224–43.

49. Milnes, *Signs, Cures and Witchery,* 157; Wigginton, *Foxfire Book,* 31–37.

50. Gainer, *Witches, Ghosts and Signs,* 112–20; Wigginton, *Foxfire Book,* 208–11.

51. Richard Chase, ed., *Grandfather Tales: American-English Folk Tales* (Boston: Houghton Mifflin Co., 1948), 52–64, 235.

52. Ibid. Consistent with an oral tradition tale, this may also refer to the numerous regional peaks named Pine Mountain, including those in Virginia, Kentucky, North Carolina, Tennessee, and Georgia. "Whitebear Wigginton" also includes the Indian-like naturalistic elements observed by Albanese of various Jack Tales.

53. Ibid., 186–94, 238. Besides the obvious agricultural component, "Tall Cornstalk" also reflects a "tale of plenty" that, unfortunately, would likely find special appeal in an area of limited agricultural productivity.

54. Mar. 29, 2008, correspondence, on file with the author.

55. Ibid. For a very similar account, see Robert Coles, "Appalachia's Moral Life," in *Mountain People,* ed. Michael Tobias (Norman: Univ. of Oklahoma Press, 1986), 19.

56. Circa 1991, during field work in Washington County, Virginia.

57. Coles, "Appalachia's Moral Life," 21.

58. Ibid.

59. Albanese, *America: Religion and Religions,* 226–27.

60. See J. Trent Alexander, "'They're Never Here More Than a Year': Return Migration in the Southern Exodus, 1940–1970," *Journal of Social History* 38 (Spring 2005): 653–71 and J. Trent Alexander, "Defining the Diaspora: Appalachians in the Great Migration," *Journal of Interdisciplinary History* 37 (Autumn 2006): 219–47.

61. In addition to the sources cited here, see other examples in Angie Cheek, Lacy Hunter Nix, and Foxfire students, eds., *The Foxfire 40th Anniversary*

Book: Faith, Family, and the Land (New York: Anchor Books, 2006), 480; Coles, "Appalachia's Moral Life," 19, 21; Gable and Davis, *Oral History of Konnarock,* 76, 81, 86, 87; Linda Garland Page and Eliot Wigginton, eds., *Aunt Arie: A Foxfire Portrait* (Chapel Hill: Univ. of North Carolina Press, 1992), 22.

62. Albanese, *America: Religion and Religions,* 228; Gable and Davis, *Oral History of Konnarock,* 88.

63. See http://archiver.rootsweb.ancestry.com/th/read/AMERIND-US-SE/2001–11/1005923156, attributed to a printed article in *Kentucky Explore* 11 (Mar. 1997).

64. Gable and Davis, *Oral History of Konnarock,* 88.

65. Alice Hayes interview, 4, Konnarock Oral History, archived at Emory and Henry College, available at http://library.ehc.edu/konnarock.html.

66. Earl V. Shaffer, *Walking with Spring: The First Thru-Hike of the Appalachian Trail* (Harpers Ferry, WV: Appalachian Trail Conference, 1995). People dispute whether or not Shaffer's hike was actually the first from end to end, but certainly it was among the earliest.

67. Christopher P. Collier, "Good Times at a Backpackers' Paradise, *New York Times,* May 12, 2006; Edward B. Garvey, *The New Appalachian Trail* (Birmingham, AL: Menasha Ridge Press, 1997), 53; Luxenberg, *Walking the Appalachian Trail,* 139.

68. Luxenberg, *Walking the Appalachian Trail,* 140. Also see Sharla Bardin, "A Hiker Hostel Steeped in History," *Roanoke Times and World News,* Nov. 26, 2009.

69. Pearl Dawson, *Appalachian Trail Journal,* May 23, 2002, at http://www.trail journals.com/entry.cfm?id=16754.

70. Mark Hughes, *Appalachian Trail Journal,* May 13, 2006, available at http://www.trailjournals.com/entry.cfm?id=137845.

71. Literally thousands of hikers' journal entries and their photographs (from the mid-1990s to the present) describing the above can be found at http://www.trailjournals.com/ and http://www.fred.net/kathy/at/journal.html.

72. Ken Berry, *Appalachian Trail Journal,* Apr. 10, 2006, at http://www.trail journals.com/entry.cfm?id=130993.

73. Nate Olive, *Appalachian Trail Journal,* May 7, 2001, at http://www.trailjournals.com/entry.cfm?id=5844.

74. Simon Coleman and John Elsner, *Pilgrimage: Past and Present in the World Religions* (Cambridge: Harvard Univ. Press, 1995), 213.

75. Ibid., 214.

76. Ibid., 216–20. Also see Victor Turner and Edith Turner, *Image and Pilgrimage in Christian Culture: Anthropological Perspectives* (New York: Columbia Univ. Press, 1978), 20.

77. An interesting short essay addressing the overview of this topic might be found in Stuart Thompson, "Physical or Metaphysical," *Appalachian Trailway News* 61 (Nov.–Dec., 2000), 21–22.

78. Turner and Turner, *Image and Pilgrimage,* 7, 33–34.

79. In the foreword to Larry Luxenberg, *Walking the Appalachian Trail* (Mechanicsburg, PA: Stackpole Books, 1994), ix.

80. Flager Films Presentation, *Appalachian Impressions* (2004), opening minutes.

81. Jan. 19, 2009, correspondence on file with the author. Also see Thomas Crutcher, "Life after Katahdin?" *Appalachian Trailway News* 62 (Sept.–Oct. 2001), 26–27.

82. A good indication of spiritual experiences may be found in the "Spiritual Awakenings" chapter in Larry Luxenberg, *Walking the Appalachian Trail,* 127–31, 132–34. As an example of a religious association, a recent editorial comment in the *Appalachian Trailway News* read, "Moses on Mount Sinai, Jesus wandering forty days in the desert, the Buddha meditating for years under a bodhi tree, Mohammed in the hills outside Mecca receiving the Koran from the Angel Gabriel—the world's religions testify to the fact that something about going out into the wilderness opens people up to the big lessons. Along the Appalachian Trail, the wilderness isn't as big, and the lessons are usually not announced with a trumpet or a burning bush, but they're worth learning nevertheless." See *Appalachian Trailway News* 65 (July–Aug. 2004): 26.

83. Luxenberg, *Walking the Appalachian Trail,* 126, 207–12.

84. E. Jeffrey Coons, "Keeping Company with the Avatar Tree," *Appalachian Trailway News* 66 (Mar.–Apr. 2005), 36–37.

85. Tony Barrett, "Hail Dan," *Appalachian Trailway News* 62 (May–June 201): 28; Luxenberg, *Walking the Appalachian Trail,* 130, 136–37, 183.

86. Melissa Norman, *Appalachian Trail Journal,* May 22, 2002, available at http://www.trailjournals.com/entry.cfm?id=16115.

87. Luxenberg, *Walking the Appalachian Trail,* 48, 136–42, 178, 179, 180, 195; Turner and Turner, *Image and Pilgrimage,* 2, 8–9, 14, 31, 35, 37.

88. Luxenberg, *Walking the Appalachian Trail,* 169.

89. Coleman and Elsner, *Pilgrimage,* 6.

90. John Scott, "The Appellation Trail," *Appalachian Trailway News* 62 (Nov.–Dec. 2001), 23–24. Also see, Luxenberg, *Walking the Appalachian Trail,* ix.

91. Luxenberg, *Walking the Appalachian Trail,* 119, 135, 140, 177, 198.

92. Ibid., 178.

93. Trail diaries sent as e-mails to friends and family, generously shared with the author by Anthony J. Barrett.

94. Feb. 14, 2009, correspondence, on file with the author.

95. Albanese, *America: Religion and Religions,* 241, 243.

12. Old Commons Meets the New

1. Cheek et al., *The Foxfire 40th Anniversary Book,* 461–69; Everett Dick, *The Lure of the Land: A Social History of the Public Lands from the Articles of Confederation to the New Deal* (Lincoln: Univ. of Nebraska Press, 1970), 5; Steven Hahn,

The Roots of Southern Populism: Yeoman Farmers and the Transformation of the Georgia Upcountry, 1850–1890 (New York: Oxford Univ. Press, 1983), 58, 243, 245–46, 252, 253, 282; Shawn E. Kantor, *Politics and Property Rights: The Closing of the Open Range in the Postbellum South* (Chicago: Univ. of Chicago Press, 1998), 2, 17–18.

2. See Richard William Judd, *Common Lands, Common People: The Origins of Conservation in Northern New England* (Cambridge: Harvard Univ. Press, 1997). For Old World precursors of commons practices, see Tom Williamson and Liz Bellamy, *Property and Landscape: A Social History of Land Ownership and the English Countryside* (London: George Philip, 1987), 101–3, 115–16, 137–39; Marc Bloch, *French Rural History: An Essay on Its Basic Characteristics*, trans. Janet Sondheimer (1931; Berkeley: Univ. of California Press, 1966), 135, 186–87.

3. Gifford Pinchot, *Breaking New Ground* (New York: Harcourt, Brace & Co., 1947), 61; Thomas, *Southern Appalachia, 1885–1915*, 24, 29, 35–36, 39, 41.

4. Hicks, *Appalachian Valley*, 53. For a similar evocation of "God-given" or unwitting natural law sentiment regarding common property, see Hahn, *Roots of Southern Populism*, 252.

5. Donald S. Lutz, "The Relative Influence of European Writers on Eighteenth-Century American Political Thought," *American Political Science Review* 78 (Mar. 1984), 190, 192–96.

6. John Locke, *Two Treatises of Government*, ed. Peter Laslett (New York: Cambridge Univ. Press, 1988) 1, 4, 102, 271, 283–84, 285–302, and chap. 5.

7. Declaration of Independence (1776) preamble, available at http://www.archives.gov/exhibits/charters/declaration_transcript.html. The constitutional clause is found in the Fifth Amendment in the Bill of Rights, available at Yale Law School's Avalon Project: http://avalon.law.yale.edu/18th_century/rights1.asp.

8. James W. Ely Jr., *The Guardian of Every Other Right: A Constitutional History of Property Rights*, 2nd ed. (New York: Oxford Univ. Press, 1998), 17.

9. Garrett Hardin, "Tragedy of the Commons," *Science* 162 (1968): 1243–48.

10. John A. Baden and Douglas S. Noonan, eds., *Managing the Commons* (Bloomington: Indiana Univ. Press, 1998), xvii.

11. For a very small sampling, see S. V. Ciriacy-Wantrup and Richard C. Bishop, "'Common Property' as a Concept in Natural Resources Policy," *Natural Resources Journal* 15 (Oct. 1975): 713–27; Susan Jane Buck Cox, "No Tragedy on the Commons," *Environmental Ethics* 7 (Spring 1985): 49–61; F. Berkes, D. Feeney, B. J. McCay, and J. M. Acheson, "Benefits of the Commons," *Nature* 340 (July 13, 1989): 91–93; Bonnie J. McCay, "Culture and the Commons," in Hoage and Moran, *Culture*, 65–76. For an enormous list gathered by Charlotte Hess at Indiana Univ.'s Digital Library of the Commons, see http://www.indiana.edu/~workshop/wsl/tragedy.html.

12. For a radical proposal regarding a return to selective timber management, see the appendix regarding Leo Drey's Pioneer Forest in Missouri.

13. This is the question that goes begging in the work of Kathryn Newfont and Donald Edward Davis. See Newfont, "Moving Mountains," 91, 92, 225, 259; Davis, *Where There Are Mountains,* 174, 175, 179, 181.

14. Ronald Eller, *Uneven Ground: Appalachia since 1945* (Lexington: Univ. Press of Kentucky, 2008), 7, 10.

15. A recent survey of national forests in the South found that Virginia residents were among the highest levels (59 percent) of people who continue to use the forest for gathering herbs, mushrooms, berries, and the like. See David N. Wear and John G. Greis, *Southern Forest Resource Assessment: Summary Report,* Gen. Tech. Rep. SRS-54 (Asheville, NC: USDA, Forest Service, Southern Research Station, 2002), 270–72.

16. This generally dichotomous scenario informs Newfont's "Moving Mountains." See Newfont, "Moving Mountains," esp. iii, 105, 107, 110, 120. Also see Michael Ann Williams, "'When I Can Read My Title Clear': Anti-Environmentalism and Sense of Place in the Great Smoky Mountains," in *Culture, Environment, and Conservation in the Appalachian South,* ed. Benita J. Howell (Urbana: Univ. of Illinois Press, 2002), 87, 94.

17. Newfont, "Moving Mountains," iii, 14–15, 53, 64–82.

18. For a massive example involving the old 1,700-acre Black Diamond Ranch abutting the former New Castle Ranger District, see Cody Lowe, "Craig County Ranch: A Diamond in the Rough," *Roanoke Times,* Dec. 14, 2008.

19. Dick Austin, e-mail to the author, May 8, 2009.

20. For example, Virginia Forest Watch, Southern Appalachian Forest Coalition, Southern Environmental Law Center, the Clinch Coalition, and the Citizens Task Force for National Forest Management. Also see Hays, *Wars in the Woods,* 30–31, 84–85; Hirt, *A Conspiracy of Optimism,* 281–92.

21. A Virginia Department of Forestry ranger who wished to remain anonymous informed the author of this "pervasive cultural attitude" after working in southwestern Virginia for part of the 1980s and most of the 1990s. Retired forester Jim Willis, who has worked with Appalachia Sustainable Development, said compliance has improved in recent years (e-mail to author, May 12, 2009).

22. *Annotated Code of Virginia,* Title 10.1 Conservation, chap. 11, Forest Resources and the Department of Forestry, sec. 10.1-1100 through sec. 10.1-1181.12; Hays, *Wars in the Woods,* 99–100. For the recent tallying of logging violations, see the "Regional Reports" from the Virginia Forest Watch at http://www.virginiaforestwatch.org/badlogging.htm#western.

23. Robert Jackson Allen, "Sustainable Forestry in Virginia: Opportunities for Overdue Legislation and Options for Private Landowners," *Appalachian Journal of Law* 7 (Winter 2007): 5, 22–23, 28–33.

24. See Appalachian Sustainable Development at http://www.asdevelop.org/.

25. Nancy J. Manring, "The Politics of Accountability in National Forest Planning," *Administration and Society* 37 (Mar. 2005): 65–75.

26. For example, during the early and mid 2000s the Clinch Coalition was able to influence the JNF cut in the High Knob area, reducing it by half. See John Mongle, "Coalition Celebrates Seven Years," *Coalfield Progress,* Oct. 26, 2004.

27. These were journalistically documented, to some extent, in Chris Bolgiano, *Living in the Appalachian Forest: True Tales of Sustainable Forestry* (Mechanicsburg, PA: Stackpole Books, 2002).

28. Chris Bolgiano, *The Appalachian Forest: A Search for Roots and Renewal* (Mechanicsburg, PA: Stackpole Books, 1998), 62, 181.

29. Numerous EPA and USDA studies and congressional testimonials attest as much. In U.S. Congress, Senate, *Water Quality: Hearing before the Committee on Agriculture, Nutrition, and Forestry,* 106th Cong., 2nd sess., 2000, see the comments of Dan Glickman, secretary of the USDA (p. 12), Carol Browner, EPA Administrator (p. 120), National Association of State Foresters (pp. 287, 288, 289). In U.S. Congress, House, *Nonpoint Source Pollution: Atmospheric Deposition and Water Quality: Hearing before the Subcommittee on Water Resources and Environment of the Committee on Transportation and Infrastructure,* 110th Cong., 1st sess., 2007, see references to the EPA's *National Water Quality Inventory—2000 Report* (p. xx) and the comments of Craig Hooks, EPA director of Office of Wetlands, Oceans, and Watersheds (p. 239). Also see EPA, *Managing Nonpoint Source Pollution: Final Report to Congress on Section 319 of the Clean Water Act* (Washington, DC: EPA, 1992).

30. J. S. Shortle, D. Abler, and M. Ribaudo, "Agriculture and Water Quality: The Issues," in *Agricultural Nonpoint Source Pollution: Watershed Management and Hydrology,* ed. William F. Ritter and Adel Shirmohammadi (Boca Raton, FL: Lewis, 2001), 3; M. Ribaudo, "Non-point Source Pollution Control Policy in the USA," in *Agricultural Nonpoint Source Pollution: Watershed Management and Hydrology,* William F. Ritter and Adel Shirmohammadi, eds. (Boca Raton, FL: Lewis, 2001), 147.

31. M. Ribaudo, "Non-point Source Pollution Control Policy," 142.

32. Retired sustainable forester Jim Willis (e-mail to author, May 12, 2009) pointed out that loggers in southwestern Virginia have often asked "why us?" regarding targeted criticism over water quality. Willis also indicated that loggers have become more and more receptive to "best practices" when logging, especially after visiting old logging sites and seeing the consequences of erosion around logging roads.

33. See the woodland acreage statistics gathered by the U.S. Forest Service and Virginia Department of Forestry at http://www.fia.fs.fed.us/. For the broader national picture regarding national land use, see the USDA data at http://www.ers.usda.gov/Publications/EIB14/.

34. For an example, see Coles, "Appalachia's Moral Life," 21.

35. For examples spanning over a century, see Cheek et al., *Foxfire 40th Anniversary Book,* 415–16, 420–24, 480, 482–83, 485, 486; Gable and Davis, *Oral History of Konnarock,* 76; Miles, *Spirit of the Mountains,* 190–96; Rev. A. Rufus

Morgan interview in Eliot Wigginton, ed., *Foxfire 4* (Garden City, NY: Anchor Books, 1977), 439.

36. Al Fritsch and Kristin Johannsen, *Ecotourism in Appalachia: Marketing the Mountains* (Lexington: Univ. Press of Kentucky, 2004); see esp. chap. 7.

37. Fritsch and Johannsen, *Ecotourism in Appalachia,* 185.

38. Wagner, "It May Not Be Heaven, But It's Close," 19–20.

39. Wear and Greis, *Southern Forest Resource Assessment,* 185, 269–70.

40. Mohai et al., *Change in the USDA Forest Service,* 1, 7, 8, 9, 11, 13.

41. Ibid., 13.

42. Anthony J. Barrett trail diary, May 16, 1999.

43. Ibid.

44. Fritsch, *Appalachia: A Meditation,* 6.

45. Ibid.

46. Cheek et al., *Foxfire 40th Anniversary Book,* 480.

47. Ibid., 418, 461–69. Also see Dick, *Lure of the Land,* 5, and Kantor, *Politics and Property Rights,* 2, 17–18.

48. Melinda Wagner, ed. and comp., "It May Not Be Heaven, But It's Close: Land and People in Craig County, Virginia," Mar. 1995, position paper, Appalachian Regional Study Center, Radford Univ., Radford, VA, 8, 10, 11, 15, 16.

49. Feb. 3, 2009, correspondence, on file with author.

50. Jan. 6, 2009, correspondence, on file with author.

51. Wagner, "It May Not Be Heaven," 13. Barbara Rasmussen perpetuates this misunderstanding in the wider literature, citing Si Kahn's long-obsolete work, *The Forest Service and Appalachia* (Morganton, GA: John Hay Whitney Foundation, 1974). See Rasmussen, *Absentee Landowning and Exploitation in West Virginia,* 150. Regarding PILT, see appendix B.

52. In general, see Sarvis, "Land and Home in the American Mind," 107–37.

53. Liza Field, "A Century in the Life of the Forest," *Blue Ridge Country Magazine* (Nov.–Dec. 2005), at http://www.blueridgecountry.com/FavoriteArticles/CenturyOld_ND05/index.html.

54. Ibid.

55. Public Broadcasting System, "Appalachians: A History of Mountains and People" (2009), pt. 1, opening minutes.

Epilogue

1. See Milton S. Heath Jr., "The North Carolina Mountain Ridge Protection Act," *North Carolina Law Review* (Nov. 1984): 183–97.

2. The major exception being the coalfields abutting Kentucky and West Virginia, where intensive industrial exploitation continues.

3. The overview of this evolution is ably and concisely handled in Hays, *The American People and the National Forests.*

4. National Environmental Policy Act of 1969, Public Law 91-190, 91st Cong., 1st sess., 1970, sec. 2.

5. From Shands and Healy, *The Lands Nobody Wanted.*

6. For some interesting observations on island biogeography, see Ecological Society of America, *Conserving Biological Diversity in Our National Forests* (Washington, DC: Wilderness Society, 1986), 18, 19, 22, 28, 38, 47, 50.

7. For a lengthy depiction of the sort of range in points of view that defines modern JNF multiple use management, see *Roanoke Times and World News,* Sept. 5, 1993.

8. Loesel, interview by the author, Apr. 22, 1997. Also see Joseph L. Sax, "Parks, Wilderness, and Recreation," in Lacey, *Government and Environmental Politics,* 135.

9. See Silas House, *Something's Rising: Appalachians Fighting Mountaintop Removal* (Lexington: Univ. Press of Kentucky, 2009) and Shirley Stewart Burns, Mari-Lynn Evans, and Silas House, eds., *Coal Country: Rising up Against Mountaintop Removal Mining* (San Francisco: Sierra Club Books, 2009). This astonishing environmental change seems legal only through how courts have interpreted section 404 of the Clean Water Act. See Mark Baller and Leor Joseph Pantilat, "Defenders of Appalachia: The Campaign to Eliminate Mountaintop Removal Coal Mining and the Role of Public Justice," *Environmental Law* (Summer 2007): 629–64.

Appendix A

1. This 1998 essay and subsequent versions were based on the author's interviews with Leo A. Drey, Jan. 27, 1998, and Clint Trammel, May 20, 1998; tapes and transcripts deposited with the Western Historical Manuscript Collection, Columbia, Missouri, Environment Oral History Project, Collection No. 3966. Also see Susan Flader, "Missouri's Pioneer in Sustainable Forestry," *Forest History Today* (Spring–Fall 2004): 2–15. For an important work documenting one of the oldest sustainable forestry efforts in North America, see Thomas Davis's work pertaining to the Menominee Indians of Wisconsin: Thomas Davis, *Sustaining the Forest, the People, and the Spirit* (Albany: State Univ. of New York Press, 2000).

2. Davis, *Sustaining the Forest, the People, and the Spirit,* 12, 18, 20, 21. Interestingly, the Menominee example also contradicts the inevitability of a "tragedy of the commons" as discussed in chapter 12.

3. These include the Allegheny (Pennsylvania), Monongahela (West Virginia), Cherokee (Tennessee), Pisgah and Nantahala (North Carolina), Daniel Boone (Kentucky)—and of course the Pioneer's nearest neighbors, the Mark Twain (Missouri) and Ozark–St. Francis (Arkansas).

4. 58th Cong., 3rd sess., 1905, Public Laws chap. 1405; McSweeney-McNary Act of 1928, Public Law 70-466, 70th Cong., 1st sess., 1928, secs. 1, 2; Cooperative Farm Forestry Act of 1937, Public Law 75-95, 75th Cong., 1st sess., 1937; 81st

Cong., 2nd sess., 1950, Public Laws chap. 896; National Forest Management Act of 1976, Public Law 94-588, 94th Cong., 2nd sess., 1976, secs. 2(5), 2(6).

Appendix B

1. Fred Rogers Fairchild, *Forest Taxation in the United States,* USDA Miscellaneous Publication no. 218 (Washington, DC: GPO, 1935), 503, 517, 519, 639–40.

2. Dana and Fairfax, *Forest and Range Policy,* 90.

3. Ise, *The United States Forest Policy,* 279–80.

4. Dana and Fairfax, *Forest and Range Policy,* 289; Frome, *The Forest Service,* 233.

5. Kneipp, "Uncle Sam Buys Some Forests," 483.

6. Fairchild et al., *Forest Taxation in the United States,* 197.

7. *New Castle Record,* Nov. 27, 1958.

8. Shands and Healy, *The Lands Nobody Wanted,* 226.

9. Ibid.

10. Public Land Law Review Commission, *One Third of the Nation's Land: A Report to the President and to the Congress* (Washington, DC: GPO, 1970), chap. 14, "Tax Immunity."

11. Payment in Lieu of Taxes Act, sec. 2(a)(1).

12. Ms. Ann Huebner, Forest Service economist, telephone interview, Feb. 3, 1992.

13. Appalachian Land Ownership Task Force, *Land Ownership Patterns and Their Impacts on Appalachian Communities,* vol. 6, *Virginia* (Washington, DC: Appalachian Regional Commission, 1980–81). Twenty-five percent of the fund and PILT data was supplied by Bill Howell of the BLM.

Source Materials

A Note on Primary Sources: By the time I reached the National Archives in 1992, they had already destroyed the eighty-some boxes of Jefferson National Forest records. Only a few files remained. The correspondence, memorandums, and other documents would have made for a far different history, had they been available. Some primary documents have survived, and all are (or were) located at the supervisor's office in Roanoke, Virginia, or at the individual ranger district offices throughout southwest Virginia. The nature of these documents range from well-kept, orderly materials still used in daily operations to haphazardly stored documents kept in boxes or file cabinets at the whim of a staff person out of historical curiosity or for personal reasons. Researchers should remember that the JNF is a land management institution, however, and not an archive.

The land files, located at JNF headquarters, constitute many hundreds of cubic feet and represent all parcels acquired by the National Forest Reservation Commission beginning and following 1911. These documents remain viable and regularly consulted by JNF staff when dealing with acquisitions, land trades, boundary surveys, and such. They are not archived as historic documents would be but generally receive the same kind of care in terms of their retrieval, use, and refiling. They contain invaluable historic material and proved absolutely essential for ascertaining certain aspects of acquisition history (particularly surrounding the Mount Rogers NRA) related in these pages.

Other primary documents located in the Roanoke supervisor's office were in a much more precarious state when I last saw them. The Mount Rogers NRA documents were crucial to this study's most complicated chapter and had survived far longer than typically so, mostly because of Charles Blankenship's interest. After his retirement (which was before I began working for the JNF in 1991), they continued to reside in a file cabinet in the recreation staff officer's quarters. Timber, mineral, wildlife, botanical, and air-quality documents were much more recent and existed more as working references than archived, historical documents. Finally, the oral history audio material remains the most fragile of resources, given the temporary nature of magnetic tape. The majority of these tapes were boxed in the archaeology office. A few interviews are deposited in Missouri with the Western Historic Manuscript Collection at University of Missouri–Columbia.

Archival Sources

Appalachian Regional Study Center, Radford Univ., Radford, VA.

> Wagner, Melinda Bollar. "Appalachian Attitude toward Land." Paper presented at Appalachian Studies Conference, 1996 (194.99.d).
>
> ———., ed. and comp. "It May Not Be Heaven, But It's Close: Land and People in Craig County, Virginia." Position paper, Mar. 1995 (195.8.a).

Citizens Task Force for National Forest Management, Roanoke, VA.

> JNF budget documents. Deposited with Jim Loesel.

Emory and Henry College, Emory, VA.

> Saltworks lawsuit, printed manuscripts.
>
> > *N. K. White and Others vs. Stuart, Buchanan and Co.* Abingdon, VA: Standard Print, n.d. [after Feb. 17, 1879].
> >
> > *Stuart and Others vs. White and Others.* South West Virginia Enterprise Office, Wytheville, 1872.
> >
> > *Stuart, Buchanan and Co. vs. N.K. White et al.* Virginia office, Abingdon, 1867.
> >
> > *White, Robertson and Others vs. Stuart, Buchanan and Co.* October 18, 1877.

Forest History Society, Durham, NC.

> Mustian, Albert. Papers. "Holston Working Circle Development Plan, 1941–1950."

Jefferson National Forest, Roanoke, VA.

> Air Resource Management.
>
> Cultural Resource.
>
> Glenwood Ranger District History.
>
> > Robens, Ward. "Historical Data, Glenwood Working Circle, Jefferson National Forest." September 1949.
>
> Land Status Atlases and Individual Land.
>
> Mount Rogers National Recreation Area Headquarters Land (2100 File).
>
> Recreation and Planning.
>
> Timber Management.
>
> Watershed Protection and Management.
>
> Wildlife Management.

National Archives, Washington, DC.

> U.S. Forest Service, Record Group 95.

Univ. of Virginia, Alderman Library Archives, Charlottesville.

> Preston-Davis Papers.

Virginia Historical Society, Richmond.

Campbell-Preston-Floyd Papers.

Virginia Polytechnic Institute and State Univ., Special Collections, Blacksburg.

Preston Family Papers.

Norfolk & Western Archives.

Sarvis, Will. "The Salt Industry of Nineteenth Century Saltville, Virginia."
MS.

Virginia State Library and Archives, Richmond.

Board of Public Works—Turnpike Companies.

Laws Cited

Acquired Lands Act of 1947. Public Law 80-382. 80th Cong., 1st sess., Aug. 7, 1947.

Annotated Code of Virginia, Title 10.1. Conservation, Chapter 11, Forest Resources and the Dept. of Forestry, sec. 10.1-1100 through sec. 10.1-1181.12.

Appalachian Regional Development Act of 1965. Public Law 89-4. 89th Cong., 3rd sess., Mar. 9, 1965.

Archaeological and Historic Preservation Act of 1974. Public Law 93-291. 93rd Cong., 2nd sess., May 24, 1974.

Archaeological Resources Protection Act of 1979. Public Law 96-95. 96th Cong., 1st sess., Oct. 31, 1979.

Clark-McNary Act of 1924. Public Law 108-198. 108th Cong., 1st sess., June 7, 1924.

Eastern Wilderness Act of 1975. Public Law 93-622. 93rd Cong., 2nd sess., Jan. 3, 1975.

Endangered Species Act of 1973. Public Law 93-205. 93rd Cong., 1st sess., Dec. 28, 1973.

Federal Onshore Oil and Gas Leasing Act of 1987. In Title V: Energy and Environmental Programs, Omnibus Budget Reconciliation Act of 1987. Public Law 100-203. 100th Congress, 1st sess., Dec. 22, 1987, sec. 5101.

Historic Sites Act of 1935. Public Law 74-292. 74th Cong., 1st sess., Aug. 21, 1935.

Land and Water Conservation Fund Act of 1964. Public Law 88-578. 88th Cong., 2nd sess., Sept. 3, 1964.

Materials Disposal Act of 1947. Public Law 80-291. 80th Cong., 1st sess., July 31, 1947.

Minerals Leasing Act of 1920. Public Law 66-146. 66th Cong., 2nd sess., Feb. 25, 1920.

Mount Rogers National Recreation Act of 1966. Public Law 89-438. 89th Cong., 2nd sess., May 31, 1966.

Multiple-Use and Sustained Yield Act of 1960. Public Law 86-517. 86th Cong., 2nd sess., June 12, 1960.

National Forest Management Act of 1976. Public Law 94-588. 94th Cong., 2nd sess., Oct. 22, 1976.

National Historic Preservation Act of 1966. Public Law 89-665. 89th Cong., 2nd sess., Oct. 15, 1966.

Omnibus Public Land Management Act of 2009. Public Law 111-11. 111th Cong., 1st sess., Mar. 30, 2009, Subtitle B, "Virginia Ridge and Valley Wilderness."

Organic Administration Act of 1897. 55th Cong., 1st sess., June 4, 1897.

Payment in Lieu of Taxes Act of 1976. Public Law 94-565. 94th Cong., 2nd sess., Oct. 20, 1976.

Pittman-Robertson Act of 1937. Public Law 75-415. 75th Congress, 1st sess., Sept. 2, 1937.

Surface Mining Control and Reclamation Act of 1977. Public Law 95-87. 95th Cong., 1st sess., Aug. 3, 1977.

Uniform Relocation Assistance and Real Property Acquisition Policies Act of 1971. Public Law 91-646. 91st Cong., 1st sess., Jan. 2, 1971.

Virginia Wilderness Act of 1984. Public Law 98-586. 98th Cong., 2nd sess., Oct. 30, 1984.

Virginia Wilderness Act of 1988. Public Law 100-326. 100th Cong., 2nd sess., June 7, 1988.

Watershed Protection and Flood Prevention Act of 1954. Public Law 83-566. 83rd Cong., 2nd sess., Aug. 4, 1954.

Weeks Act of 1911. Public Law 61-435. 61st Cong., 3rd sess., Mar. 1, 1911.

White Pine Blister Rust Control Act of 1940. Public Law 76-486. 76th Cong., 3rd sess., Apr. 26, 1940.

Wilderness Act of 1964. Public Law 88-577. 88th Cong., 2nd sess., Sept. 3, 1964.

Published Primary Sources

Note: Many of the following are on file with the JNF Cultural Resources Division.

Glenwood Working Circle, Jefferson National Forest. *Streamliner Timber Management Plan.* Natural Bridge Station, VA: U.S. Forest Service. Approved between Apr. 27 and May 28, 1951.

———. *Timber Sale Report of Chestnut Mt. Chance.* Natural Bridge Station, VA: U.S. Forest Service, Apr. 17, 1948.

Homans, G. M. *Report on a Working Plan for the Timberlands of the Interstate Investment Company, Lee County, Virginia, and Harlan County, Kentucky.* N.p.: U.S. Forest Service, 1905.

Jefferson News Letter. Vol. 1 (May 25, 1936): 1–17.

Message from the President of the United States Transmitting a Report of the Secretary of Agriculture in Relation to the Forests, Rivers, and Mountains of the Southern Appalachian Region. Washington, DC: GPO, 1902.

National Forest Reservation Commission. *Progress of Purchase of Eastern National Forests.* Washington, DC: GPO, 1920.

———. Annual reports. Washington, DC: GPO, 1911–65.

Natural Bridge National Forest. Map. USDA MF-20 R.7. Washington, DC: GPO, 1930.

Sargent, Charles S. *Report on the Forests of North America*. Washington, DC: GPO, 1884.

Sears, H. M. *Management Plan for the Arcadia Working Circle*. Natural Bridge Station, VA: Natural Bridge National Forest, 1928.

Shrauder, Paul. "Extensive Forest-Wide Wildlife Habitat Survey." Paper presented at the Timber-Wildlife Coordination Workshop, Asheville, NC, Oct.–Nov., 1973.

Tabbutt, D. W. *Stoney Creek Unit*. Timber sale summary. Bedford Working Circle, Natural Bridge National Forest, Sept. 17, 1927.

U.S. Dept. of Agriculture, Forest Service. *National Forest Manual: Regulations and Instructions*. Field manual. Washington, DC: GPO, 1928–33.

U.S. Forest Service. *Fire Handbook—Region Seven*. Washington, DC: GPO, 1931.

———. *Service Bulletin* 18 (May 7, 1934).

Virginia State Chamber of Commerce. *National Forests in Virginia: Vacation Opportunities*. Richmond: Virginia State Chamber of Commerce, 1926.

Taped Interviews

Note: All interviews conducted by the author, except where noted. Tapes on file with the JNF Cultural Resources, Roanoke, Virginia, unless otherwise indicated.

Arthur, Elmore. Resident of Potts Creek Valley, next to New Castle Ranger District. Interviewed in Monroe County, WV, Aug. 20, 1992.

Barber, Mike. Forest Service archaeologist since 1977, covering national forests in Virginia, Kentucky, Tennessee, and North Carolina. Interviewed in Roanoke, VA, June 3, 1992.

Blackburn, William D. Began his career in the Forest Service during the mid-1950s and worked in Regions 5, 3, and 1. He came to the JNF in 1977, where he worked in minerals, as the State and Private Forestry liaison, and other capacities. Interviewed in Roanoke, VA, July 10, 1992.

Blankenship, Charles. Worked for the JNF August 1966 to January 1990 in such areas as Mt. Rogers NRA planning, recreation, land management, fire control, and law enforcement. Interviewed in Roanoke, VA, Dec. 12, 1991.

Campbell, William. Worked as a CCC supervisor at the Natural Bridge Camp in early 1936, then transferred to the old Holston District in 1937, where he worked in timber survey and general district work, which included maintenance of trails, telephones, and fire control. He moved to the Newport CCC camp in 1940 and was laid off when that camp closed in 1941. He returned to the Holston District in 1948 as general district assistant, then transferred to Roanoke in 1959, where his survey work coincided with the increased land acquisition program. Mr. Campbell retired in 1977. Interviewed in Roanoke, VA, Nov. 14, 1991, and Dec. 13, 1991.

Clark, Blanch. Retired Green Cove Station employee and longtime resident of Green Cove. Interviewed in Green Cove, VA, Oct. 7, 1991.

Clark, C. B. Worked as a fire lookout on the Clinch Purchase Unit and early Clinch Ranger District, 1931–42. Once Mr. Clark transferred into the district office, he worked in fire control, timber management, budget and finance, and just about everything else, from 1942 to 1965. Interviewed in Jasper, VA, Jan. 7, 1992.

Cordell, Cecil. Began working for the Forest Service in 1954 on the Pisgah National Forest. He also worked on the Cherokee and Ouachita before coming to the Jefferson in 1969. From 1969 to 1984 he worked as administrative officer, including deputy job corps coordinator. Interviewed in Bent Mountain, VA, Dec. 18, 1991.

Dickerman, Ernest M. One of the charter members in the founding of the Wilderness Society in 1936, an organization he has been a member of ever since. His work helped bring about the 1964 Wilderness Act and the 1975 Eastern Wilderness Act. He was a formal employee of the Wilderness Society from 1966 to 1976 involved in working with the Smoky Mountain National Park. He served as president of the Virginia Wilderness Committee from 1976 to circa 1979 and has been involved in the establishment of all national forest wilderness areas in Virginia. Interviewed in Buffalo Gap, VA, Apr. 24, 1997. Interview on file in the Western Historical Manuscript Collection, Columbia, MO.

Edwards, Adele. Retired schoolteacher and surviving daughter of W. M. Buchanan, Green Cove Station owner and operator. Interviewed in Green Cove, VA, Sept. 17, 1991.

Gable, William. Led the opposition to the Mount Rogers NRA development, beginning in the early 1970s as a Lutheran minister out of Konnarock, Virginia. Interviewed in Abingdon, VA, Jan. 8, 1992.

Hadacek, Art. Came to the Jefferson in 1966 and worked until 1983 in timber management, manpower programs (including Mainstream America), safety, law enforcement, civil defense, communications, Job Corps coordination, and fire control. Interviewed in Bent Mountain, VA, Dec. 16, 1991.

Hassinger, Martin. Son of timber industrialist Luther Hassinger. Interviewed by Mary Kegley and Lionel Melancon in Bristol, VA, July 24, 1979.

Huffman, Lucy. Longtime resident of Johns Creek Valley, near the New Castle Ranger District. Interviewed in Craig County, VA, June 2, 1992.

Jefferson, James N. Worked as a CCC supervisor on the Glenwood and Clinch Ranger Districts from 1933 to 1939. During the late 1930s he worked in the supervisor's office as a CCC administrator. From 1943 to 1949 he worked as a forest engineer on the Jefferson, then transferred to the George Washington, where he finished his engineer and Forest Service career in 1969. Interviewed in Salem, VA, Jan. 17, 1992.

Jennings, Pat. Born in Camp, Virginia, in 1919, and grew up with the old Unaka National Forest. He worked with Holston District personnel as Smyth County sheriff, 1947–54, then became instrumental in the Mount Rogers NRA legislation as U.S. congressman, 1954–66, and clerk of the U.S. Congress, 1966–76. Interviewed in Marion, VA, Jan. 6, 1992.

Kibler, Eleanor. Helped get the Blacksburg Ranger District started in 1958–59. Her late husband, Bruce Kibler, began working with the WPA near or on JNF territory during the 1930s and worked with the Jefferson until 1970 in fire control and as construction superintendent. Interviewed in New Castle, VA, Jan. 23, 1992.

Kinman, Reginald. Worked as a landscape architect and in planning for the U.S. Military and the Army Corps of Engineers before joining the Forest Service in 1966. He worked in these fields on the Jefferson from 1966 to 1989, with three years on the Ozark National Forest, 1974–78. Interviewed in Roanoke, VA, Feb. 4, 1992.

Loesel, James E. Founding member of the Citizens Task Force in 1982, he has worked for the organization ever since. The Roanoke-based Citizens Task Force functions as a watchdog group monitoring Forest Service land management practices, particularly those of Region 8, and especially the Jefferson and George Washington national forests. Interviewed in Roanoke, VA, Apr. 22, 1997. Interview on file in the Western Historical Manuscript Collection, Columbia, MO.

Lowe, John C. Represented some of the landowners involved in land condemnation cases surrounding the Mount Rogers NRA during the mid-1970s. Interviewed in Charlottesville, VA, May 5, 1992.

McGinnes, Burd. Worked for the U.S. Fish and Wildlife Service as leader of the Virginia Cooperative Wildlife Research Unit at Virginia Tech, 1958–82. Interviewed in Blacksburg, VA, Jan. 29, 1992.

Medley, Hermie B. Remembered early Forest Service land acquisitions near his home in Goldbond, Virginia. He worked on the New Castle District during the early 1940s in road construction, fire suppression, and wildlife habitat creation. He was a frequent "202C" customer of Ranger J. N. Van Alstine and W. W. Taylor. Interviewed in Goldbond, VA, Jan. 22, 1992.

Melancon, Lionel. Came to the Jefferson in 1969 as interpretive planner for the Mount Rogers project. In that capacity, and as public information officer, Mr. Melancon worked until 1989. Interviewed in Roanoke, VA, Dec. 17, 1991.

Morrison, Bill. Worked on North Carolina national forests before coming to the Jefferson in 1967, where he worked in timber management, at the Flatwoods CCC, and in watershed, air quality, and soil management until 1986. Interviewed in Hardy, VA, Jan. 16, 1992.

Penfold, Michael. Son of the famous conservationist Joe Penfold, Michael began working for the Forest Service in 1957. For the next twelve years he worked as a timber cruiser, smoke jumper, recreation manager, and other occupations in the western United States. He served as supervisor of the Jefferson National Forest from 1969 to 1976, subsequently taking a job with the Council for Environmental Quality. He was assistant director of the Bureau of Land Management at the time of the interview. Interviewed in Washington, DC, Oct. 26, 1993. Interview on file at the Western Historical Manuscript Collection, Columbia, MO.

Runyan, Francis. Personnel clerk, 1961–91. According to Cecil Cordell, Runyan filled the capacity of personnel management specialist for a year during the late 1970s and did a "bang-up job" even though she never received the title or salary appropriate for her accomplishments. Interviewed in Roanoke, VA, Dec. 9, 1991.

Saboites, Charles. New Castle district ranger, 1961–70, and Clinch District ranger, 1976–91, also worked on the Allegheny, Monongahela, Sabine, and Kisatchie National Forests. Interviewed in Wise, VA, Feb. 20, 1992.

Sexton, Charles. Worked in TSI, fire control, timber sales administration, and other jobs on the Clinch District, 1962–80. Interviewed in Wise County, VA, Feb. 20, 1992.

Short, Jim. Flatwoods CCC director, January 1968 to October 1971, and deputy forest supervisor, October 1971 to September 1977. Interviewed in Roanoke, VA, Dec. 12, 1991.

Shrauder, Paul. JNF's first full-time wildlife biologist, 1960–65; region eight wildlife biologist, 1965–72; JNF wildlife biologist and timber staff officer, 1972–88. Interviewed in Roanoke, VA, Dec. 10, 1991.

Smith, J.H.W. Born in 1910 in Craig County, where, except for five years in Covington, he has lived all his life. He was an early timber buyer on the New Castle District and worked in the timber and lumber industry for more than thirty years. He was a volunteer firefighter for fifty-one years and helped fight many fires on national forest land. He continues to use the national forest for hunting every year. Interviewed in New Castle, VA, Jan. 30, 1992.

Stamper, Marjorie. Worked in the New Castle Ranger District office from 1957 to 1978. Her late husband, Paul Stamper, worked for the Jefferson from the 1930s until 1964 in fire control, road construction, and heavy equipment operation. Interviewed in New Castle, VA, Jan. 23, 1992.

Timko, Paul. Began his Forest Service career in 1951 and worked on national forests in Texas, Colorado, Arkansas, Oklahoma, and Georgia before coming to the Jefferson in 1974, where he worked in Lands until 1984. Interviewed in Roanoke County, VA, May 19, 1992.

Untaped Interviews and Other Personal Communications

Anonymous. Virginia Tech forestry professor who wished to remain anonymous. Interviewed in Blacksburg, VA, Jan. 28, 1992.

Barker, Jim. Mount Rogers National Recreation Area office. Personal communications, 1991–92.

Beck, Don. Forest scientist, Southeast Experiment Station. Interview by telephone, Feb. 25, 1992.

Blevins, Lucille. Retired schoolteacher and lifelong resident of Green Cove. Interviewed in Green Cove, VA, Oct. 15, 1991.

Boardwine, Bob. New Castle Ranger District. Personal communication, Jan. 29, 1992.

Branham, Ken. Clinch Ranger District. Interview in person, Jan. 7, 1992.

Brossey, Louis. Wythe Ranger District. Interview by telephone, Feb. 14, 1992.

Collins, David. Blacksburg Ranger District. Interview by telephone, Feb. 24, 1992.

Colona, Bob. JNF silviculturist, 1967–88. Interviewed in Roanoke, VA, Mar. 2, 1992.

Harris, Megaera. Research Historian, Office of the Postmaster General. Letter, Oct. 17, 1991.

Harry, Herman. Longtime resident of Potts Creek Valley. Interviewed in Monroe County, WV, Aug. 20, 1992.

Hedrick, Joe. Glenwood Ranger District. Personal communication, Feb. 1992.

Huebner, Ann. Washington Office Forest Service Economist. Interview by telephone, Feb. 3, 1992.

Karger, Ernie. Began working with the CCC in 1933 on the Allegheny National Forest. He moved down to the Natural Bridge Station in January 1936 and worked on the Jefferson until 1944 as assistant ranger, ranger, and assistant supervisor. Interview by telephone, Mar. 12, 17, 1992.

Miller, Ken. Roanoke chapter president, National Railway Historical Society. Interview by phone, Oct. 21, 1991.

Paradzinski, Paul. New Castle Ranger District. Personal communication, Jan. 29, 1992.

Piedmont, Don. N&S public relations director. Interview by phone, Sept. 16, 1991.

Purcell, Lewis. Mount Rogers NRA office. Personal communications, 1991, 1992.

Sloan, Hank. JNF logging engineer. Personal communication, Feb. 23, 1993.

Spiller, Donald. Wythe Ranger District. Personal communication, Feb. 18, 1992.

Walters, Ray. Wythe Ranger District. Personal communication, Feb. 18, 1992.

Wilbur, Henry M. Director, Mountain Lake Biological Station. Interviewed at Mountain Lake, Mar. 30, 1992.

Unpublished Sources

Note: The following materials are on file at JNF Cultural Resources Division, unless noted otherwise.

Barber, Michael B. "Human Prehistory beyond the Blue Ridge: A Brief Introduction." Revised 1989.

Blankenship, Charles A. "Bureaucracy and the Appalachian Trail." Nov. 1989.

———. "Mount Rogers National Recreation Plan." Apr. 1967.

Citizens for Southwest Virginia. "Response of the Citizens for Southwest Virginia to the Draft Environmental Impact Statement for the Mt. Rogers National Recreation Area and Mt. Rogers Scenic Highway." Aug. 1, 1978. Virginia State Library, Richmond.

Clark, C. B. "First Thirty Years on the Clinch, 1935–1965." 1965. Also on file at Virginia State Library, Richmond.

Cochran, A. R. Memoirs. Undated.

Colona, Robert T., and Louise Givens-Reynolds. "The Jefferson National Forest: A Stressed and Aging Condition." Undated [after 1983].

Commission of Game and Inland Fisheries et al. "Cooperative Wildlife Management on Virginia National Forests." June 1963.

Conrad, David. "The Return of the Forests." History of Region 7/Region 9. American Resources Group, 1988.

"Jefferson National Forest Public Involvement Handbook." Forest Service [after Aug. 1983].

Jefferson National Forest. "A Proposal: Mill Creek Wilderness." Draft Environmental Impact Statement, May 22, 1981, Roanoke, VA.

Morrison, Bill. "Writings on History of Forestry on the Jefferson." Nov. 1975.

Paxton, Percy J. "National Forests and Purchase Units of Region Eight." USDA Forest Service, Region Eight Office, Atlanta, Jan. 1, 1955.

Sloan, Hank, and Karen Goode. "Timber Sale Program Information Reporting System Report (TSPIRS), Fiscal Year 1992."

Smith, Lewis. "Glenwood District Ranger, August 1931–September 1958." Memoirs. 1979.

Wilbur, Henry M. "Facilities Support for Mountain Lake Biological Station." 1992.

Miscellaneous Governmental Publications

Assessment of Regional Priority for the Mount Rogers N.R.A. Washington, DC: GPO, 1977.

Ayers, H. B., and W. W. Ashe. The Southern Appalachian Forests. U.S. Geological Survey, Professional Paper No. 37. Washington, DC: GPO, 1905.

Blankenship, Charles A. The Mount Rogers National Recreation Area: History of Land Use Planning. Washington, DC: GPO, 1976.

Bureau of Agricultural Economics. Economic and Social Problems and Conditions of the Southern Appalachians. Washington, DC: USDA, 1935.

Catton, Theodore, and Lisa Mighetto. The Fish and Wildlife Job on the National Forests: A Century of Game and Fish Conservation, Habitat Protection, and Ecosystem Management. Washington, DC: USDA, Forest Service, 1998.

Cherokee National Forest. Final Environmental Impact Statement. Cleveland, TN: USDA, Forest Service, 1986.

Craig, Ronald B. Virginia Forest Resources and Industries. Miscellaneous Publication No. 681. Washington, DC: USDA, 1949.

Daniel Boone National Forest. Draft Land and Resource Management Plan. Winchester, KY: USDA, Forest Service, 1985.

Environmental Protection Agency. Managing Nonpoint Source Pollution: Final Report to Congress on Section 319 of the Clean Water Act. Washington, DC: EPA, 1992.

Fedkiw, John. Managing Multiple Uses on National Forests, 1905–1995: A 90-Year Learning Experience and It Isn't Finished Yet. Washington, DC: USDA, Forest Service, 1998.

Glenn, Leonidas C. *Denudation and Erosion in the Southern Appalachian Region and the Monongahela Basin.* U.S. Geological Survey, Professional Paper No. 72. Washington, DC: GPO, 1911.

Hoots, Thomas A. "Some Facts about Timber Management and Roads on the Jefferson National Forest, Virginia." Forest Service MS, Apr. 6, 1983.

Hughes, Jay M., and R. Duane Lloyd, comps. *Outdoor Recreation: Advances in Application of Economics.* General Technical Report WO-2. Washington, DC: USDA, Forest Service, 1977.

Jefferson National Forest. *Jefferson National Forest General Report to the Public for 1991.* Roanoke, VA: Jefferson National Forest, 1991-1993.

———. *Mount Rogers National Recreation Area: A Review.* Roanoke, VA: 1969.

———. *Mount Rogers National Recreation Area Final Management Plan and Environmental Impact Statement.* Washington, DC: GPO, 1978.

———. *Mount Rogers National Recreation Area, Jefferson National Forest, Virginia.* Information pamphlet, 1966.

———. *Mount Rogers NRA Story.* Public information publication, circa 1978–79.

———. *Wildlife Management Handbook.* Roanoke, VA: Jefferson National Forest, 1949.

Kaufmann, Merrill R. *An Ecological Basis for Ecosystem Management.* General technical report RM-246. Fort Collins, CO: USDA Forest Service, Rocky Mountain Forest and Range Experiment Station, 1994.

Knight, Herbert A., and Joe P. McClure. *Virginia's Timber, 1966.* U.S. Forest Service Resource Bulletin SE8. Asheville, NC: USDA, Forest Service, Southeastern Forest Experiment Station, 1967.

———. *Virginia's Timber, 1977.* Forest Service resource bulletin SE-44. Asheville, NC: USDA Forest Service, Southeastern Forest Experiment Station, 1978.

Larson, Robert W., and Mackay B. Bryan. *Virginia's Timber.* Forest Survey Release No. 54. Asheville, NC: USDA, Forest Service, Southeastern Forest Experiment Station, 1959.

Leighton, M. O., and A. H. Horton. *Relation of the Southern Appalachian Mountains to Inland Water Navigation.* USDA, Forest Service Circular No. 143, 1908.

Mohai, Paul. *Change in the USDA Forest Service: Are We Heading in the Right Direction?* USDA Forest Service General Technical Report NC-172. Washington, DC: GPO, 1994.

Outdoor Recreation Resources Review Commission. *Economic Studies of Outdoor Recreation.* Report No. 24. Washington, DC: GPO, 1962.

Public Land Law Review Commission. *One Third of the Nation's Land: A Report to the President and to the Congress.* Washington, DC: GPO, 1970.

Rhyne, Edwin H. "Recreation, Continuity, and Change: The Impact of the Socio-Cultural Environment of the Development of a NRA in the Vicinity of Mount Rogers, Virginia." Unpublished report, Oct. 1974, 16, 18, 2100 File. Roanoke, VA: Jefferson National Forest.

Rose, Anita Kristine. *Virginia's Forests, 2001.* Resource Bulletin SRS-120. Asheville, NC: USDA FS Southern Research Station, 2007.

U.S. Congress. House. *Economics of Federal Timber Sales: Hearings before the Subcommittee on Forests, Family Farms, and Energy of the Committee on Agriculture.* 99th Cong., 1st sess., 1985.

———. *Internal Commerce of the United States: Virginia,* by J. D. Imboden. 49th Cong., 2nd sess., 1886. H. Doc., vol. 17, no. 7, pt 2.

———. *Nonpoint Source Pollution: Atmospheric Deposition and Water Quality: Hearing before the Subcommittee on Water Resources and Environment of the Committee on Transportation and Infrastructure.* 110th Cong., 1st sess., 2007.

———. *Salt Works—United States,* by S. D. Ingham. 21st Cong., 1st sess., 1830. H. Doc. 55.

———. Senate. *Water Quality: Hearing before the Committee on Agriculture, Nutrition, and Forestry.* 106th Cong., 2nd sess., 2000.

U.S. Dept. of Agriculture, Forest Service. *Facts about Region 7.* Upper Darby, PA: USDA, Forest Service, 1961–62.

———. *The Fairest One of All.* Milwaukee: USDA, Forest Service, 1973.

———. "Forest Service Roles in Outdoor Recreation." Program Aid 1205, Jan. 1978.

———. *Guide for Managing the National Forests in the Appalachians.* Washington, DC: GPO, 1971.

———. *Jefferson National Forest, Virginia: Facts 1981.* Roanoke, VA: Jefferson National Forest, 1981.

———. *Story of the Jefferson National Forest.* Washington, DC: GPO, 1964.

———. *System for Managing the National Forests in the East.* Washington, DC: GPO, 1970.

———. *Wildlife Habitat Management Handbook.* FSH 2609.23R. Washington, DC: GPO, 1971.

U.S. Dept. of Conservation and Recreation, Division of State Parks. *Virginia State Parks, Facts and Figures, 1990.* Richmond: Commonwealth of Virginia, 1990.

Virginia's Forests. USDA FS Research Bulletin SE-95, Aug. 1987.

Wear, David N., and John G. Greis. *Southern Forest Resource Assessment: Summary Report.* Gen. Tech. Rep. SRS-54. Asheville, NC: USDA, Forest Service, Southern Research Station, 2002.

Wilson, James. *Report on the Forests and Forest Conditions of the Southern Appalachian Region.* Washington, DC: GPO, 1902.

Newspapers

Abingdon Virginian

Bland Messenger

Bluefield Daily Telegraph

Bristol Herald Courier

Bristol Virginia–Tennessean

Clinch Valley News

Coalfield Progress

Cumberland Times

Dickenson Star

Galax Gazette

Giles Virginian-Leader

Montgomery News Messenger

Mount Rogers Citizen

Mountain Eagle Newspaper

New Castle Record

The (Lynchburg, VA) News

The (Abingdon, VA) Plow (defunct; back issues available at Kelly Library, Emory and
  Henry College, Emory, VA)

Roanoke Times

Roanoke Times and World News

Roanoke World News

Saltville News–Messenger

Scott County Herald Virginian

Southwest Virginia Enterprise

Sunset (Princeton-Bluefield) News Observer

Smyth County News

Washington County News

Films

Flager, Mark, dir. *Appalachian Impressions.* Austin, TX: Flagler Films Presentation,
  2004.

Spears, Ross, dir. *Appalachians: A History of Mountains and People.* Arlington, VA:
  Public Broadcasting System, 2009.

Dissertations and Theses

Adams, Harold S. "Analysis of Vegetation on the South Slopes of Peters Mountain,
  Virginia." Ph.D. diss., Virginia Polytechnic Institute and State Univ., 1974.

Atkins, Charlotte Lou. "Rockbridge Alum Springs: A History of the Spa: 1790–
  1974." Master's thesis, Virginia Polytechnic Institute and State Univ., 1974.

Bachant, Joseph Peter. "The Distribution and Abundance of the Eastern Wild Turkey as Influenced by Forest Wildlife Management Practices. " Master's thesis,
  Virginia Polytechnic Institute and State Univ., 1963.

Blymyer, Michael J. "Impact of Clearcutting on Indigenous Mammals of Southwest
  Virginia." Master's thesis, Virginia Polytechnic Institute and State Univ., 1976.

Curtis, Robert L. Jr. "Climatic Factors Influencing Hunter Sightings of Deer on the Broad Run Research Area." Master's thesis, Virginia Polytechnic Institute and State Univ., 1971.

Diebel, Kenneth Edward. "Isozyme Variation within the Fraser Fir Population on Mt. Rogers, Virginia." Ph.D. diss., Virginia Polytechnic Institute and State Univ., 1989.

Edmunds, Paul C. "Seasonal Fluctuations of Rotifer Populations Related to Selected Biological, Chemical and Physical Parameters in a Small Mountain Pond, Jefferson National Forest, Virginia." Master's thesis, Virginia Polytechnic Institute and State Univ., 1974.

England, Russell H. "Some Effects of Abandoned Manganese Strip Mines in Smyth County, Virginia on Stream Ecology." Master's thesis, Virginia Polytechnic Institute and State Univ., 1968.

Holmes, Oliver Wendell. "Stagecoach and Mail: From Colonial Days to 1820." Ph.D. diss., Columbia Univ., 1956.

Hooper, Robert Guy. "The Influence of Habitat Disturbance on Bird Populations." Master's thesis, Virginia Polytechnic Institute and State Univ., 1967.

Hunter, Robert F. "The Turnpike Movement in Virginia, 1816–1860." Ph.D. diss., Columbia Univ., 1957.

Klyza, Christopher James. "The Political Economy of Natural Resource Development: A Case Study of Mount Rogers National Recreation Area." Master's thesis, Duke Univ., 1983.

Larson, J. S. "Wildlife Forage Clearings on Forest Lands: A Critical Appraisal and Research Needs." Ph.D. diss., Virginia Polytechnic Institute and State Univ., 1966.

Mays, Thomas Davidson. "The Price of Freedom: The Battle of Saltville and the Massacre of the Fifth United States Colored Cavalry." Master's thesis, Virginia Polytechnic Institute and State Univ., 1992.

McTague, S. B., Jr. "Petrography and Petrology of 'Unakites' Located in the Mount Rogers Area, Southwestern Virginia." Master's thesis, Virginia Polytechnic Institute and State Univ., 1967.

Newfont, Kathryn. "Moving Mountains: Forest Politics and Commons Culture in Western North Carolina. Ph.D. diss., Univ. of North Carolina, Chapel Hill, 2001.

Nicholas, Niki S. "Stand Structure, Growth, and Mortality in Southern Appalachian Spruce-Fir." Ph.D. diss., Virginia Polytechnic Institute and State Univ., 1992.

Nicholls, Leland Leon. "A Geographical Analysis of Selected Ski Resorts in the South." Ph.D. diss., Univ. of Tennessee, 1972.

Piper, Jacquelyn G. "An Interpretation of Mount Rogers National Recreation Area." Master's thesis, Univ. of South Florida, 1977.

Reddick, Genevieve S. "Ozone Effects on Forest Vegetation." Master's thesis, Virginia Polytechnic Institute and State Univ., 1983.

Reeves, John Henry, Jr. "The History and Development of Wildlife Conservation in Virginia: A Critical Review." Ph.D. diss., Virginia Polytechnic Institute and State Univ., 1960.

Roberts, Edward Graham. "The Roads of Virginia, 1607–1840." Ph.D. diss., Univ. of Virginia, 1951.

Schaeffer, Mary Alice. "Effects of Land Use on Oxygen Uptake by Microorganisms on Fine Benthic Organic Matter in Two Appalachian Mountain Streams." Master's thesis, Virginia Polytechnic Institute and State Univ., 1993.

Schutz, Noel W., Jr. "The Study of Shawnee Myth in an Ethnographic and Ethnohistorical Perspective." Ph.D. diss., Indiana Univ., 1975.

Seneca, Earnest Davis. "Deer Distribution, Usage and Population on the Broad Run Management Area, Craig County, Virginia." Master's thesis, Virginia Polytechnic Institute and State Univ., 1961.

Shields, Arthur Randolph. "The Isolated Spruce and Spruce-Fir Forests of Southwestern Virginia: A Biotic Study." Ph.D. diss., Univ. of Tennessee, 1962.

Snyder, Ernest Cline. "Fincastle, Virginia, 1772–1942: A Study in Small Town Life." Master's thesis, Virginia Polytechnic Institute, 1942.

Turner, Jeffrey C. "Cloverdale Furnace: A Century of Iron Manufacture in Botetourt County, Virginia, 1789–1889." Master's thesis, Virginia Polytechnic Institute and State Univ., 1984.

Weekes, William Dickey. "A 15-Year Research Summary and Hunting Harvest Data Evaluation of the Broad Run Management Area." Master's thesis, Virginia Polytechnic Institute, 1974.

Books, Articles, and Papers

Abdy, E. S. *Journal of a Residence and Tour in the United States of North America.* Vol. 2. London, 1855.

Abernethy, Thomas Perkins. *Three Virginia Frontiers.* Baton Rouge: Louisiana State Univ. Press, 1940.

Albanese, Catherine L. *America: Religions and Religion.* Belmont, CA: Wadsworth Pub. Co., 1981.

Albion, Robert G., ed. *Philip Vickers Fithian: Journal, 1775–1776: Written on the Virginia-Pennsylvania Frontier and in the Army around New York.* Princeton: Princeton Univ. Press, 1934.

Alderman, John P. *Carroll, 1765–1815.* Roanoke: Alderman Books, 1985.

Alexander, Edward P. *The Journal of John Fontaine: An Irish Huguenot Son in Spain and Virginia 1710–1719.* Charlottesville: Univ. Press of Virginia, 1972.

Alexander, J. Trent. "Defining the Diaspora: Appalachians in the Great Migration." *Journal of Interdisciplinary History* 37 (Autumn 2006): 219–47.

———. "'They're Never Here More Than a Year': Return Migration in the Southern Exodus, 1940–1970." *Journal of Social History* 38 (Spring 2005): 653–71.

Allen, Joel A. *The American Bisons: Living and Extinct.* Cambridge: Cambridge Univ. Press, 1876. Reprint, New York: Arno Press, 1974.

Allen, Robert Jackson. "Sustainable Forestry in Virginia: Opportunities for Over-due Legislation and Options for Private Landowners." *Appalachian Journal of Law* 7 (Winter 2007): 1–34.

Alvord, Clarence W., and Lee Bidgood. *The First Explorations of the Trans-Allegheny Region by the Virginians, 1650–1674.* Cleveland: Arthur H. Clark Co., 1912.

American Society of Planning Officials. *Subdividing Rural America: Impacts of Recreational Lot and Second Home Development.* Washington, DC: GPO, 1976.

Anderson, Edgar. *Plants, Man and Life.* Berkeley: Univ. of California Press, 1969.

Appalachian Land Ownership Task Force. *Land Ownership Patterns and Their Impacts on Appalachian Communities.* 7 vols. Washington, DC: Appalachian Regional Commission, 1980–81.

Arese, Count Francesco. *A Trip to the Prairies and in the Interior of North America.* New York: Harbor Press, 1934.

Ashe, W. W. "The Creation of the Eastern National Forests." *American Forestry* 28 (Sept. 1922): 521–25.

Atkinson, Edward. *The Future Sites of the Principle Iron Production of the World.* Baltimore: Manufacturers' Record Co., 1890.

Baden, John A., and Douglas S. Noonan, eds. *Managing the Commons.* Bloomington: Indiana Univ. Press, 1998.

Baden, John, and Richard L. Stroup, eds. *Bureaucracy vs. Environment: The Environmental Costs of Bureaucratic Governance.* Ann Arbor: Univ. of Michigan Press, 1981.

Badgett, C. S. "Equipment for Incline Logging." *American Lumberman* 144 (Oct. 8, 1921): 54.

Bailey, Kenneth P. *The Ohio Company of Virginia and the Westward Movement 1748–1792: A Chapter in the History of the Colonial Frontier.* Glendale, CA: Arthur H. Clark Co., 1939.

Baker, Harry Lee. "The Forests of Lee County, Virginia." In *The Geology and Coal Resources of the Coal-Bearing Portion of Lee County, Virginia.* Virginia Geological Survey Bulletin No. 26. Charlottesville: Univ. of Virginia, 1925.

Baldwin, Donald N. *The Quiet Revolution: Grass Roots of Today's Wilderness Preservation Movement.* Boulder, CO: Pruett, 1972.

Baller, Mark, and Leor Joseph Pantilat. "Defenders of Appalachia: The Campaign to Eliminate Mountaintop Removal Coal Mining and the Role of Public Justice." *Environmental Law* (Summer 2007): 629–64.

Barrett, Tony. "Hail Dan." *Appalachian Trailway News* 62 (May–June 2001): 28.

Bauxar, J. Joseph. "Yuchi Ethnoarchaeology, Part I: Some Yuchi Identifications Reconsidered." *Ethnohistory* 4 (Summer 1957): 279–301.

Bechtold, William A., Mark J. Brown, and John B. Tansey. *Virginia's Forests.* USDA Forest Service Research Bulletin SE-95, Aug. 1987.

Beck, Donald E. *Polymorphic Site Index Curves for White Pine in the Southern Appalachians.* USDA Forest Research Paper No. 80, Apr. 1971.

Beck, Donald E., and Lino Della-Bianca. *Board-Foot and Diameter Growth of Yellow Poplar after Thinning.* USDA Forest Research Paper SE-123, Jan. 1975.

———. *Growth and Yield of Thinned Yellow-Poplar.* USDA Forest Research Paper SE-101, Nov. 1972.

———. *Yield of Unthinned Yellow Poplar.* USDA Forest Research Paper SE-58, Mar. 1970.

Bennett, Hugh H. "Thomas Jefferson, Soil Conservationist." In *Jefferson Reader,* ed. Francis Coleman Rosenberger, 184–92. New York: E. P. Dutton & Co., 1953.

Benthall, Joseph L. *Archeological Investigation of the Shannon Site.* Richmond: Virginia State Library, 1969.

———. "The Litten Site: A Late Woodland Village Complex, Washington County, Virginia." *Quarterly Bulletin of the Archeological Society of Virginia* 26 (Sept. 1971): 1–34.

Berkebile, Don H. *Carriage Terminology: An Historical Dictionary.* Washington, DC: Smithsonian Institution Press, 1978.

Berkes, F., D. Feeney, B. J. McCay, J. M. Acheson. "Benefits of the Commons." *Nature* 340 (July 13, 1989): 91–93.

Bernbaum, Edwin. *Sacred Mountains of the World.* San Francisco: Sierra Club Books, 1990.

Berry, Frederick H. "Chestnut Breeding in the U.S. Department of Agriculture." In *Proceedings of the American Chestnut Symposium.* Morgantown: West Virginia Univ. Books, 1978.

Billings, Dwight B., Gurney Norman, and Katherine Ledford, eds. *Confronting Appalachian Stereotypes: Back Talk from an American Region.* Lexington: Univ. Press of Kentucky, 1999.

Billington, Ray Allen. *Westward Expansion: A History of the American Frontier.* 4th ed. New York: Macmillan Pub. Co., 1974.

Bining, Arthur Cecil. *Pennsylvania Iron Manufacture in the Eighteenth Century.* 2nd ed. Harrisburg: Pennsylvania Historical & Museum Commission, 1973.

Bland County Centennial Corporation. *History of Bland County (Virginia).* Radford, VA: Commonwealth Press, 1961.

Bodell, Dorothy H. *Montgomery White Sulphur Springs: A History of the Resort, Hospital, Cemeteries, Markers, and Monument.* Blacksburg, VA: Pocahontas Press, 1993.

Boivin, Nicole, and Mary Ann Owoc, eds. *Soils, Stones and Symbols: Cultural Perceptions of the Mineral World.* London: University College London, 2004.

Botkin, Daniel B. *Discordant Harmonies: A New Ecology for the Twenty-first Century.* New York: Oxford Univ. Press, 1990.

Bott, Keith Edward. *44RU7: Archaeological Test Excavations at a Late Woodland Village in the Lower Uplands of Southwest Virginia.* Richmond: Virginia Division of Historic Landmarks, 1981.

Boyd, Charles Rufus. *Resources of South-West Virginia.* New York: John Wiley & Sons, 1881.

Bradford, S. Sydney. "The Negro Ironworker in Ante Bellum Virginia." *Journal of Southern History* 25 (May 1959): 194–206.

Branch, Michael P., and Daniel J. Philippon, eds. *The Height of Our Mountains: Nature Writing from Virginia's Blue Ridge Mountains and Shenandoah Valley.* Baltimore: Johns Hopkins Univ. Press, 1998.

Braund, Kathryn E. Holland. *Deerskins and Duffels: The Creek Indian Trade with Anglo-America, 1685–1815.* Lincoln: Univ. of Nebraska Press, 1993.

Brenneman, Ron. "Impact of Deer on Forest Regeneration." In *Proceedings: Guidelines for Regenerating Appalachian Hardwood Stands.* Morgantown: West Virginia Univ. Books, 1988.

Briceland, Alan V. *Westward from Virginia: The Exploration of the Virginia-Carolina Frontier 1650–1710.* Charlottesville: Univ. Press of Virginia, 1987.

Brooks, A. B. *West Virginia Geological Survey, Vol. 5: Forest and Wood Industries.* Morgantown, WV: Acme Publishing Co., 1910.

Brooks, Leslie. "Incline Logging." *Southern Lumberman* 114 (May 10, 1924): 49–50.

Brooks, Maurice. *The Appalachians.* Boston: Houghton Mifflin Co., 1965.

Brown, Dona. *Inventing New England: Regional Tourism in the Nineteenth Century.* Washington, DC: Smithsonian Institution Press, 1995.

Brown, Margaret Lynn. *The Wild East: A Biography of the Great Smoky Mountains.* Gainesville: Univ. Press of Florida, 2000.

Brown, Nelson Courtlandt. *Logging—Transportation: The Principles and Methods of Log Transportation in the United States and Canada.* New York: John Wiley & Sons, 1936.

Brubaker, Sterling, ed. *Rethinking the Federal Lands.* Washington, DC: Resources for the Future, 1984.

Bruce, Kathleen. *Virginia Iron Manufacture in the Slave Era.* New York: Century Co., 1931.

Bruce, Thomas. *Southwest Virginia and Shenandoah Valley.* Richmond, VA: J. L. Hill Publishing Co., 1891.

Bryan, Mackay B. *Forest Statistics for the Mountain Region of Virginia, 1957.* Forest Survey Release No. 52. Asheville, NC: USDA Forest Service, Southeast Forest Experiment Station, 1958.

Bryant, Ralph Clement. *Logging: The Principles and General Methods of Operation in the United States.* 2nd ed. New York: John Wiley & Sons, 1923.

Buchanan, William T., Jr. "The Hall Site, Montgomery County, Virginia." *Quarterly Bulletin of the Archeological Society of Virginia* 35 (Dec. 1980): 72–99.

———. *The Trigg Site, City of Radford, Virginia.* Richmond: Archeological Society of Virginia, 1984.

Burke, William. *Mineral Springs of Western Virginia with Remarks on Their Use.* 2nd ed. New York: Wiley & Putnam, 1846.

Burns, Shirley Stewart, Mari-Lynn Evans, and Silas House, eds. *Coal Country: Rising up against Mountaintop Removal Mining.* San Francisco: Sierra Club Books, 2009.

Burton, Charles T. *Fincastle Virginia, Past: People with Places.* N.p: n.d. Copy in Virginia Polytechnic Institute and State Univ.'s Newman Library, Blacksburg.

Bushnell, David I., Jr. "The Five Monacan Towns in Virginia, 1607." *Smithsonian Misc. Collection,* vol. 82, no. 12. Washington, DC: GPO, 1930.

Butler, Ovid, ed. *The Biltmore Story: Recollections of the Beginning of Forestry in the United States.* Saint Paul: American Forest History Foundation, 1955.

————, ed. "The National Forests of the East." *American Forests* 41 (Sept. 1935): 508–9.

Butterfield, K. P. "Forests and Game." *Virginia Wildlife* 7 (Nov. 1946): 9–11.

Caldwell, Joseph R. "Eastern North America." In *Prehistoric Agriculture,* ed. Stuart Struever, 361–82. Garden City, NY: American Museum of Natural History, 1971.

Callicott, J. Baird. "American Indian Land Wisdom? Sorting out the Issues." *Journal of Forest History* 33 (Jan. 1989): 35–42.

————. "Traditional American Indian and Western European Attitudes toward Nature: An Overview." *Environmental Ethics* 4 (Winter 1982): 293–318.

Campbell, John C. *The Southern Highlander and His Homeland.* 1921. Lexington: Univ. Press of Kentucky, 1969.

Campbell, Joseph. *Transformations of Myth through Time.* New York: Harper & Row, 1990.

Carr, Lucien. "Report on the Exploration of a Mound in Lee County, Virginia." In *Tenth Annual Report of the Peabody Museum,* 75–94. Cambridge: Salem Press, 1877.

Carrier, Lyman, and Katharine S. Bort. "The History of Kentucky Bluegrass and White Clover in the United States." *Journal of the American Society of Agronomy* 8 (July–Aug. 1916): 256–66.

Cavender, Anthony P. *Folk Medicine in Southern Appalachia.* Chapel Hill: Univ. of North Carolina Press, 2003.

Chase, Richard, ed. *Grandfather Tales: American-English Folk Tales.* Boston: Houghton Mifflin Co., 1948.

————, ed. *The Jack Tales.* Cambridge, MA: Houghton Mifflin, 1943.

Cheek, Angie, Lacy Hunter Nix, and Foxfire students, eds. *The Foxfire 40th Anniversary Book: Faith, Family, and the Land.* New York: Anchor Books, 2006.

Ciriacy-Wantrup, S. V., and Richard C. Bishop. "'Common Property' as a Concept in Natural Resources Policy." *Natural Resources Journal* 15 (Oct. 1975): 713–27.

Clark, Thomas D. *The Greening of the South: The Recovery of Land and Forest.* Lexington: Univ. Press of Kentucky, 1984.

Clary, David A. *Timber and the Forest Service.* Lawrence: Univ. Press of Kansas, 1986.

Clawson, Marion. *Economics of National Forest Management.* Baltimore: Johns Hopkins Univ. Press, 1976.

————. *Federal Lands Revisited.* Washington, DC: Resources for the Future, 1983.

———. *Federal Lands since 1956: Recent Trends in Use and Management.* Washington, DC: Resources for the Future, 1967.

———. *Forests for Whom and for What?* Baltimore: Resources for the Future, by the Johns Hopkins Univ. Press, 1975.

Clawson, Marion, and Jack L. Knetsch. *Economics of Outdoor Recreation.* Baltimore: Johns Hopkins Univ. Press, 1966.

Coal, Charles B. *The Life and Adventures of Wilburn Waters: The Famous Hunter and Trapper of White Top Mountain.* 1878. Richmond, VA: M. D. Hart, 1960.

Cochran, A. R. "Business Aspects of Wildlife Management." *Virginia Wildlife* 11 (Dec. 1950): 10–12.

———. "Jefferson National Forest." *Virginia Wildlife* 10 (May 1949): 6–8, 24.

Cohen, Stan. *Historic Springs of the Virginias: A Pictorial History.* Charleston, WV: Pictorial Histories, 1981.

Coleman, Simon, and John Elsner. *Pilgrimage: Past and Present in the World Religions.* Cambridge: Harvard Univ. Press, 1995.

Coles, Robert. "Appalachia's Moral Life." In *Mountain People,* ed. Michael Tobias (Norman: Univ. of Oklahoma Press, 1986), 19–21.

———. *Migrants, Sharecroppers, Mountaineers.* Boston: Little, Brown, 1971.

Collier, Christopher P. "Good Times at a Backpackers' Paradise." *New York Times,* May 12, 2006.

Conn, Charles W. *Like a Mighty Army: A History of the Church of God, 1886–1976.* Rev. ed. Cleveland, TN: Pathway Press, 1977.

Connelley, William Elsey. *The Founding of Harman's Station.* New York: Torch Press, 1910.

Cooke, John Esten. *The Last of the Foresters.* New York: Derby and Jackson, 1856.

———. "The White Sulphur Springs." *Harper's New Monthly Magazine* 62 (Aug. 1878): 337–56.

Coons, E. Jeffrey. "Keeping Company with the Avatar Tree." *Appalachian Trailway News* 66 (Mar.–Apr. 2005): 36–37.

Cooper, Thomas. "What Is the Problem of Mountain Agriculture?" *Mountain Life and Work* 3 (July 1927): 13–15.

Corkran, David H. *The Creek Frontier 1540–1783.* Norman: Univ. of Oklahoma Press, 1967.

Cowdrey, Albert E. *This Land, This South: An Environmental History.* Lexington: Univ. Press of Kentucky, 1983.

Cox, Susan Jane Buck. "No Tragedy on the Commons." *Environmental Ethics* 7 (Spring 1985): 49–61.

Cox, Thomas R., Phillip Drennon Thomas, and Joseph J. Malone. *This Well-Wooded Land: Americans and Their Forests from Colonial Times to the Present.* Lincoln: Univ. of Nebraska Press, 1985.

Crane, Eva. *A Book of Honey.* New York: Charles Scribner's Sons, 1980.

Crane, Verner W. *The Southern Frontier 1670–1732.* 1929. Ann Arbor: Univ. of Michigan Press, 1956.

Crawford, Hewlette. "Deer Range Potential in Selective and Clearcut Oak-Pine Stands in Southwestern Virginia." USDA FS Research Publication SE-134, June 1975.

Crayon, Porte (David Hunter Strother). "Virginia Illustrated." *Harper's New Monthly Magazine* 12 (Jan. 1856): 159.

Crosby, Alfred W. *The Columbian Exchange: Biological and Cultural Consequences of 1492.* New ed. Westport, CT: Praeger, 2003.

———. *Ecological Imperialism: The Biological Expansion of Europe, 900–1900.* New ed. New York: Cambridge Univ. Press, 2004.

Crutcher, Thomas. "Life after Katahdin?" *Appalachian Trailway News* 62 (Sept.– Oct. 2001): 26–27.

Culhane, Paul J. *Public Lands Politics: Interest Group Influence on the Forest Service and the Bureau of Land Management.* Baltimore: Johns Hopkins Univ. Press for Resources for the Future, 1981.

Custer, Jay F., and Dennis C. Curry. "Prehistoric Settlement-Subsistence Systems in Grayson County, Virginia." *Quarterly Bulletin of the Archeological Society of Virginia* 41 (Sept. 1986): 113–41.

Dana, Samuel Trask. *Forest and Range Policy: Its Development in the United States.* New York: McGraw-Hill, 1956.

Dana, Samuel Trask, and Sally K. Fairfax. *Forest and Range Policy: Its Development in the United States.* 2nd ed. New York: McGraw-Hill, 1980.

Darlington, William M. *Christopher Gist's Journals: With Historical, Geographical and Ethnological Notes and Biographies of His Contemporaries.* Pittsburg: J. R. Weldin & Co., 1893.

Davis, Donald Edward. *Homeplace Geography: Essays for Appalachia.* Macon, GA: Mercer Univ. Press, 2006.

———. *Where There Are Mountains: An Environmental History of the Southern Appalachians.* Athens: Univ. of Georgia Press, 2000.

Davis, Thomas. *Sustaining the Forest, the People, and the Spirit.* Albany, NY: State Univ. of New York Press, 2000.

Deacon, Robert T., and M. Bruce Johnson, eds. *Forestlands: Public and Private.* San Francisco: Pacific Institute for Public Policy Research, 1985.

Description of the Peaks of Otter. Lynchburg: Virginia Job Office, 1853.

Dew, Charles B. *Iron-Maker to the Confederacy.* New Haven: Yale Univ. Press, 1966.

Dick, Everett. *The Lure of the Land: A Social History of the Public Lands from the Articles of Confederation to the New Deal.* Lincoln: Univ. of Nebraska Press, 1970.

Dickens, Roy S., Jr. *Cherokee Prehistory: The Pisgah Phase in the Appalachian Summit Region.* Knoxville: Univ. of Tennessee Press, 1976.

Dragoo, Don W. "Some Aspects of Eastern North American Prehistory: A Review 1975." *American Antiquity* 41 (Jan. 1976): 3–27.

Duerr, William A. *The Economic Problems of Forestry in the Appalachian Region.* Cambridge: Harvard Univ. Press, 1949.

Dunaway, Wilma A. *The First American Frontier: Transition to Capitalism in Southern Appalachia, 1700–1860.* Chapel Hill: Univ. of North Carolina Press, 1996.

Dunn, Durwood. *Cades Cove: The Life and Death of a Southern Appalachian Community 1818–1937.* Knoxville: Univ. of Tennessee Press, 1988.

Dunnell, R. C. "The Prehistory of Fishtrap, Kentucky." *Yale University Publications in Anthropology,* no. 75. New Haven: Yale Univ., 1972.

Durrenberger, Joseph A. *Turnpikes: A Study of the Toll Road Movement in the Middle Atlantic States and Maryland.* New York: Columbia Univ., 1931.

Earle, Alice Morse. *Stage-coach and Tavern Days.* New York: Benjamin Bloom, 1900.

Ecological Society of America. *Conserving Biological Diversity in Our National Forests.* Washington, DC: Wilderness Society, 1986.

Egloff, Keith T. *Ceramic Study of Woodland Occupation Along the Clinch and Powell Rivers in Southwest Virginia.* Research Report Series No. 3. Richmond, VA: Dept. of Conservation and Historic Resources, Division of Historic Landmarks, 1987.

———. "The Late Woodland Period in Southwestern Virginia." In *Middle and Late Woodland Research in Virginia: A Synthesis,* ed. Theodore R. Reinhart and Mary Ellen N. Hodges, 187–224. Richmond: Archaeological Society of Virginia, 1992.

Egloff, Keith T., and Celia Reed. "Crab Orchard Site: A Late Woodland Palisaded Village." *Quarterly Bulletin of the Archeological Society of Virginia* 34 (Mar. 1980): 130–48.

Egloff, Keith T., and Deborah Woodward. *First People: The Early Indians of Virginia.* Richmond, VA: Dept. of Historic Resources, 1992.

Eliade, Mircea. *The Myth of the Final Return.* Trans. Willard R. Trask. New York: Pantheon Books, 1954.

———. *Patterns in Comparative Religion.* Trans. Rosemary Sheed. New York: Meridian Books, 1966.

Eller, Ronald D. *Miners, Millhands and Mountaineers: The Industrialization of the Appalachian South, 1880–1930.* Knoxville: Univ. of Tennessee Press, 1982.

———. *Uneven Ground: Appalachia since 1945.* Lexington: Univ. Press of Kentucky, 2008.

Ely, James W., Jr. *The Guardian of Every Other Right: A Constitutional History of Property Rights.* 2nd ed. New York: Oxford Univ. Press, 1998.

Ergood, Bruce, and Bruce E. Kuhre, eds. *Appalachia: Social Context Past and Present.* 2nd ed. Dubuque, IA: Kendall Hunt, 1983.

Evans, Thomas C., G. E. Morrill, and John Carow. *Virginia: Preliminary Tables of Forest Area, Timber Volumes and Utilization.* Asheville, NC: USDA, Forest Service, Appalachian Forest Experiment Station, 1941.

Evers, Alf. *The Catskills: From Wilderness to Woodstock.* Garden City: Doubleday & Co., 1972.

Ewan, Joseph, and Nesta Ewan, eds. *John Banister and His Natural History of Virginia 1678–1692.* Urbana: Univ. of Illinois Press, 1970.

Fairchild, Fred Rogers. *Forest Taxation in the United States.* USDA Miscellaneous Publication 218. Washington, DC: GPO, 1935.

Faulkner, Charles H. "Industrial Archaeology of the 'Peavine Railroad': An Archaeological and Historical Study of an Abandoned Railroad in East Tennessee." *Tennessee Historical Quarterly* 44 (Spring 1985): 40–58.

Fearnow, Theodore, and I. T. Quinn. "Action on the Blue Ridge." In *The Yearbook of Agriculture, 1949,* 586–92. Washington, DC: GPO.

Foerster, Robert F. *The Italian Emigration of Our Times.* Cambridge: Harvard Univ. Press, 1924.

Foss, Phillip O., comp. *Recreation.* New York: Chelsea House, 1971.

Foster, Edward Halsey. *The Civilized Wilderness: Backgrounds to American Romantic Literature, 1817–1860.* New York: Free Press, 1975.

Foster, Stephen W. *The Past Is Another Country: Representation, Historical Consciousness, and Resistance in the Blue Ridge.* Berkeley: Univ. of California Press, 1988.

Franklin, W. Neil, "Virginia and the Cherokee Indian Trade, 1753–1755." *East Tennessee Historical Society Publications* 5 (1933): 22–38.

Freeman, Richard. "The Ecofactory: The United States Forest Service and the Political Construction of Ecosystem Management." *Environmental History* 7 (Oct. 2002): 632–58.

Freinkel, Susan. *American Chestnut: The Life, Death, and Rebirth of a Perfect Tree* (Berkeley: Univ. of California Press, 2007).

Fries, Adelaide L., ed. *Records of the Moravians in North Carolina.* Vol. 1. Raleigh: State Dept. of Archives and History, 1968.

Fritsch, Al, and Kristin Johannsen. *Ecotourism in Appalachia: Marketing the Mountains.* Lexington: Univ. Press of Kentucky, 2004.

Fritsch, Albert J. *Appalachia: A Meditation.* Chicago: Loyola Univ. Press, 1986.

Frome, Michael. *The Forest Service.* 2nd ed. Boulder, CO: Westview Press, 1983.

Frothingham, E. H. "Ecology and Silviculture in the Southern Appalachians: Old Cuttings as a Guide to Future Practice." *Journal of Forestry* 15 (Mar. 1917): 343–49.

———. "Forest Research in the Southern Appalachians." *Southern Lumberman* 108 (Dec. 23, 1922): 122–27.

———. "Scientific Research and Southern Appalachian Forests." *Lumber World Review* 45 (Nov. 10, 1923): 47–52.

Gable, William B., and Edward H. Davis, eds. *An Oral History of Konnarock, Virginia.* Charlottesville: Virginia Foundation for the Humanities, 1997.

Gainer, Patrick W. *Witches, Ghosts and Signs: Folklore of the Southern Appalachians.* Grantsville, WV: Seneca Books, 1975.

Galloway, Patricia, ed. *The Southeastern Ceremonial Complex: Artifacts and Analysis.* Lincoln: Univ. of Nebraska Press, 1989.

Gardner, William M., and Lauralee Rappleye. "Test Excavations at 44WS28, Clinch Ranger District, Jefferson National Forest, Wise County, Virginia." Front Royal, VA: Thunderbird Research Corp., 1979.

Garren, Kenneth H. "Effects of Fire on Vegetation of the Southeastern United States." *Botanical Review* 9 (1943): 617–54.

Garvey, Edward B. *The New Appalachian Trail.* Birmingham, AL: Menasha Ridge Press, 1997.

Geier, Clarence R. "Development and Diversification: Cultural Directions during the Late Woodland/Mississippian Period in Eastern North America." In *Middle and Late Woodland Research in Virginia: A Synthesis,* ed. Theodore R. Reinhart and Mary Ellen N. Hodges, 277–302. Richmond: Archaeological Society of Virginia, 1992.

Gersmehl, Phil. "Factors Leading to Mountaintop Grazing in the Southern Appalachians." *Southeastern Geographer* 10 (Apr. 1970): 67–72.

Giles County Historical Society. *Giles County, Virginia History—Families.* Pearisburg, VA: Giles County Historical Society, 1983.

Glacken, Clarence J. *Traces on the Rhodian Shore: Nature and Culture in Western Thought from Ancient Times to the End of the Eighteenth Century.* Berkeley: Univ. of California Press, 1967.

Glassie, Henry. "The Appalachian Log Cabin." *Mountain Life and Work* 39 (Winter 1963): 5–14.

———. "The Types of the Southern Mountain Cabin." In *The Study of American Folklore,* ed. Jan H. Brunwand, 529–62. 3rd ed. New York: Norton & Co., 1986.

Goodwin, Gary C. *Cherokees in Transition: A Study of Changing Culture and Environment Prior to 1775.* Chicago: Univ. of Chicago Press, 1977.

Gottfried, Robin R. "Observations on Recreation-Led Growth in Appalachia." *American Economist* 21 (Spring 1977): 44–50.

Graves, Henry. "A Crisis in National Recreation." *American Forestry* 26 (July 1920): 391–400.

Gray, L. C. "Economic Conditions and Tendencies in the Southern Appalachians as Indicated by the Cooperative Survey." *Mountain Life and Work* 9 (July 1933): 7–12.

Gregg, Sara M. "Uncovering the Subsistence Economy in the Twentieth-Century South: Blue Ridge Mountain Farms." *Agricultural History* 78 (Autumn 2004): 417–37.

Griffin, James B. "Eastern North American Archaeology." *Science* 156 (Apr. 14, 1967): 175–91.

Guerrant, Edward O. *The Galax Gatherers.* 1910. Knoxville: Univ. of Tennessee Press, 2005.

Haan, Richard L. "The 'Trade Do's Not Flourish as Formerly': The Ecological Origins of the Yamassee War of 1715." *Ethnohistory* 28 (Fall 1981): 341–58.

Hahn, Steven. *The Roots of Southern Populism: Yeoman Farmers and the Transformation of the Georgia Upcountry, 1850–1890.* New York: Oxford Univ. Press, 1983.

Hale, Horatio. "The Tutelo Tribe and Language." *Proceedings of the American Philosophical Society* 21 (Mar. 2, 1883): 1–47.

Hale, J. P. *History of the Great Kanawha Valley.* Vol. 1. Madison, WI: Brant, Fuller & Co., 1891.

Hall, George R. "The Myth and Reality of Multiple Use Forestry." In *Politics, Policy, and Natural Resources,* ed. Dennis L. Thompson, 363–75. New York: Free Press, 1972.

Hall, William L. "The Appalachian Work." *American Forestry* 18 (Mar. 1912): 192.

———. "Influence of the National Forests in the Southern Appalachians." *Journal of Forestry* 17 (1919): 402–7.

———. "To Remake the Appalachians: A New Order in the Mountains that Is Founded on Forestry—What the Government's Appalachian Forests Mean to the People in the Mountains and to the Millions Who Want Recreation." *World's Work* 28 (July 1914): 321–38.

———. *The Waning Hardwood Supply and the Appalachian Forests.* USDA Forest Service, Circular No. 116, Sept. 24, 1907.

Hamel, Paul B., and Mary U. Chiltoskey. *Cherokee Plants and Their Uses: A 400 Year History.* Sylva, NC: Herald, 1975.

Hampton, O. W. *Culture of Stone: Sacred and Profane Use of Stone among the Dani.* College Station: Texas A&M Univ. Press, 1999.

Hantman, Jeffrey L. "Between Powhatan and Quirank: Reconstructing Monacan Culture and History in the Context of Jamestown." *American Anthropologist* 92 (Sept. 1990): 676–90.

Hart, John F. "Land Rotation in Appalachia." *Geographical Review* 67 (Apr. 1977): 148–66.

———. "Loss and Abandonment of Cleared Farm Land in the Eastern United States." *Annals of the Association of American Geographers* 58 (Sept. 1968): 417–40.

Hassinger, Luther C. "The Lumber Industry of Southwest Virginia." *Historical Society of Washington County, Virginia Bulletin,* ser. 2, no. 4 (Spring 1967): 1–16.

Hawley, Willis C. "Buying National Forests." *American Forests* 31 (May 1925): 293–95, 304.

Hayden, C. B. "On the Rock Salt and Salines of the Holston." *American Journal of Science* 44 (1843): 173–79.

Hays, Samuel P. *The American People and the National Forests: The First Century of the U.S. Forest Service.* Pittsburgh: Univ. of Pittsburgh Press, 2009.

———. *Beauty, Health, and Permanence: Environmental Politics in the United States, 1955–1985.* New York: Cambridge Univ. Press, 1987.

———. *Conservation and the Gospel of Efficiency: The Progressive Conservation Movement, 1890–1920.* Rev. ed. Cambridge: Harvard Univ. Press, 1969.

———. *Wars in the Woods: The Rise of Ecological Forestry in America.* Pittsburgh: Univ. of Pittsburgh Press, 2007.

Heath, Milton S., Jr. "The North Carolina Mountain Ridge Protection Act." *North Carolina Law Review* (Nov. 1984): 183–97.

Heizer, Robert F. "Primitive Man as an Ecological Factor." *Kroeber Anthropological Society Papers 13* (Fall 1955): 1–31.

Henderson, Archibald. *Conquest of the Old Southwest.* New York: Century, 1920.

Hendricks, Walter H. "Daniel Boone as a Virginian." *Historical Society of Washington County, Virginia, Bulletin* 24 (1987): 1–8.

Hicks, George L. *Appalachian Valley.* New York: Holt, Rinehart, & Winston, 1976.

Hirt, Paul W. *A Conspiracy of Optimism: Management of the National Forests since World War Two.* Lincoln: Univ. of Nebraska Press, 1994.

Hoage, R. J., and Katy Moran, eds. *Culture: The Missing Element in Conservation and Development.* Dubuque, IA: Kendall Hunt, 2000.

Hobart, Seth G., George W. Dean, Edwin E. Rodger. *The History of the Virginia Division of Forestry: 1914–1981.* Charlottesville: Virginia Dept. of Forestry, 1981.

Holbrook, Stewart H. *Burning an Empire: The Story of American Forest Fires.* New York: Macmillan Co., 1944.

Horner, William Edmonds. *Observations on the Mineral Waters in the South Western Part of Virginia.* Philadelphia: J. Thompson, 1834.

House, Silas. *Something's Rising: Appalachians Fighting Mountaintop Removal.* Lexington: Univ. Press of Kentucky, 2009.

Howell, Benita J., ed. *Culture, Environment, and Conservation in the Appalachian South.* Urbana: Univ. of Illinois Press, 2002.

Hsiung, David C. "How Isolated Was Appalachia? Upper East Tennessee, 1780–1835." *Appalachian Journal* 16 (Summer 1989): 336–49.

Hudson, Charles. "Cherokee Concept of Natural Balance." *Indian Historian* 3 (1970): 51–54.

———. *The Southeastern Indians.* Knoxville: Univ. of Tennessee Press, 1976.

Hulbert, Archer Butler. *Historic Highways of America.* Vol. 6, *Boone's Wilderness Road.* Cleveland: Arthur H. Clark Co., 1903.

Humbert, R. L., C. C. Beggs et al. *Industrial Survey, Craig County, Virginia.* Blacksburg: Engineering Extension Division, Virginia Polytechnic Institute, 1930.

Ingram, Helen M., and R. Kenneth Godwin. "Conservation and the Forces of Change." In *Public Policy and the Natural Environment,* vol. 4, *Public Policy Studies.* Greenwich, CT: JAI Press, 1985.

Inscoe, John C. *Mountain Masters, Slavery, and the Sectional Crisis in Western North Carolina.* Knoxville: Univ. of Tennessee Press, 1989.

Ise, John. *The United States Forest Policy.* New Haven: Yale Univ. Press, 1920.

James River Project Committee. *The James River Basin: Past, Present and Future.* Richmond: Virginia Academy of Science, 1950.

Johnson, Kenneth A. "Origins of Tourism in the Catskill Mountains." *Journal of Cultural Geography* 11, no. 1 (1990): 5–16.

Johnston, David E. *A History of Middle New River Settlements and Contiguous Territory.* 1906. Radford: Commonwealth Press, 1969.

Jolliffe, William. *The Buchanan and Clifton Forge Railroad: A Necessity to Virginia: Developing the Coal and Iron of the James and Kanawha River Valleys.* Lynchburg, VA: Bell, Browne & Co., 1875.

Jones, Emory Eugene, Jr. "The Gilbert Site, Tazewell County, Virginia." *Quarterly Bulletin of the Archeological Society of Virginia* 44 (Dec. 1989): 201–29.

Jones, Loyal. "Appalachian Values." In *Voices from the Hills: Selected Readings of Southern Appalachia,* ed. Robert J. Higgs and Ambrose N. Manning, 507-17. New York: Frederick Ungar, 1975.

Kantor, Shawn E. *Politics and Property Rights: The Closing of the Open Range in the Postbellum South.* Chicago: Univ. of Chicago Press, 1998.

Karamanski, Theodore J. "Logging, History, and the National Forests: A Case Study of Cultural Resource Management." *Public Historian* 7, no. 2 (1985): 27–40.

Kaufman, Herbert. *The Forest Ranger: A Study in Administrative Behavior.* Baltimore: Johns Hopkins Univ. Press, 1960.

Keel, Bennie C. *Cherokee Archaeology: A Study of the Appalachian Summit.* Knoxville: Univ. of Tennessee Press, 1976.

Keever, Catherine. "Present Composition of Some Stands of the Former Oak–Chestnut Forest in the Southern Blue Ridge Mountains." *Ecology* 34 (Jan. 1953): 44–54.

Kegley, F. B. *Kegley's Virginia Frontier.* Roanoke: Southwest Virginia Historical Society, 1938.

Kimmins, J. P. *Forest Ecology: A Foundation for Sustainable Forest Management and Environmental Ethics in Forestry.* Upper Saddle River, NJ: Prentice Hall, 2004.

Klein, Michael J., and Thomas Klatka. "Late Archaic and Early Woodland Demography and Settlement Patterns." In *Late Archaic and Early Woodland Research in Virginia: A Synthesis,* ed. Theodore R. Reinhart and Mary Ellen N. Hodges, 139–84. Richmond: Archaeological Society of Virginia, 1991.

Kneipp, L. F. "Uncle Sam Buys Some Forests." *American Forests* 42 (Oct. 1936): 443–46, 483.

Krech, Shepard. *The Ecological Indian: Myth and History.* New York: Norton, 1999.

Krochmal, Arnold, and Connie Krochmal. "Checklist of Common Plants of the Appalachian Trail in the Jefferson National Forest, Virginia." USDA, FS Research Note SE-305, Feb. 1981.

Kurath, Gertrude P. "The Tutelo Fourth Night Spirit Release Singing." *Midwest Folklore* 4, no. 2 (1954): 87–105.

———. "The Tutelo Harvest Rites: A Musical and Choreographic Analysis." *Scientific Monthly* 76 (Mar. 1953): 153–62.

Lacey, Michael J., ed. *Government and Environmental Politics: Essays on Historical Developments since World War Two.* Baltimore: Johns Hopkins Univ. Press, 1989.

Lambert, Robert S. "Logging the Little River, 1890–1940." *East Tennessee Historical Society's Publications* 33 (1961): 32–42.

Lanman, Charles. *Letters from the Alleghany Mountains.* New York, 1848.

———. "Our Landscape Painters." *Southern Literary Messenger* 16 (May 1850): 272–80.

Larkin, Bob. "A Winter in the South." *Harper's New Monthly Magazine* 15 (Sept. 1857): 447–50.

Lawson, John. *A New Voyage to Carolina.* 1709. Chapel Hill: Univ. of North Carolina Press, 1967.

Lawson, Karol Ann Peard. "An Inexhaustible Abundance: The National Landscape Depicted in American Magazines, 1780–1820." *Journal of the Early Republic* 12 (Fall 1992): 303–30.

Leonard, Bill J., ed. *Christianity in Appalachia: Profiles in Regional Pluralism.* Knoxville: Univ. of Tennessee Press, 1999.

Leopold, Aldo. "The Wilderness and Its Place in Forest Recreation Policy." *Journal of Forestry* 19 (Nov. 1921): 718–21.

Lewis, Charles D. "Government Forests and the Mountain Problem." *Mountain Life and Work* 6 (Jan. 1931): 2–7.

Lewis, John B. "Success Story: Wild Turkey." In *Restoring America's Wildlife 1937–1987: The First 50 Years of the Federal Aid in Wildlife Restoration (Pittman-Robertson Act).* Washington, DC: U.S. Dept. of Interior, 1987.

Lewis, Lloyd D. "An Abingdon Branch Accolade." *Trains* 44 (June 1984): 21–28.

Lewis, R. Barry, ed. *Kentucky Archaeology.* Lexington: Univ. Press of Kentucky, 1996.

Lewis, Ronald L. *Coal, Iron, and Slaves: Industrial Slavery in Maryland and Virginia, 1715–1865.* Westport, CT: Greenwood Press, 1979.

Lewis, Terry. "Great Game Preserve Proposed in Virginia." *Virginia Forests* 43 (Spring 1988): 21–22.

Lewis, Thomas M. N., and Madeline Kneberg. *Hiwassee Island: An Archaeological Account of Four Tennessee Indian Peoples.* Knoxville: Univ. of Tennessee Press, 1946.

Lillard, Richard G. *The Great Forest.* New York: Alfred A. Knopf, 1947.

Link, O. Winston, and Tim Hensley. *Steam, Steel and Stars: America's Last Steam Railroad.* New York: Harry N. Abrams, 1987.

Linzey, Donald W., ed. *Proceedings of the Symposium on Endangered and Threatened Plants and Animals of Virginia.* Blacksburg: Virginia Polytechnic Institute and State Univ., 1979.

Lonn, Ella. *Salt as a Factor in the Confederacy.* New York: Walter Neale, 1933.

Loucks, Orie. "In Changing Forests: A Search for Answers." In *An Appalachian Tragedy: Air Pollution and Tree Death in the Eastern Forests of North America,* ed. Harvard Ayers Jenny Hager, and Charles E. Little, 85–97. San Francisco: Sierra Club Books, 1998.

Lucia, Ellis. *The Big Woods: Logging and Lumbering—from Bull Teams to Helicopters—in the Pacific Northwest.* New York: Doubleday, 1975.

Lutts, Ralph H. "Like Manna from God: The American Chestnut Trade in Southwestern Virginia." *Environmental History* 9 (July 2004): 497–525.

MacCord, Howard A., Sr. "The Brown Johnson Site—Bland County, Virginia." *Quarterly Bulletin of the Archeological Society of Virginia* 25 (June 1971): 230–71.

———. "The Flannery Site, Scott County, Virginia." *Quarterly Bulletin of the Archeological Society of Virginia* 34 (Sept. 1979): 1–32.

———. "The Sullins Site, Washington County, Virginia." *Quarterly Bulletin of the Archeological Society of Virginia* 36 (Dec. 1981): 94–121.

MacDonald, William L., and Mark L. Double. "Frequency of Vegetative Compatibility Types of *Endothia parasitica* in Two Areas of West Virginia." In *Proceedings of the American Chestnut Symposium.* Morgantown, WV: West Virginia Univ. Books, 1978.

Mann, Ralph. "Mountains, the Land, and Kin Networks: Burkes Garden, Virginia, in the 1840s and 1850s." *Journal of Southern History* 58 (1992): 411–34.

Manring, Nancy J. "The Politics of Accountability in National Forest Planning." *Administration and Society* 37 (Mar. 2005): 57–88.

Martin, Calvin, ed. *The American Indian and the Problem of History.* New York: Oxford Univ. Press, 1987.

———. "Fire and Forest Structure in the Aboriginal Eastern Forest." *Indian Historian* 6 (Fall 1973): 38–42, 54.

———. *Keepers of the Game: Indian-Animal Relations and the Fur Trade.* Berkeley: Univ. of California Press, 1978.

Martineau, Harriet. *Society in America.* New York: Saunders and Otley, 1837. Reprint, New Brunswick: Transaction Books, 1981.

Mastran, Shelley Smith, and Nan Lowerre. *Mountaineers and Rangers.* Washington, DC: GPO, 1983.

Matton, M. A. "Appalachian Comeback." In *Trees, Yearbook of Agriculture 1949,* 304–9. Washington, DC: USDA, 1949.

Maxwell, Hu. "The Use and Abuse of Forests by the Virginia Indians." *William and Mary Quarterly* 19 (Oct. 1910): 73–103.

McCauley, Deborah V. *Appalachian Mountain Religion: A History.* Urbana: Univ. of Illinois Press, 1995.

McCleod, William C. "Conservation among Primitive Hunting Peoples." *Scientific Monthly* 43 (Dec. 1936): 562–66.

McClure, Joe P., Noel D. Cost, and Herbert A Knight. *Multiresource Inventories—A New Concept for Forest Survey.* USDA Forest Service Research Paper SE-191, Apr. 1979.

McCormick, J. Frank, and Robert B. Platt. "Recovery of an Appalachian Forest Following the Chestnut Blight." *American Midland Naturalist* 104 (Oct. 1980): 264–73.

McDonald, Jerry N. *North American Bison: Their Classification and Evolution.* Berkeley: Univ. of California Press, 1981.

McDonald, Michael J., and John Muldowny. *TVA and the Dispossessed: The Resettlement of Population in the Norris Dam Area.* Knoxville: Univ. of Tennessee Press, 1982.

McGee, Charles E. *Heavy Mortality and Succession in a Virgin Mixed Mesophytic Forest.* USDA Forest Service Research Paper SO-209, Aug. 1984.

McGee, Charles E., and Lino Della-Bianca. *Diameter Distributions in Natural Yellow Poplar Stands.* USDA Forest Research Paper SE-25, Feb. 1967.

McLearen, Douglas C. "Late Archaic and Early Woodland Material Culture in Virginia." In *Late Archaic and Early Woodland Research in Virginia: A Synthesis,* ed. Theodore R. Reinhart and Mary Ellen Hodges, 89–138. Richmond: Archaeological Society of Virginia, 1991.

———. "Virginia's Middle Woodland Period: A Regional Perspective." In *Middle and Late Woodland Research in Virginia: A Synthesis,* ed. Theodore R. Reinhart and Mary Ellen N. Hodges, 39–64. Richmond: Archaeological Society of Virginia, 1992.

McLoughlin, William G. *Cherokee Renascence in the New Republic.* Princeton: Princeton Univ. Press, 1986.

McMichael, Edward V. "Environment and Culture in West Virginia." *Proceedings of the West Virginia Academy of Science* 33 (1961): 146–50.

Mead, C. P. "Forests and Game." *Virginia Wildlife* 9 (Jan. 1948): 16–18.

Meyers, Maureen S. "The Mississippian Frontier in Southwestern Virginia." *Southeastern Archaeology* 21 (Winter 2002): 178–92.

Michael, Edwin D. "Effects of White-Tailed Deer on Appalachian Hardwood Regeneration." In *Proceedings: Guidelines for Regenerating Appalachian Hardwood Stands.* Morgantown: West Virginia Univ. Books, 1988.

Miles, Emma Bell. *The Spirit of the Mountains.* 1905. Knoxville: Univ. of Tennessee Press, 1975.

Miller, Carl F. "Revaluation of the Eastern Siouan Problem with Particular Emphasis on the Virginia Branches—The Occaneechi, the Saponi, and the Tutelo." Bureau of American Ethnology Bulletin 164. *Anthropological Papers,* no. 52. Washington, DC: GPO, 1957.

Miller, Char, and Hal Rothman, eds. *Out of the Woods: Essays in Environmental History.* Pittsburgh: Univ. of Pittsburgh Press, 1997.

Miller, Jim Wayne. *The Mountains Have Come Closer.* Boone, NC: Appalachian Consortium Press, 1980.

Milnes, Gerald. *Signs, Cures and Witchery: German Appalachian Folklore.* Knoxville: Univ. of Tennessee Press, 2007.

Mitchell, Robert D. *Appalachian Frontiers: Settlement, Society, and Development in the Preindustrial Era.* Lexington: Univ. Press of Kentucky, 1991.

———. *Commercialism and Frontier: Perspectives on the Early Shenandoah Valley.* Charlottesville: Univ. Press of Virginia, 1977.

Mohlenbrock, Robert H. "Mount Rogers, Virginia." *Natural History* 99 (Dec. 1990): 72–74.

Moir, Esther. *The Discovery of Britain: The English Tourists.* London: Routledge and Kegan Paul, 1964.

Mooney, James. "The Cherokee River Cult." *Journal of American Folklore* 13 (Jan.–Mar. 1900): 1–10.

———. *Myths of the Cherokee*. Washington, DC: Bureau of American Ethnology, 1902.

———. *The Sacred Formulas of the Cherokees* Washington, DC: Bureau of American Ethnology, 1891.

———. *The Siouan Tribes of the East*. Bureau of American Ethnology, Bulletin 22. Washington, DC: GPO, 1894.

Moorman, John J. *The Mineral Springs of Western Virginia*. New York: Wiley and Putnam, 1846.

———. *The Virginia Springs and Springs of the South and West*. Philadelphia: J. B. Lippincott & Co., 1859.

Moquin, Wayne, ed. *A Documentary History of the Italian American*. New York: Praeger, 1974.

Morse, David. *American Romanticism*. Vol. 1, *From Cooper to Hawthorne, Excessive America*. Totowa, NJ: Barnes & Noble Books, 1987.

Morton, Oren F. *A History of Monroe County, West Virginia*. 1916. Baltimore: Regional Publishing Co., 1980.

Mosby, Henry, and Charles Handley. *The Wild Turkey in Virginia*. Richmond: Virginia Commission of Game and Inland Fisheries, 1942.

Muller, Jon. "The Southern Cult." In *The Southeastern Ceremonial Complex: Artifacts and Analysis*, ed. Patricia Galloway, 11–26. Lincoln: Univ. of Nebraska Press, 1989.

Mullin, George B. P. "The Growing Recreational Use on the Jefferson National Forest." *Virginia Wildlife* 16 (Apr. 1955): 18–20.

Murphy, J. P. "Advantages of Incline and Mechanical Logging." *Southern Lumberman* 119 (Apr. 25, 1925): 35–36.

Myers, Andrew H. "The Creation of Shenandoah National Park: Albemarle County Cultures in Conflict." *Magazine of Albemarle County History* 51 (1993): 52–89.

Myers, Kenneth. *The Catskills: Painters, Writers, and Tourists in the Mountains 1820–1895*. Yonkers: Hudson River Museum of Westchester, 1987.

Nash, Roderick. *The American Environment: Readings in the History of Conservation*. Reading, MA: Addison-Wesley, 1968.

———. *Wilderness and the American Mind*. 4th ed. New Haven: Yale Univ. Press, 2001.

National Parks and Conservation Association. "Assault on Mount Rogers National Recreation Area." *National Parks and Conservation Magazine* 52 (June 1978): 24–25.

Neel, L. R. "Agriculture in the Southern Mountains." *Mountain Life and Work* 3 (Apr. 1927): 4–6, 10.

Nelson, Robert H. "The Future of Federal Forest Management: Options for Use of Market Methods." In *Federal Lands Policy*, ed. Phillip O. Foss, 159–76. Westport, CT: Greenwood Press, 1987.

Newell, Alan S. "Identification and Interpretation: Managing Cultural Resources in the U.S. Forest Service." *Public Historian* 9 (1987): 147–53.

Nicholls, W. D. "Families on Submarginal Land." *Mountain Life and Work* 10 (Apr. 1934): 26–28.

Nichols, William H. "Economic Development in Upper East Tennessee Valley." *Journal of Political Economy* 64 (Aug. 1956): 277–302.

Nicolson, Marjorie Hope. *Mountain Gloom and Mountain Glory: The Development of the Aesthetics of the Infinite.* 1959. Seattle: Univ. of Washington Press, 1997.

Noe, Kenneth W. *Southwest Virginia's Railroad: Modernization and the Sectional Crisis.* Urbana: Univ. of Illinois Press, 1994.

Nolt, John, ed. *A Land Imperiled: The Declining Health of the Southern Appalachian Bioregion.* Knoxville: Univ. of Tennessee Press, 2005.

Novak, Barbara. *Nature and Culture: American Landscape Painting 1825–1875.* New York: Oxford Univ. Press, 1980.

Oak, Steven W., Cindy M. Huber, Raymond M. Sheffield. *Incidence and Impact of Oak Decline in Western Virginia, 1986.* USDA Resource Bulletin SE-123, Nov. 1991.

Obermiller, Phillip J., Thomas E. Wagner, and E. Bruce Tucker, eds. *Appalachian Odyssey: Historical Perspectives on the Great Migration.* Westport, CT: Praeger, 2000.

O'Byrne, J. W. "The Forests of Russell County, Virginia." In *The Geology and Coal Resources of Russell County, Virginia.* Virginia Geological Survey Bulletin No. 22. Charlottesville: Univ. of Virginia, 1922.

O'Donnell, Kevin E., and Helen Hollingsworth, eds. *Seekers of Scenery: Travel Writing from Southern Appalachia, 1840–1900.* Knoxville: Univ. of Tennessee Press, 2004.

Oertel, Everett. "Bicentennial Bees: Early Records of Honey Bees in the Eastern United States." *American Bee Journal* 116 (5 parts, Feb.–June 1976).

Otis, Alison T., William D. Honey, Thomas C. Hogg, and Kimberly K. Lakin. *The Forest Service and the Civilian Conservation Corps: 1933–42.* Washington, DC: USDA, Forest Service, 1986.

O'Toole, Randal. *Reforming the Forest Service.* Washington, DC: Island Press, 1988.

Otto, John S. "Forest Fallowing among the Appalachian Mountain Folk: An Ethnohistorical Study." *Anthropologica* 30, no. 1 (1988): 3–22.

———. "Forest Fallowing in the Southern Appalachian Mountains: A Problem in Comparative Agricultural History." *Proceedings of the American Philosophical Society* 133 (Mar. 1989): 51–63.

Ousby, Ian. *The Englishman's England: Taste, Travel and the Rise of Tourism.* Cambridge: Cambridge Univ. Press, 1990.

Owen, A. L. Riesch. *Conservation under F.D.R.* New York: Praeger, 1983.

Page, Linda Garland, and Eliot Wigginton, eds. *Aunt Arie: A Foxfire Portrait.* Chapel Hill: Univ. of North Carolina Press, 1992.

Page, Thomas Nelson. *Social Life in Old Virginia before the War.* New York: Charles Scribner's Sons, 1898.

Paulding, James Kirke. *Letters from the South, Written during an Excursion in the Summer of 1816.* Vol. 1. New York: James Eastburn & Co., 1817.

Peaks of Otter, A Monograph of the Religious Experience of a Young Man. Philadelphia: Presbyterian Board of Publication, 1859.

Pederson, Fred C. *The Forests of the Valley Coal Fields of Virginia.* Charlottesville: Virginia Geological Commission, 1925.

———. "The Forests of Wise County." In *Geology and Mineral Resources of Wise County and the Coal-Bearing Portion of Scott County, Virginia.* Virginia Geological Survey Bulletin No. 24, 1923.

Pencil, Mark. *The White Sulphur Papers.* New York: Samuel Colman, 1839.

Pendleton, William C. *History of Tazewell County and Southwest Virginia, 1748–1920.* Richmond, VA: W. C. Hill Printing Co., 1920.

Perdue, Charles L., Jr., and Nancy J. Martin-Perdue. "Appalachian Fables and Facts: A Case Study of the Shenandoah National Park Removals." *Appalachian Journal* 7 (Autumn–Winter 1979–80): 84–104.

———. "'To Build a Wall around These Mountains': The Displaced People of Shenandoah." *Magazine of Albemarle Country History* 49 (1991): 48–71.

———, eds. *Talk about Trouble: A New Deal Portrait of Virginians in the Great Depression.* Chapel Hill: Univ. of North Carolina Press, 1996.

Philpot, J. H. *The Sacred Tree.* New York: Macmillan Co., 1897.

Pierce, Daniel S. *The Great Smokies: From Natural Habitat to National Park.* Knoxville: Univ. of Tennessee Press, 2000.

Pinchot, Gifford. *Breaking New Ground.* New York: Harcourt, Brace & Co., 1947.

Pisani, Donald J. "Forests and Conservation, 1865–1890." *Journal of American History* 72, no. 2 (1985): 340–59.

Pollard, Edward A. *Virginia Tourist: Sketches of the Springs and Mountains of Virginia.* Philadelphia: J. B. Lippincott & Co., 1870.

Porteous, Alexander. *The Forest in Folklore and Mythology.* New York: Macmillan Co., 1928.

Porterfield, Richard L., and John B. Crist, eds. *Impacts of the Changing Quality of Timber Resources.* Madison, WI: Forest Products Research Society, 1978.

Powell, Katrina M. "Writing the Geography of the Blue Ridge Mountains: How Displacement Recorded the Land." *Biography* 25, no. 1 (2002): 73–94.

Preister, Kevin, and James A. Kent. "Using Social Ecology to Meet the Productive Harmony Intent of the National Environmental Policy Act." *Hastings West-Northwest Journal of Environmental Law and Policy* 7 (Spring 2001): 235–51.

Preston, Thomas L. *Historical Sketches and Reminiscences of an Octogenarian.* Richmond, VA: B. F. Johnson, 1900.

Price, Overton W. "Practical Forestry in the Southern Appalachians." In *Yearbook of the United States Department of Agriculture, 1900,* 357–68. Washington, DC: GPO, 1901.

Prolix, Peregrine [Philip Houlbrooke Nicklin]. *Letters Descriptive of the Virginia Springs.* Philadelphia: H. S. Tanner, 1837. Reprint, Austin: AAR/Tantalus, 1978.

Pudup, Mary Beth, Dwight B. Billings, and Altina L. Waller, eds. *Appalachia in the Making: The Mountain South in the Nineteenth Century.* Chapel Hill: Univ. of North Carolina Press, 1995.

Purrington, Burton L. "Ancient Mountaineers: An Overview of the Prehistoric Archaeology of North Carolina's Western Mountain Region." In *The Prehistory of North Carolina,* ed. Mark A. Mathis and Jeffrey J. Crow. Raleigh: North Carolina Division of Archives and History, 1983.

Pyle, Charlotte, and Michael P. Schafale. "Land Use History of Three Spruce-Fir Forest Sites in Southern Appalachia." *Journal of Forest History* 32, no. 1 (1988): 4–21.

Pyne, Stephen J. *Fire in America: A Cultural History of Wildland and Rural Fire.* Princeton: Princeton Univ. Press, 1982.

Rachal, William M. E. "Salt the South Could Not Savor." *Virginia Cavalcade* 3 (Autumn 1953): 4–7.

Randall, Charles Edgar. "Jefferson's Forest." *American Forests* 75 (Apr. 1969): 20–22, 57–62.

Rasmussen, Barbara. *Absentee Landowning and Exploitation in West Virginia, 1760–1920.* Lexington: Univ. Press of Kentucky, 1994.

Ray, Arthur J. *Indians in the Fur Trade: Their Role as Trappers, Hunters, and Middlemen in the Lands Southwest of Hudson Bay.* Toronto: Univ. of Toronto Press, 1974.

Rehder, John B. *Appalachian Folkways.* Baltimore: Johns Hopkins Univ. Press, 2004.

Reinhart, Theodore R., and Mary E. Hodges, eds. *Middle and Late Woodland Research in Virginia: A Synthesis.* Richmond: Archaeological Society of Virginia, 1992.

Reniers, Perceval. *The Springs of Virginia: Life, Love, and Death at the Waters, 1775–1900.* Chapel Hill: Univ. of North Carolina Press, 1941.

Richardson, Elmo R. *The Politics of Conservation: Crusades and Controversies 1897–1913.* Berkeley: Univ. of California Press, 1962.

"Ride to the Peaks of Otter." *Southern Literary Messenger* 7 (Dec. 1841): 850–53.

Ritter, William F., and Adel Shirmohammadi, eds. *Agricultural Nonpoint Source Pollution: Watershed Management and Hydrology.* Boca Raton, FL: Lewis, 2001.

Rivkin, Dean Hill. "Lawyering, Power, and Reform: The Legal Campaign to Abolish the Broad Form Mineral Deed." *Tennessee Law Review* 66 (1999): 467–98.

Robbins, William G. *American Forestry: A History of National, State, and Private Cooperation.* Lincoln: Univ. of Nebraska Press, 1985.

———. *Lumberjacks and Legislators: Political Economy of the U.S. Lumber Industry, 1890–1941.* College Station: Texas A&M Univ. Press, 1982.

Robertson, Wyndham. "Some Notes on the Holstein (Va.) Salt and Gypsum." *Virginias* 3 (Feb. 1882): 20–21, 42.

Robinson, W. Stitt. *The Southern Colonial Frontier, 1607–1763.* Albuquerque: Univ. of New Mexico Press, 1979.

Roe, Frank G. *The North American Buffalo: A Critical Study of the Species in Its Wild State*. 2nd ed. Toronto: Univ. of Toronto Press, 1970.

Rogers, Anne Frazer, ed. "The Jaybird Branch Project: Report of Investigations." Cullowhee: Western Carolina Univ., 1982.

Rohrbach, Holmes, and Peter T. *Stagecoach East: Stagecoach Days in the East from the Colonial Period to the Civil War*. Washington, DC: Smithsonian Institution Press, 1983.

Rosenbaum, Walter A. *Environmental Politics and Policy*. Washington, DC: Congressional Quarterly Press, 1985.

Rostlund, Erhard. *Freshwater Fish and Fishing in Native North America*. Berkeley: Univ. of California Press, 1952.

Roth, Dennis M. *The Wilderness Movement and the National Forests: 1964–1980*. Washington, DC: USDA, Forest Service, 1984.

Rountree, Helen C. *The Powhatan Indians of Virginia*. Norman: Univ. of Oklahoma Press, 1989.

Royall, Anne Newport. *Sketches of History, Life, and Manners, in the United States*. 1826. New York: Johnson Reprint Corp., 1970.

Ruffner, H. "Notes of a Tour from Virginia to Tennessee, in the Months of July and August, 1838." *Southern Literary Messenger* 5 (Jan. 1839): 44–48.

Rutland, Robert A. "Men and Iron in the Making of Virginia—Part I (1619–1860)." *Iron Worker* 40 (Summer 1976): 2–17.

———. "Men and Iron in the Making of Virginia—Part II." *Iron Worker* 40 (Autumn 1976): 2–17.

Saggs, H.W.F. *Civilization before Greece and Rome*. New Haven: Yale Univ. Press, 1989.

Salstrom, Paul. *Appalachia's Path to Dependency: Rethinking a Region's Economic History, 1730–1940*. Lexington: Univ. Press of Kentucky, 1994.

Santoa, Carlos. "Shenandoah Park, Counties Aiming for Better Relations." *Richmond Times-Dispatch*, Dec. 7, 1992.

Sargent, C. S. "Fuel Consumption in the Virginias and U.S." *Virginias* 4 (Jan. 1883): 3.

Sarvis, Will. "Americans and Their Land: The Deep Roots of Property and Liberty." *Contemporary Review* 290 (Spring 2008): 40–46.

———. "An Appalachian Forest: Creation of the Jefferson National Forest and Its Effects on the Local Community." *Forest and Conservation History* 37 (October 1993): 169–78.

———. "A Difficult Legacy: Creation of the Ozark National Scenic Riverways." *Public Historian* 24 (Winter 2002): 31–52.

———. "Fisheries and Wildlife Management: Part of the History of the Jefferson National Forest." *Virginia Forests* 48 (Summer 1992): 6–8.

———. "The Great Anti-Fire Campaign." *American Forests* 99 (May–June 1993): 33–35, 58.

———. "Green Cove Station: An Appalachian Train Depot and Its Community." *Virginia Cavalcade* 42 (Autumn 1992): 52–61.

———. "Land and Home in the American Mind." *Journal of Natural Resources and Environmental Law* 22, no. 2 (2008–9): 107–37.

———. "The Mount Rogers National Recreation Area and the Rise of Public Involvement in Forest Service Planning." *Environmental History Review* 28 (Summer 1994): 41–65.

———. "Old Eminent Domain and New Scenic Easements: Land Acquisition for the Ozark National Scenic Riverways." *Western Legal History* 13 (Winter–Spring 2000): 1–37.

———. "The Potts Valley Branch Railroad and Tri-State Incline Lumber Operation in West Virginia and Virginia, 1892–1932." *West Virginia History* 54 (1995): 42–58.

———. "Prehistoric Southwest Virginia: Aboriginal Occupation, Land Use, and Environmental Worldview." *Smithfield Review* 5 (Apr. 2000): 125–51.

———. "Turnpike Tourism in Western Virginia." *Virginia Cavalcade* 48 (Winter 1998): 14–23.

Sauer, Carl O. "The Settlement of the Humid East." *Climate and Man, 1941 Yearbook of Agriculture,* 158–66. Washington, DC: GPO, 1941.

———. *Sixteenth Century North America: The Land and the People as Seen by the Europeans.* Berkeley: Univ. of California Press, 1971.

Schiff, Ashley L. *Fire and Water: Scientific Heresy in the Forest Service.* Cambridge: Harvard Univ. Press, 1962.

Schwab, W. G. "The Forests of Buchanan County, Virginia." In *The Geology and Coal Resources of Buchanan County, Virginia.* Virginia Geological Survey Bulletin No. 18. Charlottesville: Univ. of Virginia, 1918.

Scott, John. "The Appellation Trail." *Appalachian Trailway News* 62 (Nov.–Dec. 2001): 23–24.

Sears, John F. *Sacred Places: American Tourist Attractions in the Nineteenth Century.* New York: Oxford Univ. Press, 1989.

Sellars, Richard West. *Preserving Nature in the National Parks.* New Haven: Yale Univ. Press, 1997.

Sessoms, H. W. "Lowering Systems and Inclines." *Lumber World Review* 43 (Nov. 10, 1922): 83–84.

Seyden, Terry. "Jefferson National Forest—Rich in History." *Virginia Forests Magazine* 45 (Spring 1990): 13–15, 22.

Shackelford, Laurel, and Bill Weinberg, eds. *Our Appalachia: An Oral History.* New York: Hill & Wang, 1977.

Shaffer, Earl V. *Walking with Spring: The First Thru-Hike of the Appalachian Trail.* Harpers Ferry, WV: Appalachian Trail Conference, 1995.

Shands, William E., and Robert G. Healy. *The Lands Nobody Wanted.* Washington, DC: Conservation Foundation, 1977.

Shands, William E., and Thomas E. Waddell. *Below-Cost Timber Sales in the Broad Context of National Forest Management.* Washington, DC: Conservation Foundation, 1988.

Shands, William E., Perry R. Hagenstein, and Marissa T. Roche. *National Forest Policy: From Conflict toward Consensus.* Washington, DC: Conservation Foundation, 1979.

Shapiro, Henry D. *Appalachia on Our Mind: The Southern Mountains and Mountaineers in the American Consciousness, 1870–1920.* Chapel Hill: Univ. of North Carolina Press, 1978.

Short, C. Brant. *Ronald Reagan and the Public Lands: America's Conservation Debate, 1979–1984.* College Station: Texas A&M Univ. Press, 1989.

Simmons, Dennis E. "Conservation, Cooperation, and Controversy: The Establishment of the Shenandoah National Park, 1924–1936." *Virginia Magazine of History and Biography* 89 (1981): 387.

Sisson, Sebert L. "Pot Rock Cliff Shelter, Carroll County, Virginia." *Quarterly Bulletin of the Archeological Society of Virginia* 34 (Sept. 1979): 50–56.

Sludes, Earl R. *White Pine Provenance Study in the Southern Appalachians.* USDA Forest Research Paper SE-2, Feb. 1963.

Sludes, Earl R., and Keith W. Dorman. *Performance in the Southern Appalachians of Eastern White Pine Seedlings from Different Provinces.* USDA Forest Research Paper SE-90, Dec. 1971.

Smith, Lee. *Oral History.* New York: G. P. Putnam's Sons, 1983.

Smith, Marvin T. "Aboriginal Population Movements in the Early Historic Period of the Interior Southeast." In *Powhatan's Mantle: Indians in the Colonial Southeast,* ed. Peter H. Wood, Gregory A. Waselkov, and M. Thomas Hatley, 21–34. Lincoln: Univ. of Nebraska Press, 1989.

Solomon, Stephen D. "Chestnut Trees Return." *Scientific American Earth* 19, no. 1 (2009): 4.

Speck, Frank G. "The Ethnic Position of the Southeastern Algonkian." *American Anthropologist,* n.s. 26 (1924): 184–200.

———. *Ethnology of the Yuchi Indians.* Univ. of Pennsylvania Museum of Anthropological Publications No. 1. Philadelphia: Univ. Museum, 1909–11.

Spurr, Stephen H., and Burton V. Barnes. *Forest Ecology.* 3rd ed. Malabar, FL: Krieger, 1992.

Starkey, Edith Kimmell. "Over the Mountain: Timbering at Braucher." *Goldenseal* 13 (Summer 1987): 34–39.

Steen, Harold K., ed. *Forest and Wildlife Science in America: A History.* Durham, NC: Forest History Society, 1999.

———. *The U.S. Forest Service: A History.* Seattle: Univ. of Washington Press, 1976.

Still, James. *River of Earth.* 1940. Lexington: Univ. Press of Kentucky, 1978.

Stine, Jeffrey K. "Environmental Politics in the American South: The Fight over the Tennessee-Tombigbee Waterway." *Environmental History Review* 15 (Spring 1991): 1–24.

Stoner, Robert Douthat. *A Seed-Bed of the Republic: A Study of the Pioneers in the Upper (Southern) Valley of Virginia.* Roanoke: Roanoke Historical Society, 1962.

Stratton, Ezra M. *The World on Wheels.* New York: Benjamin Bloom, 1972.

Strother, David Hunter (aka "Porte Crayon"). *Virginia Illustrated: Containing a Visit to the Virginian Canaan.* New York: Harper & Brothers, 1871.

Stroup, Richard, and John Baden, eds. *Natural Resources: Bureaucratic Myths and Environmental Management.* San Francisco: Pacific Institute of Public Policy Research, 1983.

Sturgill, Mack H. *Hungry Mother: History and Legends.* Marion, VA: Tucker Printing, 1986.

Suddeth, Ruth E. "The Myths of the Cherokees." *Georgia Review* 10 (Spring 1956): 84–91.

Summers, Lewis Preston. *History of Southwest Virginia, 1746–1786, Washington County, 1777–1870.* 1903. Baltimore: Genealogical Publishing Co., 1966.

Swain, Donald C. *Federal Conservation Policy 1921–1933.* Berkeley: Univ. of California Press, 1963.

Swank, James M. *History of the Manufacture of Iron in All Ages.* Philadelphia: American Iron & Steel Association, 1892.

Swank, Wayne T., and D. A. Crossley Jr., eds. *Forest Hydrology and Ecology at Coweeta.* New York: Springer-Verlag, 1988.

Swanton, John R. *The Indians of the Southeastern United States.* Bureau of American Ethnology, Bulletin 137. Washington, DC: GPO, 1946.

Tanner, Helen Hornbeck. "The Land and Water Communication Systems of the Southeastern Indians." In *Powhatan's Mantle: Indians in the Colonial Southeast,* ed. Peter H. Wood, Gregory A. Waselkov, and M. Thomas Hatley, 6–20. Lincoln: Univ. of Nebraska Press, 1989.

Taylor, George Rogers. *The Transportation Revolution, 1815–1860.* New York: Harper & Row, 1951.

Terrie, Philip G. "The New York Natural History Survey in the Adirondack Wilderness, 1836–1840." *Journal of the Early Republic* 3 (Summer 1983): 186–206.

———. "Romantic Travelers in the Adirondack Wilderness." *American Studies* 24, no. 2 (1983): 59–75.

Thomas, Roy E., comp. *Southern Appalachia, 1885–1915: Oral Histories from the Residents of the State Corner Area of North Carolina, Tennessee and Virginia.* Jefferson, NC: McFarland & Co., 1991.

Thompson, Stuart. "Physical or Metaphysical." *Appalachian Trailway News* 61 (Nov.–Dec. 2000): 21–22.

Tindall, P. B. *Observations on the Mineral Waters of Western Virginia.* Richmond: Charles H. Wynne, 1858.

Titus, Warren I. *John Fox, Jr.* New York: Twayne, 1971.

Townsend, W. B. "Railroad Construction and Operation in Mountain Logging." *Lumber World Review* 43 (Nov. 10, 1922): 111.

Turner, E. Randolph. "PaleoIndian Settlement Patterns and Population Distribution in Virginia." In *PaleoIndian Research in Virginia: A Synthesis,* ed. J. Mark Wittkofski and Theodore R. Reinhart. Special Publication No. 19. Richmond: Archaeological Society of Virginia, 1989.

Turner, Victor, and Edith Turner. *Image and Pilgrimage in Christian Culture: Anthropological Perspectives.* New York: Columbia Univ. Press, 1978.

Van Lear, David H., and Thomas A. Waldrop. "Effects of Fire on Natural Regeneration in the Appalachian Mountains." In *Guidelines for Regenerating Appalachian Hardwood Stands: Workshop Proceedings,* ed. Arlyn W. Perkey et al. Morgantown: West Virginia Univ. Books, 1988.

Van Zandt, Roland. *The Catskill Mountain House.* New Brunswick, NJ: Rutgers Univ. Press, 1966.

"Viator." "The White Sulphur Springs." *New England Magazine* 3 (Sept. 1832): 222–27.

Virginia Historical Society. *First Resorts: A Visit to Virginia's Springs.* Richmond: Virginia Historical Society, 1987.

Virginia Polytechnic Institute and State Univ., Engineering Extension Division. *Industrial Survey,* nos. 1–19 (Southwest Virginia). Blacksburg: Virginia Polytechnic Institute and State Univ., 1929–30.

"Virginia Springs." *Southern Literary Messenger* 3 (May 1837): 281.

"Visit to the Virginia Springs, during the Summer of 1834, No. 1." *Southern Literary Messenger* 1 (May 1835): 474–77.

"Visit with a Forest Ranger." *Norfolk and Western Magazine* 29 (July 1951): 418–21, 474, 475.

Voegelin, C. F. *The Shawnee Female Deity.* New Haven: Yale Univ. Publications in Anthropology, 1970.

W. M. Ritter Lumber Company. *The Romance of Appalachian Hardwood Lumber.* Columbus, OH: W. M. Ritter Lumber Co., 1940.

Wallach, Bret. "The Slighted Mountains of Upper East Tennessee." *Annals of the Association of American Geographers* 71, no. 3 (1981): 359–73.

Ward, Barry J. "Italian-American Folk Poetry in the Industrialized Appalachian Mountains." *West Virginia History* 43 (Spring 1982): 285–302.

Waselkov, Gregory A. "Indian Maps of the Colonial Southeast." In *Powhatan's Mantle: Indians in the Colonial Southeast,* ed. Peter H. Wood, Gregory A. Waselkov, and M. Thomas Hatley, 292–343. Lincoln: Univ. of Nebraska Press, 1989.

Watson, Thomas L. *Mineral Resources of Virginia.* Lynchburg, VA: J. P. Bell Co., 1907.

Weidensaul, Scott. *Mountains of the Heart: A Natural History of the Appalachians.* Golden, CO: Fulcrum, 1994.

Wenke, Robert J. *Patterns in Prehistory: Humankind's First Three Million Years.* 3rd. ed. New York: Oxford Univ. Press, 1990.

Whisnant, David E. *All That Is Native and Fine: The Politics of Culture in an American Region.* Chapel Hill: Univ. of North Carolina Press, 1983.

———. *Modernizing the Mountaineer.* Boone, NC: Appalachian Consortium Press, 1980.

White, Peter S., ed. *The Southern Appalachian Spruce-Fir Ecosystem: Its Biology and Threats.* Gatlinburg, TN: National Park Service, Southeast Regional Office, Research/Resource Management Report #SER-71, Nov. 1984.

Widner, Ralph R., ed. *Forests and Forestry in the American States: A Reference Anthology.* Missoula, MT: National Association of State Foresters, 1968.

Wigginton, Eliot, ed. *The Foxfire Book.* Garden City, NY: Doubleday, 1972.

———. *Foxfire 4.* Garden City, NY: Anchor Books, 1977.

Wigginton, Eliot, and Margie Bennett, eds. *Foxfire 9.* Garden City, NY: Doubleday, 1986.

Wilhelm, Gene, Jr. "The Mullein: Plant Piscicide of the Mountain Folk Culture." *Geographical Review* 64 (Apr. 1974): 235–52.

———. "Shenandoah Resettlements." *Pioneer America* 14, no. 1 (1982): 15–40.

Wilkinson, Charles F., and H. Michael Anderson. *Land Resource Planning in the National Forests.* Washington, DC: Island Press, 1987.

Williams, John Alexander. *Appalachia: A History.* Chapel Hill: Univ. of North Carolina Press, 2002.

Williams, Michael Ann. *Great Smoky Mountains Folklife.* Jackson: Univ. Press of Mississippi, 1995.

———. "'When I Can Read My Title Clear': Anti-Environmentalism and Sense of Place in the Great Smoky Mountains." In *Culture, Environment, and Conservation in the Appalachian South,* ed. Benita J. Howell, 87–99. Urbana: Univ. of Illinois Press, 2002.

Williams, Michael. *Americans and Their Forests: A Historical Geography.* New York: Cambridge Univ. Press, 1989.

Williams, Samuel Cole. *Adair's History of the American Indians.* 1755. New York: Argonaut Press, 1966.

———. *Early Travels in the Tennessee Country, 1540–1800.* Johnson City, TN: Watauga Press, 1928.

Williams, Wellington. *Appletons' Southern and Western Travellers' Guide.* New York: D. Appleton & Co., 1850.

Wilson, Goodridge. *Smyth County History and Traditions.* Kingsport, TN: Kingsport Press, 1932.

Withey, Lynne. *Grand Tours and Cook's Tours: A History of Leisure Travel, 1750 to 1915.* New York: William Morrow & Co., 1997.

Witthoft, John, and William A. Hunter. "The Seventeenth Century Origins of the Shawnee." *Ethnohistory* 2 (Winter 1955): 42–57.

Wolfe, Charles K. *Children of the Heav'nly King: Religious Expression in the Central Blue Ridge.* Washington, DC: Library of Congress Recording Laboratory, [1982].

Wolfe, Margaret Ripley. "Aliens in Appalachia: The Construction of the Clinch-field Railroad and the Italian Experience." In *Appalachia: Family Traditions in Transition,* ed. Emmet M. Essin. Johnston City: Eastern Tennessee State Univ. Press, 1975.

Wood, Peter H. "The Changing Population of the Colonial South: An Overview by Race and Region, 1685–1790." In *Powhatan's Mantle: Indians in the Colonial Southeast,* ed. Peter H. Wood et al., 35–103. Lincoln: Univ. of Nebraska Press, 1989.

Wood, Peter H., Gregory A. Waselkov, and M. Thomas Hatley, eds. *Powhatan's Mantle: Indians in the Colonial Southeast.* Lincoln: Univ. of Nebraska Press, 1989.

Woods, Frank W., and Royal E. Shanks. "Natural Replacement of Chestnut by Other Species in the Great Smoky Mountains National Park." *Ecology* 40 (July 1959): 349–61.

Worrell, Albert C. *Principles of Forest Policy.* New York: McGraw-Hill Book Co., 1970.

Worster, Donald, ed. *The Ends of the Earth: Perspectives on Modern Environmental History.* New York: Cambridge Univ. Press, 1988.

———. *Nature's Economy: A History of Ecological Ideas.* New York: Cambridge Univ. Press, 1977.

———. *The Wealth of Nature: Environmental History and the Ecological Imagination.* New York: Oxford Univ. Press, 1993.

Worthington, Jane Tayloe. "Spring Memories among the Mountains." *Southern Literary Messenger* 12 (June 1846): 349–52.

———. "To the 'Far Blue Mountain.'" *Southern Literary Messenger* 13 (Jan. 1847): 41–42.

Wright, J. Leitch, Jr. *The Only Land They Knew: The Tragic Story of the American Indians in the Old South.* New York: Free Press, 1981.

Yahner, Richard H. *Eastern Deciduous Forest: Ecology and Wildlife Conservation.* Minneapolis: Univ. of Minnesota Press, 1995.

Yoakley, Ina C. "Wild Plant Industry of the Southern Appalachians." *Economic Geography* 8 (July 1932): 311–17.

Young, T. "False, Cheap and Degraded: When History, Economy and Environment Collided at Cades Cove, Great Smoky Mountains National Park." *Journal of Historical Geography* 32, no. 1 (2006): 169–89.

Index

Bedford County (VA), 206
beech trees, 134
Belmont Corporation, 96
Bennett Logging and Lumber Company, 132, 281nn34–36
Bentham, Jeremy, 185
Big Stone Gap (VA), 74, 163,
Biltmore Estate (NC), 54–55
bison, eastern, 13, 21, 31, 224n10
Blacksburg Ranger District, **96**, 100, 106, 182, 278n10, 279n12; manganese mining on, 98; territorial designation of, xxi, 220; wilderness areas in, 219
Bland County (VA), 16, 203, 226n22, 252n35; JNF land, 206
Blue Ridge Mountains, 30, 37, 39; prehistory associations, 17, 18, 20
Blue Ridge Parkway, 1, 105
Blue Ridge Reservoir, 121, 275n70
Boone, Daniel, 45, 109, 151
botanists and botany, 6, 100, 170; JNF staff, 104, 145–46, 149, 188
Botetourt County (VA), 66; JNF land, 206
buffalo. *See* bison, eastern
Bureau of Land Management (BLM), 137, 138, 139
Burke's Garden, 172

Cades Cove (NC), 67, 117, 253n45
Calvinism, 162, 164, 173
Calvinists, Green, 148, 190
Carroll County (VA), 16, 32, 110, 163, 252n35; JNF land, 206
Cave Mountain Lake, 84
charcoal manufacturing and use, 3, 4, 32–33, 236n27
Cherokee National Forest, 252n33, 252n34, 254n49, 301n3
Cherokee, eastern band; 2, 3, 23; buffer zone in southwestern Virginia, 27, 28, 151, 235n16; cessation of lands, 29; decimation of, 29; descendants,

164, 168; and deer skin trade, 3, 7, 24, 28–29, 234n10; prehistory, 11, 19; worldview of, 23–25
chestnut oak, 51, 79, 92, 101
chestnut trees, 11, 45, 61, 186, 207, 258n30, 259n32; blight, 79–82, 101, 144, 207, 258n29, 265n37; mast, 13, 31, 80, 155, 156, 258n30; uses of, 78, 80, 82, 154
chestnut, Chinese, 101, 258n29
Citizens for Southwest Virginia, 123, 274n49, 277n81
Citizens Task Force, 189, 199
Civil War, 4, 32, 33, 150, 221n2
Civilian Conservation Corps (CCC), 71, 74, 79, 80, 84, 86, 88
Clarke-McNary Act (1924), 61
clear-cutting, timber, x, xiii–xiv, 133, 209, 210; abuse of, 120, 129, 130; compared to timber-stripping, 129, 279n11; criticism of, 6, 54, 93, 127, 131, 147, 202–3, 278n10, 279n15; forest policy, 5, 94, 106, 128–30, 279n18; and habitat creation, 140, 141; and Pioneer Forest, 193–94, 198–99; silvicultural practice, 129, 130, 131, 194–95, 279n11, 279n18
Clinch Coalition, 179, 199, 298n20, 299n26
Clinch Purchase Unit, 89, 252n35
Clinch Ranger District, 3, 63, 66, **81**, 178, 189, **190**; fire fighting, 69–70, **73**, **77**, 78; minerals, 98, 136–39, 283n66; moonshine, 88; prehistory, 12, 17, 182, 228n43, 288n7; recreation, 84, **94**; tenants, 68, 254n56; timber, 92, 132; white pine blister rust, 80; wilderness, 142–43, 220; wildlife, 88, 100
Clinch River, 12, 17, 30, 167, 249n10, 288n12
Clinchfield Coal Corporation, 65, 74, 85
coal, 64, 136, 137, 151, 221n1, 223n11; broad form mineral deed, 137,

Leopold, Aldo, 70, 85, 86
Letcher County (KY), 65, 206
libertarianism, 176
liquor, 31. *See also* moonshine
Locke, John, 166–67, 175, 176, 182–83
log cabins, 3, 80, 117, 189, 253n45
logging, **54**; with bull dozers, 131; with
    cables, 131; and environmental
    protection, 139, 178; incline opera-
    tions, 49–54; preindustrial, 3, 179,
    195; with skidders, 131, 195–96;
    202–C permits, 83, 127, 189
Loyal Land Company, 30
Lynchburg (VA), 37, 39, 40, 79, 239n14

maple trees, 45, 134, 193
Marion (VA), 68, 69, 74, 115, 273n36
Marley Moldings Company, 132
Materials Disposal Act (1947), 283n58
McKinley, William, 58
Menominee Forest and Menominee
    Indians, 197, 301n1
mineral springs, 3, 7, 37, **40**, 39–42, 50,
    241n27
Minerals Leasing Act (1920), 136
mining and mineral rights, 136–39;
    broad form deed, 137, 252n30,
    283n62: federal, 98, 283n58;
    manganese, **97**, 96–98, 135–36;
    private, 66, 96, 97, 252n30. *See also*
    coal
Monogahela Decision, 129–30, 178,
    279n15
Monogahela National Forest, 93, 120,
    129, 262n65, 301n3
Monroe County (WV), 37, 49, 50, 51;
    JNF land, 206
Montgomery County (VA), 37, **44**, JNF
    land, 206
moonshine, 78, **89**, 88–90, 182, 189,
    262nn65–66
Mount Rogers, 69, 84, 95, 151, 164, 167,
    205; ecology of, 100–101, 141, 146;
    hiking, 169, 173

Mountain Lake Biological Station, 70,
    145–46, 286n108
Mountain Lake, 37, 44, 49, 50, 84, 101,
    141, 146
mountain lions, 29, 31
mountaintop removal, **190**, 189–90
multiple use, 178, 188, 301n7; conflicts
    regarding, 70, 98–106 passim, 176,
    177, 180, 189, 277n86; as tradi-
    tional (pre-mandated) Forest Ser-
    vice policy, 67, 82, 84, 147, 199, 202
Multiple-Use and Sustained Yield Act
    (1960), 140, 188
mysticism (and the mountains), 25, 42,
    43, 164, 171, 172, 231n63

National Forest Management Act
    (1976), 106, 130, 148
National Forest Reservation Commis-
    sion, 49, 56, 60–61, 65, 66, 84,
    252n35
National Historic Preservation Act
    (1966), 150
National Park Service, 64, 110, 183,
    189, 267n48, 270n7; in Appala-
    chia, 63, 114, 117, 253nn45–46;
    as Forest Service rival, xv, 83, 84,
    102
National Timber Supply Act, 129
Natural Bridge (VA), 37, 42, 240n20,
    243n37, 243n45
Natural Bridge Purchase Unit, **62**
natural gas, 136, 139
Natural Tunnel (VA), 37, **42**, 44, 45,
    241n28
New Castle Ranger District, xxi, 3, 69,
    223n9, 238n12; chestnut trees, 82,
    101; ecology, 70, 100, 101, 134; fire
    fighting, 77–78, 91; moonshine,
    89, 90; tenants, 68; timber, 79, 82,
    83, **93**; wilderness, 219, 220
New Deal, 63, 71, 83, 95
New River, 19, 27, 28, 30, **44**, 48, 52,
    91, 110

ridgetops, mountain, 1, 60, 67, 112, 185, 187
Ritter Lumber Company. *See* W.M. Ritter Lumber Company
Roanoke (VA), 28, 173, 199
Roanoke County (VA): JNF land, 206
Roanoke River, 2, 12, 18, 20
Rockbridge County (VA), 84; JNF land, 206
Romantic Era, **36**, **40**, 41–46
Roosevelt, Eleanor, 109, 157
Roosevelt, Franklin D., 63, 71, 79, 95, 255n1
Roosevelt, Theodore, 185
Rustin Land, Mining & Manufacturing Company, 252n35

salt manufacturing, 3, 4, 32–33, 179, 227n35
Saltville (VA), 2, 3, 12, 32–33
sawmills, 48–53 passim, 72, 82, 181, 203, 263n15
science, 6, 109, 134; applied, 55, 100–101, 145; ecology, 146–48, 196; forest, 54–56, 144–46; hydrology, 135–36, 139–40; pure, 55, 70, 100–101, 145–46, 286n113
Scott County (VA), 18, 32, 58, 66, 69, 178; JNF land, 206; Natural Tunnel, **42**, 241n28
selective cutting, timber, 46, 128–131 passim, 141, 154, 180, 279n21; German silviculture, 54, 191; JNF tradition, 79, 93; on the Pioneer Forest, 193–99
Shawnee Indians, 2, 23, 27, 28, 151, 234n10
shellfish, 13, 14, 24, 141
Shenandoah National Park, 67, 80, 110, 117, 253n46, 271n21
Sierra Club, 121, 123, 248n1
silviculture. *See* clear-cutting, selective cutting, forestry, timber management

slaves, African American, 30, 32, 236n27. *See also* African Americans
Smyth County (VA), 109, 151; fire fighting, 74; JNF land, 61, 206; and the Mount Rogers NRA, 110, 114–15; prehistory, 19, 224n8; soil reclamation, 96–97, 135; timber, 132
soil reclamation, 96–99, 135–36
Southeast Forest Experiment Station, 6
Southeast Wildlife Habitat Survey, 140
Spiritual Life Institute, 171
spruce trees, 80, 124, 171; ecology of, 100–101, 107, 109, 146, 269n2, 286n107; prehistoric, 11
stagecoach travel, nineteenth century, 35, **38**, 37–39, 45, 240n21
steam technology; donkeys, 48; incline systems, 51–52; yarders, 48. *See also* railroads
stereotypes, Appalachian. *See under* Appalachia
strip mining, 71, 96–99, 120, 135, 136, 138, 283n62
Surface Mining Control and Reclamation Act (1977), 135, 138
Sweet Springs (WV), 37, 39

taxation, forest lands, 49, 62, 201–4
Tazewell County (VA), 31; JNF land, 206; prehistory, 16, 17–18, 224n8, 225n14, 226n23
tenants, on JNF, 67–68, 254nn55–56
Tennessee River, 12, 13, 14, 18, 19, 59, 288n12
Tennessee Valley Authority (TVA), 114, 136, 249n10, 282n53
timber management; below-cost sales, 105, 127, 131–33, 198; surveys, 56, 72, 78–79, 93, 144; timber stand improvement (TSI), 79, 94, 100, 127, 128. *See also* clear-cutting, forestry, selective cutting
timber: hardwoods, 32, 59, 79, 80, **81**, 106, 107, 134, 281n39; softwoods,

13, 61, 79–80, 96, 101, 106–7, 134, 145, 163, 195, 281n39; stripping, 33, 46, 49, **54**, 54, 60, 69, 128, 129, 179, 181

tourism, 109, 110, 121, 153, 171; nineteenth century, 3, 7, 35, **38**, 37–41, 187, 241nn27–28; boosterism, 63, 84–85; economy, 65, 85, 119, 120, 125, 186; ecotourism, 180; versus extractive forest uses, 177, 198. *See also* recreation

Trail of the Lonesome Pine, 163–64

tree planting, 5, 55, 71, 79–80

tree species. *See* individual tree species names (chestnut, Douglas fir, Fraser fir, hemlock, hickory, maple, oak, pine, spruce)

Tri-State Lumber Company, 50–53

trout, 72, 85, **86**, 87, 99, 142, 181

turkey, 13, 31, 72, 85, 99, 100

turnpikes, nineteenth century, 3, 7, 35–41, 150, 238n12, 239n13

Tutelo Indians, 2, 23, 27, 28, 233n6

U.S. Army Corps of Engineers, 120, 135, 277n86

U.S. Department of Agriculture, 9, 57, 60, 198

U.S. Fish and Wildlife Management Service, 86, 99, 100, 134, 141

U.S. Forest Service: Forest Service Employees for Environmental Ethics, xv, 178, 266n47; inter-agency tensions, 107–8, 113, 118, 127, 188, 271n22 ; Region 7, 75, 91–97 passim, 103, 104, 137; Region 8, 92, 103–4, 135, 137, 140, 147

U.S. Spruce Lumber Company, 48

Uniform Relocation Assistance and Real Property Acquisition Policies Act (1971), 113

Union Manganese Company, 96, **97**, 135

utilitarianism, 1, 7, 24, 36, 46, 179; Bentham, Jeremy, 185; Forest Service tradition, 9, 55, 70, 83, 127, 198, 199; timber resource, 10, 140, 147

valley and ridge, Appalachian subregion, 1, 158, 185, 221n1

virgin timber, xiv, 47, 50, 152, 244n1

Virginia Commission of Game and Inland Fisheries, 87, 98, 99, 100

Virginia Cooperative Wildlife Research Unit, 86, 99

Virginia Department of Forestry, 178, 179, 298n21

Virginia Division of Mined Land Reclamation, 135

Virginia Iron and Coal Company, 65, 66, 74, 252n35

Virginia Natural Heritage, 145

Virginia Plan (1938), 86–87

Virginia Tech (Virginia Polytechnic Institute & State University), 6, 98, 99, 106, 134, 182

Virginia Timber Corporation, 66

Virginia Wilderness Act (1984), 143, 145

Virginia Wilderness Act (1988), 145

Virginia Wildlife Federation, 123

W.M. Ritter Lumber Company, 48, 74

Walker Mountain (VA), 66

Warm Springs (WV), 37

Washington County (VA), 289n14; JNF land, 206; prehistory, 16, 226n23; timber, 48, 49, 151, 152, 154, 246n26

water pollution, 179–80, 299n30

water, religious value, 24–25, 164, 293n25

Watershed Protection and Flood Prevention Act (1954), 96

watershed, 87; damage, xiii, 85, **96**; protection, 5, 6, 60, 61, 94, 198, 249n10; relation to mining, **97**,

The Jefferson National Forest was designed and typeset on a Macintosh OS 10.4 computer system using InDesign software. The body text is set in 9.75/13 Warnock Pro and display type is set in ITC Legacy Sans. This book was designed and typeset by Stephanie Thompson and manufactured by Thomson-Shore, Inc.